Battery Modelling, Applications, and Technology

Battery Modelling, Applications, and Technology

Editor

Simone Barcellona

MDPI

Basel • Beijing • Wuhan • Barcelona • Belgrade • Novi Sad • Cluj • Manchester

Editor
Simone Barcellona
Politecnico di Milano
Milan
Italy

Editorial Office
MDPI
St. Alban-Anlage 66
4052 Basel, Switzerland

This is a reprint of articles from the Special Issue published online in the open access journal *Energies* (ISSN 1996-1073) (available at: https://www.mdpi.com/journal/energies/special_issues/0WS8O92BP5).

For citation purposes, cite each article independently as indicated on the article page online and as indicated below:

Lastname, A.A.; Lastname, B.B. Article Title. *Journal Name* **Year**, *Volume Number*, Page Range.

ISBN 978-3-7258-0605-8 (Hbk)
ISBN 978-3-7258-0606-5 (PDF)
doi.org/10.3390/books978-3-7258-0606-5

Contents

About the Editor

Simone Barcellona

Simone Barcellona was born in Milan, Italy, in May 1985. He obtained both his M.Sc. and Ph.D. degrees in electrical engineering from Politecnico di Milano, Milan, Italy, in July 2011 and December 2014, respectively. During his Ph.D. studies, his research focused on the theoretical and mathematical topology and operation of power static converters, as well as the study of conservative functions related to the switching networks in which the converters operate. Currently, he is a researcher at Politecnico di Milano, where his research focuses on power electronic converters and energy storage systems, with a specific emphasis on modeling and parameter estimation methods for lithium-ion batteries.

Preface

In this reprint, 14 works are presented on the topic of battery modeling and estimation methods for their parameters and states. Specifically, they cover battery state of health and state of charge estimation using different strategies such as sliding interacting multiple model partial discharge data, support vector regression, and artificial neural networks. Additionally, the reprint includes studies on 3D models for electrodes in lithium-ion batteries, focusing on the interfacial detachment of active material particles, as well as evaluations of the power generation impact for mobility in electric vehicles. Finally, a review article showcases the most recent developments in battery management systems using cloud-based artificial intelligence.

Simone Barcellona
Editor

energies

MDPI

Article

Simple Loss Model of Battery Cables for Fast Transient Thermal Simulation

Emanuele Fedele, Luigi Pio Di Noia and Renato Rizzo *

Department of Electrical Engineering and Information Technology, Università di Napoli Federico II, Via Claudio 21, 80125 Napoli, Italy
* Correspondence: renato.rizzo@unina.it

Abstract: In electric vehicles, currents with high-frequency ripples flow in the power cabling system due to the switching operation of power converters. Inside the cables, a strong coupling between the thermal and electromagnetic phenomena exists, since the temperature and Alternating Current (AC) density distributions in the strands affect each other. Due to the different time scales of magnetic and heat flow problems, the computational cost of Finite Element Method (FEM) numeric solvers can be excessive. This paper derives a simple analytical model to calculate the total losses of a multi-stranded cable carrying a Direct Current (DC) affected by a high-frequency ripple. The expression of the equivalent AC cable resistance at a generic frequency and temperature is derived from the general treatment of multi-stranded multi-layer windings. When employed to predict the temperature evolution in the cable, the analytical model prevents the use of complex FEM models in which multiple heat flow and magnetic simulations have to be run iteratively. The results obtained for the heating curve of a 35 mm^2 stranded cable show that the derived model matches the output of the coupled FEM simulation with an error below 1%, whereas the simple DC loss model of the cable gives an error of 2.4%. While yielding high accuracy, the proposed model significantly reduces the computational burden of the thermal simulation by a factor of four with respect to the complete FEM routine.

Keywords: cable; thermal analysis; skin and proximity effects; battery storage; ampacity

Citation: Fedele, E.; Di Noia, L.P.; Rizzo, R. Simple Loss Model of Battery Cables for Fast Transient Thermal Simulation. *Energies* **2023**, *16*, 2963. https://doi.org/10.3390/en16072963

Academic Editor: Simone Barcellona

Received: 22 February 2023
Revised: 15 March 2023
Accepted: 22 March 2023
Published: 23 March 2023

1. Introduction

Due to the ongoing effects of climate change, a significant reduction in greenhouse gas and pollutant emissions has become crucial. Such reduction demands a rapid transition from thermal to electrical transportation [1–7]. Electric vehicles require the use of onboard energy sources and storage systems such as fuel cells [8] and electrochemical batteries [9]. Battery electric vehicles are the most promising technology, although the actual electrochemical technology still needs improvements [10]. One of the main limitations to tackle is the reduction in volume and weight of battery cells so as to achieve higher values of specific power (kWh/kg) and power density (kWh/m^3). To this aim, abundant research on promising chemistries for the electrode and electrolyte materials has been carried on in recent years [11–14]. Besides the inherent tradeoff between weight or volume and vehicle mileage, other technical issues associated with the correct management of the battery pack also exist. In fact, the use of a battery management system able to monitor the electrical and thermal behavior of the cells is mandatory, and optimized thermal management is of utmost importance to obtain a lighter and more compact battery pack for a given power rating [15]. Indeed, the removal of heat from the cells due to the flow of current and the internal entropy variation is one of the causes that lead to a reduction in the power density of the battery pack. The thermal behavior of electrochemical cells is widely studied in the literature, and numerical and model-based techniques have been proposed for the correct evaluation of the thermal state of the cells [16–19]. The correct performance of a battery electric vehicle depends also on the proper operation of external devices attached to it, such as the DC/DC converters used as an interface to the electrical drives, the protection devices, and the wiring system [20]. The power cables play an important role in guaranteeing the vehicle's performance, especially from a reliability point of view [21,22]. In fact, due to the low volume availability and the non-negligible size of the cables, redundant wiring systems cannot be implemented onboard electric vehicles. Therefore, a failure of the main power cables between the battery pack and the propulsion system can rapidly lead to an unexpected stop of the electric vehicle. The damage and complete failure of power cables occur due to rapid degradation of the insulation, which can be caused by excessive electrical stress or overheating [23].

Virtually all the battery packs employed onboard EVs are interfaced with the traction and auxiliary loads through static power conversion stages. For this reason, the battery cabling system usually operates with distorted DC currents characterized by high-frequency ripples. Evaluating the thermal behavior of the cable in this condition is nontrivial, due to the mutual correlation between heat generation and electromagnetic phenomena, such as skin and proximity effects occurring in the conductor bundles [24,25]. On the other hand, the electrical power demanded by electric propulsion usually varies with fast dynamics determining sudden changes in the cable current, making transient analyses necessary. Complete numerical approaches based on the Finite Element Method (FEM) are often employed in the literature [26–30]; however, the high computation time required by the FEM simulations on a real multi-stranded cable geometry makes the method of application difficult for transient thermal simulations. The main idea of this paper is to derive a simple analytical model that provides a good estimation of the losses in high-power battery cables due to the flow of currents affected by high-frequency ripples. The model builds upon the premises of analytical equations known in the literature for the AC losses of round conductors in multi-layer transformer windings [31,32] and matches with good accuracy the results of electromagnetic and thermal FEM simulations. Thus, it can represent a fast and affordable means to evaluate the transient thermal behavior of the battery cabling system.

The remainder of this paper is organized as follows: Section 2 briefly introduces a typical battery EV powertrain architecture and its power cabling systems. Section 3 presents analytical loss models for multi-stranded cables carrying pure AC and distorted DC currents and provides a relation for the calculation of cable resistance at a generic temperature and frequency; in Section 4, the accuracy of the analytical loss models is assessed through the use of magnetic FEM simulations; in Section 5, a transient analysis of the cable heating is carried out using the simple method proposed in the paper and compared to that obtained by a full numerical routine comprising magnetic and thermal FEM simulations. Finally, Section 6 remarks on the methods and the results of the work and draws conclusions.

2. Battery Electric Vehicles Cabling System

Electric vehicles require an intensive adoption of power electronic converters, as they enable a flexible exchange of power between the battery pack, the propulsion motor drives, the auxiliary loads, and the onboard or offboard chargers. Figure 1 gives a sketch of a typical power circuit scheme of a battery EV, comprising the propulsion system and the onboard AC charger.

The DC/DC power converters play an important role in interfacing the battery pack with external power sources and loads. In fact, one DC/DC converter is often connected between the battery pack and the traction inverter to adapt the battery voltage to the level of the DC link and provide a stabilized voltage at the inverter input terminals. One more DC/DC converter is always included in the onboard charger to control the recharge of the battery pack from the external AC supply as prescribed by the charging profile and BMS limitations. As is well known, the operation of the power converters inevitably introduces some harmonic distortion in the currents and voltages of the power system and places extra stresses on every component of the drivetrain. For its part, the vehicle wiring system must be designed to withstand a number of stress factors, namely, steep voltage transients at the motor terminals, high environmental temperatures, and non-negligible ripple content in the DC current drawn from the battery. As is clear from the circuit scheme, such high-frequency ripples are always present in the battery current, since power converters are operating during both propulsion and vehicle recharging.

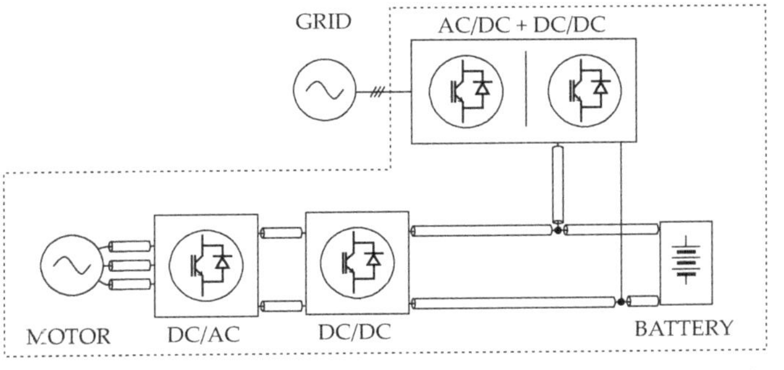

Figure 1. Typical EV power circuit including the propulsion system, the onboard charger, and the power cables.

For high-voltage EV applications, single-core multi-stranded high-temperature cables with PVC or silicon rubber insulation are usually employed in the DC stages of the propulsion system. The physical dimensions of the conductor core and insulation, as well as the nominal number of strands and their diameter for every standard cross-section, are prescribed by industry standards that cable makers are required to comply with [33,34]. Due to the multi-stranded geometry of power cables and the high-frequency harmonic of the battery current, additional Joule losses related to skin and proximity effects among strands can occur. Under particular ambient conditions, these extra Joule losses may cause excessive overheating, with a consequent fast degradation or even complete failure of the insulation layer. Therefore, an adequate evaluation of the cable's thermal behavior is of primary importance for the design and validation of the battery wiring system. In fact, a robust design can prevent accelerated aging, which can cause the intervention of protection devices or the fatal failure of the system, with consequent safety hazards [23].

3. Modeling of Cable Losses

3.1. Sinusoidal Current Waveform

The power dissipation per unit length P_0 of a round conductor subjected to an external magnetic field $\vec{H}$ and carrying a sine current of amplitude I_0 at frequency f_0 can be expressed as [35]

$$P_0 = \frac{\rho I_0^2}{\sqrt{2}\pi\delta d_0}\psi_1(\Delta) - \sqrt{2}\frac{\pi\rho d_0}{\delta}H^2\psi_2(\Delta) \tag{1}$$

where d_0 is the conductor diameter, ρ is the material resistivity, and δ is the skin depth, which is given by

$$\delta = \sqrt{\frac{\rho}{\pi\mu_0 f_0}} \tag{2}$$

In Equation (1), Δ is the normalized diameter with respect to the skin depth, i.e., $\Delta = d_0/\delta$, while $\psi_1(\Delta)$ and $\psi_2(\Delta))$ are functions of the real and imaginary parts of the k-th order Kelvin function $be_k(x)$ and their derivatives [36].

The geometry of a multi-stranded DC cable is schematized in Figure 2. Each strand is subjected to the magnetic field $\vec{H}$ produced by the currents in all the other strands inside the cable. This field can be expressed in polar coordinates as

$$\vec{H}(r,\theta) = \frac{I}{2\pi r_c^2} r\hat{i_\theta} \quad (3)$$

where I is the total current carried by the cable, and r_c is the total radius of the cable.

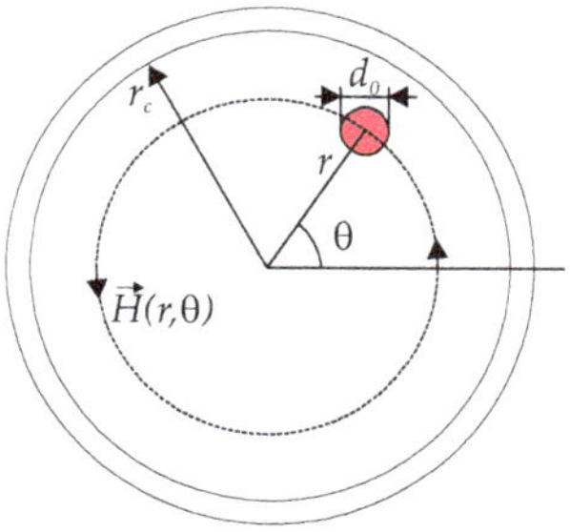

Figure 2. Multi-stranded cable geometry.

By denoting with N_0 the number of strands, the power loss density per unit volume of the cable can be expressed as

$$p = \frac{N_0 P_0}{\pi r_c^2} \quad (4)$$

By introducing the packing factor β,

$$\beta = \frac{N_0 r_0^2}{r_c^2} \quad (5)$$

and using Equations (1) and (4), the power loss density rewrites as

$$p(r,\theta) = P_0 \frac{\beta}{\pi r_0^2} = \frac{I_0^2 \rho \beta}{2\sqrt{2}\pi^2 \delta r_0^3} \psi_1(\Delta) - 2\sqrt{2}\frac{\rho\beta}{\delta r_0} H^2 \psi_2(\Delta) \quad (6)$$

By integrating Equation (6) over the entire cable cross-section, the total AC power losses per unit length of the cable can be obtained:

$$\begin{aligned} P &= \int_0^{2\pi} \int_0^{r_c} \left(\frac{I_0^2 \rho \beta}{2\sqrt{2}\pi^2 \delta r_0^3} \psi_1(\Delta) - 2\sqrt{2}\frac{\rho\beta}{\delta r_0} H^2 \psi_2(\Delta) \right) dr d\theta \\ &= \frac{\rho I^2}{\sqrt{2}\pi \delta N_0 d_0} [\psi_1(\Delta) - \beta N_0 \psi_2(\Delta)] \end{aligned} \quad (7)$$

The above result differs from the usual expression of AC losses in multi-stranded multi-layer transformer windings [32,36] and is peculiar to multi-stranded power cables in which, differently from transformer windings, the external leakage magnetic field can be considered of negligible magnitude. From Equation (7), the AC equivalent resistance of the cable can be derived straightforwardly as

$$R_{\mathrm{AC}} = \frac{P}{I^2/2} = \frac{\sqrt{2}\rho}{\pi \delta N_0 d_0} [\psi_1(\Delta) - \beta N_0 \psi_2(\Delta)] \quad (8)$$

Since the DC resistance per unit length of the cable is simply equal to the ratio between the resistivity and the cross-section,

$$R_{\mathrm{DC}} = \frac{4\rho}{N_0 \pi d_0^2} \tag{9}$$

the AC to DC resistance ratio k_R is equal to

$$k_R = \frac{R_{\mathrm{AC}}}{R_{\mathrm{DC}}} = \frac{\Delta}{2\sqrt{2}}[\psi_1(\Delta) - \beta N_0 \psi_2(\Delta)] \tag{10}$$

By recalling that the following Taylor series expansions hold true for $\psi_1(\Delta)$ and $\psi_2(\Delta)$,

$$\psi_1(\Delta) = 2\sqrt{2}\left(\frac{1}{\Delta} + \frac{1}{3\,2^8}\Delta^3 - \frac{1}{3\,2^{14}}\Delta^5 + \dots\right) \tag{11a}$$

$$\psi_2(\Delta) = \frac{1}{\sqrt{2}}\left(-\frac{1}{2^5}\Delta^3 + \frac{1}{2^{12}}\Delta^7 + \dots\right) \tag{11b}$$

and considering the terms up to the third power of Δ, the ratio k_R can be ultimately rewritten as

$$k_R = \frac{\Delta}{2\sqrt{2}}\left[2\sqrt{2}\left(\frac{1}{\Delta} + \frac{1}{3\,2^8}\Delta^3\right) + \beta N_0 \frac{1}{\sqrt{2}}\frac{1}{2^5}\Delta^3\right] = 1 + \frac{1 + 6\beta N_0}{3\,2^8}\Delta^4 \tag{12}$$

Substituting Equation (12) into Equation (10) yields the ultimate expression for the AC cable resistance at frequency f_0:

$$R_{\mathrm{AC}} = \frac{4\rho}{N_0 \pi d_0^2}\left[1 + \frac{1 + 6\beta N_0}{3\,2^8}\Delta^4\right] \tag{13}$$

where the dependency on frequency is not explicit but contained in the normalized skin depth Δ.

3.2. Generic Current Waveform

The above analysis can be extended to a generic periodic current waveform with DC and AC components, as typically encountered in electrical drive systems such as an EV powertrain. An arbitrary periodic current waveform of fundamental frequency f_0 can be expressed by means of its Fourier series expansion as

$$i(t) = I_{\mathrm{DC}} + \sum_{h=1}^{+\infty} \sqrt{2} I_h \sin(2\pi h f_0 t - \phi_h) \tag{14}$$

where the closed-form expression of I_h and ϕ_h for numerous waveforms often encountered in power electronic applications can be found in the literature. The Joule losses corresponding to such arbitrary current waveforms are then given by

$$P = R_{\mathrm{DC}} I_{\mathrm{DC}}^2 + \sum_{h=1}^{+\infty} R_{\mathrm{AC}}^{(h)} I_h^2 = R_{\mathrm{DC}}\left(I_{\mathrm{DC}}^2 + \sum_{h=1}^{+\infty} k_R^{(h)} I_h^2\right) \tag{15}$$

where $R_{\mathrm{AC}}^{(h)}$ and $k_R^{(h)}$ represent the AC cable resistance at frequency hf_0 and its normalized value, respectively. The normalized strand diameter at frequency hf_0 is equal to

$$\Delta_h = \frac{d_0}{\delta_h} = \frac{d_0}{\sqrt{1/\rho\pi\mu_0 h f_0}} = \sqrt{h}\Delta \tag{16}$$

from which the following expression for $k_R^{(h)}$ is derived:

$$k_R^{(h)} = \frac{\sqrt{h}\Delta}{2\sqrt{2}}\left[\psi_1(\sqrt{h}\Delta) - \beta N_0 \psi_2(\sqrt{h}\Delta)\right] \tag{17}$$

By recalling the Taylor series expansion provided in Equation (11a,b), Equation (17) rewrites as

$$k_R^{(h)} = 1 + \frac{1+6\beta N_0}{3\,2^8}h^2\Delta^4 \tag{18}$$

The above expression is useful because it only requires the computation of the normalized skin depth at fundamental frequency f_0. Equation (18) can be substituted into Equation (15) to calculate the total AC losses in the cable for a generic current waveform:

$$P = R_{\mathrm{DC}}\left[I_{\mathrm{DC}}^2 + \sum_{h=1}^{+\infty}\left(1 + \frac{1+6\beta N_0}{3\,2^8}\Delta^4 h^2\right)I_h^2\right] \tag{19}$$

4. Validation with Magnetic FEM Solver

The analytical expression of the AC cable resistance given by Equation (13) is validated against the results of the 2D Finite Element Method (FEM) software FEMM 4.2. The software solves the following phasor equation for the complete multi-stranded wire geometry:

$$\nabla \times \left(\frac{1}{\mu(B)}\nabla \times \mathbf{A}\right) = -\mathrm{j}\frac{\omega}{\rho}\mathbf{A} + \mathbf{J}_s - \frac{1}{\rho}\nabla V \tag{20}$$

where $\mathbf{A}$ and $\mathbf{J}_s$ are the phasors of the magnetic induction vector potential and source current density, respectively. Several simulations are carried out for different frequencies and temperatures. This is necessary because the temperature affects the conductor resistivity ρ, which in turn influences the AC resistance, both directly through the resistivity temperature coefficient and indirectly through the skin depth. The comparison is carried out on a 35 mm^2 high-temperature DC cable for automotive applications whose main parameters are reported in Table 1.

Table 1. Cable parameters.

Parameter	Value
Conductor material	Copper
Nominal section	35 mm^2
Strand diameter	0.41 mm
Number of strands	276
Ampacity	280 A at 125 °C
Insulation material	Silicon rubber
Insulation thickness	1.04 mm
Temperature range	−40 °C to +180 °C

The current density distribution inside the individual cable strands as computed by the FEM solver at 280 A, 20 °C, and for a frequency of 5 kHz and 50 kHz is shown in Figure 3.

The FEM results show how the impact of skin and proximity effects on the current density distribution differs between the two cases. At a frequency of 5 kHz, the skin depth of copper at 20 °C is 0.92 mm, approximately two times the diameter of the strands. For this reason, the current density distribution is not far from being uniform and has peaks of 9 A/mm^2. On the other hand, the skin depth at 50 kHz lowers to 0.29 mm, approximately 0.71 times the strand diameter. Hence, the distribution of $\mathbf{J}$ becomes much more nonuniform with maximum values of 48 A/mm^2, and higher Joule losses occur in the cable.

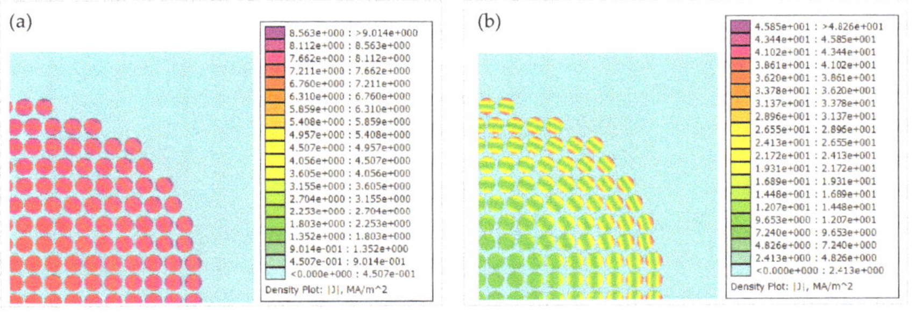

Figure 3. Current density in the cable at 20 °C and at a frequency of 5 kHz (**a**) and 50 kHz (**b**).

The AC equivalent resistance of the cable is evaluated in the post-processing phase based on the overall losses occurring in the conductor volume and the square of the current:

$$R_{\mathrm{AC}} = \frac{\int_{V_c} \rho \|\mathbf{J}\|^2 \mathrm{d}V}{1/2\left(\int_{S_c} \|\mathbf{J}\| \mathrm{d}S\right)^2} \tag{21}$$

where S_c and V_c are the cable cross section and volume, respectively. In Figure 4, the results of the FEM post-processing given by Equation (21) are compared with those given by the analytical Equation (13) for different frequencies and temperatures.

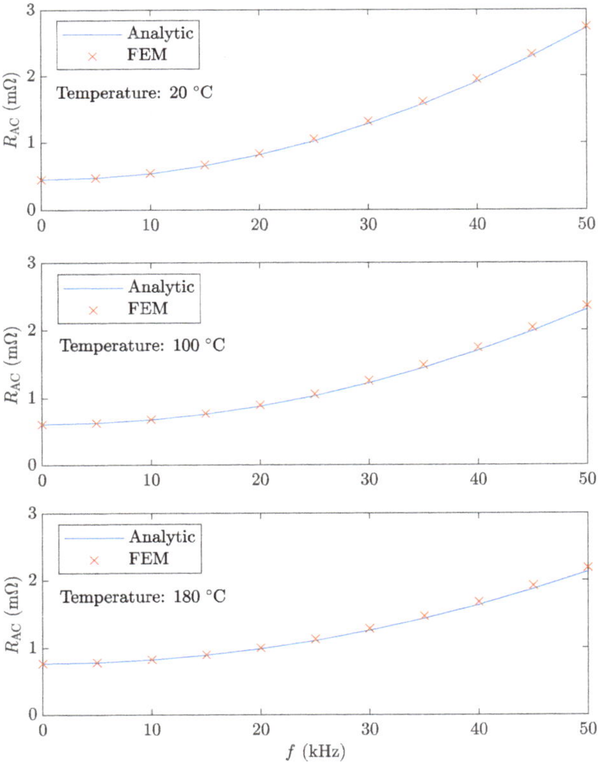

Figure 4. Comparison between analytic and FEM-calculated AC resistances at different frequencies and temperatures.

As can be seen, the analytical AC resistance equation matches with high accuracy the FEM results and proves effective in modeling the high-frequency skin and proximity effects in the strands and their complex dependency on temperature. Indeed, it is worth remarking that a higher temperature determines an increase in resistance at DC and low frequencies, but a reduction in resistance at high frequencies due to the increase in the skin depth. This is the reason why the cable has a higher resistance at 20 °C than at 180 °C at 50 kHz. The analytical model properly captures this opposite behavior at low and high frequencies.

5. Application to Thermal Transient Analysis

Thermal calculations of multi-stranded cables in free air require careful consideration of the mutual couplings between electromagnetic and thermal field problems [37]. In fact, the temperature inside the strands influences the resistivity and thus the magnetic field and current density distributions, which in turn determine the power losses inside the cable and its temperature evolution. While FEM analysis also proves effective in the thermal domain, the coupling between electromagnetic and thermal fields adds considerable computational costs, as the time scales of the two problems differ significantly [38]. If a generic current waveform is considered with multiple harmonic frequency magnetic problems, the computational burden of a fully coupled FEM model can become excessive. The practical benefit of using the analytical model of the total cable losses is therefore evaluated with application to the transient thermal analysis of the 35 mm^2 cable introduced in the previous section. A triangular current waveform is considered, which corresponds to the operation of a DC/DC converter attached to the battery. With reference to the Fourier series expansion in Equation (14), the following relations apply for a triangular shape [39]:

$$I_h = \frac{\Delta i}{d(1-d)\pi^2} \frac{\sin(h\pi d)}{h^2} \tag{22a}$$

$$\phi_h = -h\pi d \tag{22b}$$

where Δi is the peak-to-peak current ripple and d is the duty ratio.

The flowchart of the cross-coupled FEM simulation of the thermal transient is depicted in Figure 5.

For the triangular current at fundamental frequency f_0, the first N_h harmonics shall be considered. At time step k, the temperature distribution inside the strands is used to solve N_h FEM magnetic problems at peak current $\sqrt{2}I_h$ and frequency hf_0. The losses due to the harmonic currents are added to the DC losses and passed as input to the FEM thermal model that solves the heat flow equations and updates the temperature distribution, which is then used as the initial state of the subsequent iteration. Given N_t steps over the entire thermal simulation, $N_t \times N_h$ magnetic problems plus N_t thermal problems must be solved by two dedicated FEM routines.

On the other hand, the analytical loss model does not need N_h magnetic problems to be solved at each time step of the thermal simulation, reducing the total number of FEM calculations to N_t. In fact, at each time step, temperature θ is employed to calculate the DC resistance R_{DC} and normalized strand diameter Δ. These values are used by Equation (19) to evaluate the total Joule losses through a single computation. The total losses are then input to the thermal FEM solver to calculate the evolution of the temperature distribution.

The analysis was carried out for a current of 420 A of DC value (+50% with respect to the nominal ampacity) and 10% ripple. Natural convection on the external insulation surface at an air temperature of 60 °C is assumed. Matlab was employed to manage the iterative execution of electromagnetic and thermal simulations performed by FEMM 4.2 and for all the required pre- and post-processing calculations and plotting. Figure 6 shows the temperature distribution inside the cable as computed by the cross-coupled FEM routine after 10, 20, 30, and 40 min. As dictated by the circular symmetry of the problem, the heat flows from the core to the outer surface of the cable, and a temperature gradient of 10 to 20 °C is observed between the inner strands and the outer surface of the insulation layer.

After 40 min, the temperature in the silicon rubber has reached 193 °C, which is above the maximum continuous temperature rating due to a current higher than the rated ampacity.

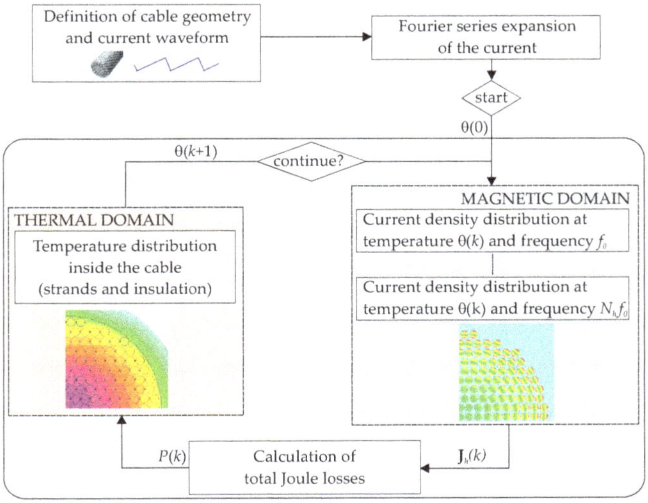

Figure 5. Flowchart for cross-coupled numerical evaluation of the cable thermal transient evolution.

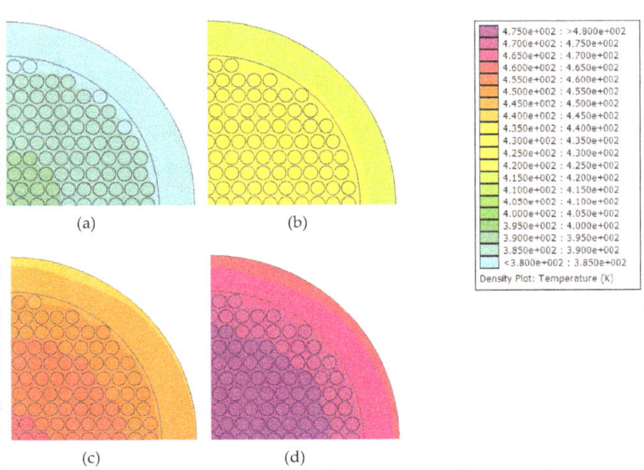

Figure 6. Temperature distribution inside the cable as computed by the coupled FEM routine after 10 min (**a**), 20 min (**b**), 30 min (**c**), and 40 min (**d**).

The variations in the average temperature of the strands and insulation as computed by the coupled FEM routine and analytical loss model are compared in Figure 7. In the plots, a third curve is also reported which represents the temperature evolution when only DC losses are considered, i.e., for $P = R_{\mathrm{DC}} I_{\mathrm{DC}}^2$.

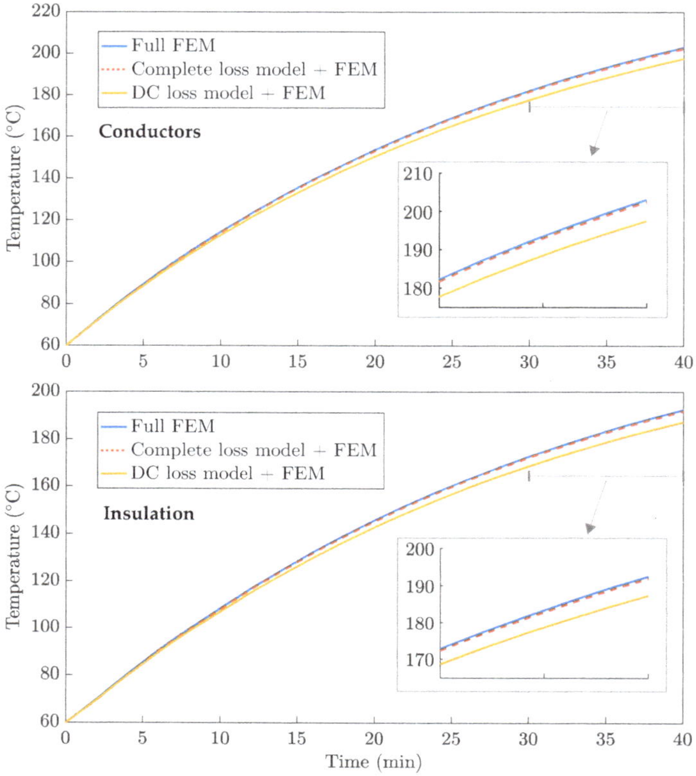

Figure 7. Temperature evolution in the conductors and insulations computed through different methods.

The comparison shows that the complete loss model accurately estimates the losses inside the cable and matches the thermal transient curve yielded by the complete FEM routine. However, it should be remarked that the coupled FEM simulation lasted approximately 17 h, while the one relying on the total loss analytical model lasted 4 h (both simulations were run on the same machine with 16 GB RAM and a six-core 3.10 GHz processor). On the other hand, the temperature profile obtained through the simple DC loss model is less accurate and presents an error that increases with the temperature. This increasing deviation can be explained as follows: The DC loss model underestimates the total cable resistance, which results in lower Joule losses and a consequently slower temperature increase. On the other hand, since the cable resistance increases with temperature (both in the DC and complete loss model), a slower temperature increase determines a slower increase in time of the internal losses. This inherent bidirectional coupling between temperature and power losses results in a discrepancy between the complete and DC loss model that increases with time. Based on the thermal constant of the cable here considered, this discrepancy becomes visible after 10 min. At the end of the simulation, the difference in the insulation temperature yielded by the two models amounts to 4.6 °C (2.4%). This discrepancy, while not excessive, suggests that the complete loss model given by Equation (19) may be preferred when a more precise evaluation of the cable temperature evolution under specific load or ambient conditions has to be performed. It is worth remarking that this higher accuracy is obtained at no additional cost, as the DC and complete loss models share the same negligible computational burden.

6. Conclusions

The DC cables connecting the battery pack to the traction system of an electric vehicle play an important role in guaranteeing the performance, reliability, and safety of the vehicle. The lifetime of a wiring system strongly depends on the temperature reached during operation. Many factors can affect the temperature of the cable, including the ambient conditions and the harmonic distortion in the current caused by the switching operation of power electronic converters attached to the battery. To evaluate the temperature evolution of DC cables under defined ambient and operating conditions, numerical approaches based on the Finite Element Method (FEM) represent the prevailing technique due to the versatility and accuracy of FEM solvers. In a multi-stranded cable geometry, which is usually found in battery power connections, a strong coupling between the thermal and electromagnetic phenomena exists, that is, they constitute a multi-physics problem. In fact, the temperature reached by the cable primarily depends on the Joule losses in the conductors, i.e., on the current density distribution within the strands. On the other hand, the distribution of harmonic AC current densities is affected by skin and proximity effects, whose impacts vary with frequency and temperature. Due to the very different time scales of magnetic and heat flow problems and the many harmonic components found in the current, a numerical solution to the problem can often be computationally hard.

This paper presented a simple analytical model to calculate the total losses occurring in a multi-stranded cable carrying a DC current that is affected by a high-frequency ripple. By considering the stranded geometry of the cable and the magnetic field inside it, the model can capture the joint effects of temperature and skin and proximity effects on the total Joule losses per unit length of the cable. The analytical equations describing the equivalent AC resistance of the cable at a fixed frequency and temperature were derived from the general treatment of high-frequency effects in multi-stranded multi-layer winding geometries and were validated against the results of time-harmonic FEM magnetic models. The model proved effective in matching with high accuracy the resistance values obtained through FEM analysis in the entire range of variation considered for frequency and temperature.

The effective advantage of using the proposed loss model to predict the temperature evolution inside the cable with a reduced computational burden was also assessed. Specifically, the output of a coupled electromagnetic–thermal FEM simulation comprising electromagnetic and heat flow calculations was compared to a simpler calculation routine, in which the loss density input to the heat flow FEM model was evaluated by means of the derived analytical loss model. The results showed that the temperature curve yielded by the simplified routine almost overlapped with that produced by the coupled FEM simulation. However, the simplified routine required only one-fourth of the computation time with respect to the full FEM simulation. On the other hand, the temperature evolution given by the simpler DC loss model was found to be affected by an error that increased with time and reached 2.4% at the end of the simulated time window. These results suggest that the proposed analytical modeling of total Joule losses in the multi-stranded cable can be of use when a more detailed evaluation of the temperature is to be performed at a largely reduced computation cost.

Author Contributions: Conceptualization, E.F., L.P.D.N. and R.R.; methodology, E.F., L.P.D.N. and R.R.; software, E.F. and L.P.D.N.; validation, E.F. and L.P.D.N.; formal analysis, E.F.; investigation, E.F. and L.P.D.N.; resources, E.F. and L.P.D.N.; data curation, E.F.; writing—original draft preparation, E.F.; writing—review and editing, L.P.D.N. and R.R.; visualization, E.F.; supervision, L.P.D.N. and R.R.; project administration, R.R. All authors have read and agreed to the published version of the manuscript.

Funding: This research received no external funding.

Data Availability Statement: The data presented in this study are available on request from the corresponding author.

Conflicts of Interest: The authors declare no conflict of interest.

References

1. Husain, I.; Ozpineci, B.; Islam, M.S.; Gurpinar, E.; Su, G.J.; Yu, W.; Chowdhury, S.; Xue, L.; Rahman, D.; Sahu, R. Electric Drive Technology Trends, Challenges, and Opportunities for Future Electric Vehicles. *Proc. IEEE* **2021**, *109*, 1039–1059. [CrossRef]
2. Sanguesa, J.A.; Torres-Sanz, V.; Garrido, P.; Martinez, F.J.; Marquez-Barja, J.M. A Review on Electric Vehicles: Technologies and Challenges. *Smart Cities* **2021**, *4*, 372–404. [CrossRef]
3. Fedele, E.; Iannuzzi, D.; Del Pizzo, A. Onboard energy storage in rail transport: Review of real applications and techno-economic assessments. *IET Electr. Syst. Transp.* **2021**, *11*, 279–309. [CrossRef]
4. Sun, X.; Li, Z.; Wang, X.; Li, C. Technology Development of Electric Vehicles: A Review. *Energies* **2020**, *13*, 90. [CrossRef]
5. Brelje, B.J.; Martins, J.R. Electric, hybrid, and turboelectric fixed-wing aircraft: A review of concepts, models, and design approaches. *Prog. Aerosp. Sci.* **2019**, *104*, 1–19. [CrossRef]
6. Fang, S.; Wang, Y.; Gou, B.; Xu, Y. Toward Future Green Maritime Transportation: An Overview of Seaport Microgrids and All-Electric Ships. *IEEE Trans. Veh. Technol.* **2020**, *69*, 207–219. [CrossRef]
7. Del Pizzo, A.; Polito, R.M.; Rizzo, R.; Tricoli, P. Design criteria of on-board propulsion for hybrid electric boats. In Proceedings of the The XIX International Conference on Electrical Machines—ICEM 2010, Rome, Italy, 6–8 September 2010; pp. 1–6. [CrossRef]
8. Sorlei, I.S.; Bizon, N.; Thounthong, P.; Varlam, M.; Carcadea, E.; Culcer, M.; Iliescu, M.; Raceanu, M. Fuel Cell Electric Vehicles—A Brief Review of Current Topologies and Energy Management Strategies. *Energies* **2021**, *14*, 252. [CrossRef]
9. Deng, J.; Bae, C.; Denlinger, A.; Miller, T. Electric Vehicles Batteries: Requirements and Challenges. *Joule* **2020**, *4*, 511–515. [CrossRef]
10. Zhao, G.; Wang, X.; Negnevitsky, M. Connecting battery technologies for electric vehicles from battery materials to management. *iScience* **2022**, *25*, 103744. [CrossRef]
11. Nitta, N.; Wu, F.; Lee, J.T.; Yushin, G. Li-ion battery materials: Present and future. *Mater. Today* **2015**, *18*, 252–264. [CrossRef]
12. Balasankar, A.; Arthiya, S.E.; Ramasundaram, S.; Sumathi, P.; Arokiyaraj, S.; Oh, T.; Aruchamy, K.; Sriram, G.; Kurkuri, M.D. Recent Advances in the Preparation and Performance of Porous Titanium-Based Anode Materials for Sodium-Ion Batteries. *Energies* **2022**, *15*, 495. [CrossRef]
13. Julien, C.M.; Mauger, A.; Zaghib, K.; Groult, H. Comparative Issues of Cathode Materials for Li-Ion Batteries. *Inorganics* **2014**, *2*, 132–154. [CrossRef]
14. Li, Q.; Xu, C.; Huang, B.; Yin, X. Rhombohedral Li1+xYxZr2-x(PO4)3 Solid Electrolyte Prepared by Hot-Pressing for All-Solid-State Li-Metal Batteries. *Materials* **2020**, *13*, 1719. [CrossRef] [PubMed]
15. Hamed, M.M.; El-Tayeb, A.; Moukhtar, I.; El Dein, A.; Abdelhameed, E.H. A review on recent key technologies of lithium-ion battery thermal management: External cooling systems. *Results Eng.* **2022**, *16*, 100703. [CrossRef]
16. Kleiner, J.; Stuckenberger, M.; Komsiyska, L.; Endisch, C. Advanced Monitoring and Prediction of the Thermal State of Intelligent Battery Cells in Electric Vehicles by Physics-Based and Data-Driven Modeling. *Batteries* **2021**, *7*, 31. [CrossRef]
17. Piccirillo, F.; Bianco, N.; Di Noia, L.P.; Guerriero, P.; Iasiello, M. Analysis of the Thermal Behavior of Li-Ion Pouch Battery Cell—Part I: Finite Element Simulations Including the Entropic Coefficient. In Proceedings of the 2022 28th International Workshop on Thermal Investigations of ICs and Systems (THERMINIC), Dublin, Ireland, 28–30 September 2022; pp. 1–6. [CrossRef]
18. Hu, Y.; Yurkovich, S.; Guezennec, Y.; Yurkovich, B. Electro-thermal battery model identification for automotive applications. *J. Power Sources* **2011**, *196*, 449–457. [CrossRef]
19. Morganti, M.V.; Longo, S.; Tirovic, M.; Blaise, C.Y.; Forostovsky, G. Multi-Scale, Electro-Thermal Model of NMC Battery Cell. *IEEE Trans. Veh. Technol.* **2019**, *68*, 10594–10606. [CrossRef]
20. Aghabali, I.; Bauman, J.; Kollmeyer, P.J.; Wang, Y.; Bilgin, B.; Emadi, A. 800-V Electric Vehicle Powertrains: Review and Analysis of Benefits, Challenges, and Future Trends. *IEEE Trans. Transp. Electrif.* **2021**, *7*, 927–948. [CrossRef]
21. Wang, C.; Liu, L.; Hou, X.; Wang, J.; Qiu, B. Failure Analysis and Improvement of High Voltage Cable of Electric Vehicle Based on Pseudo Damage Equivalence. In *Proceedings of the China SAE Congress 2020: Selected Papers*; Springer: Singapore, 2022; pp. 467–480.
22. Grandvuillemin, J.; Chamagne, D.; Tiraby, C.; Glises, R. Ampacity of power bus bars for Hybrid-Electric or Electric Vehicles. In Proceedings of the 2008 IEEE Vehicle Power and Propulsion Conference, Harbin, China, 3–5 September 2008; pp. 1–6. [CrossRef]
23. Bukya, M.; Kumar, R.; Gupta, R.K. A study on safety issues and analytical evaluation of stresses for HVDC cable in electrical vehicle. *AIP Conf. Proc.* **2020**, *2294*, 040001. [CrossRef]
24. Dias, R.A.; Lira, G.R.S.; Costa, E.G.; Ferreira, R.S.; Andrade, A.F. Skin effect comparative analysis in electric cables using computational simulations. In Proceedings of the 2018 Simposio Brasileiro de Sistemas Eletricos (SBSE), Niteroi, Brazil, 12–16 May 2018; pp. 1–6. [CrossRef]
25. Rozegnał, B.; Albrechtowicz, P.; Mamcarz, D.; Rerak, M.; Skaza, M. The Power Losses in Cable Lines Supplying Nonlinear Loads. *Energies* **2021**, *14*, 1374. [CrossRef]
26. Karahan, M.; Kalenderli, O. Coupled Electrical and Thermal Analysis of Power Cables Using Finite Element Method. In *Heat Transfer*; Vikhrenko, V.S., Ed.; IntechOpen: Rijeka, Croatia, 2011; Chapter 9. [CrossRef]
27. Di Noia, L.P.; Rizzo, R. Thermal Analysis of Battery Cables for Electric Vehicles. In Proceedings of the 2020 2nd IEEE International Conference on Industrial Electronics for Sustainable Energy Systems (IESES), Cagliari, Italy, 1–3 September 2020; Volume 1, pp. 298–302. [CrossRef]

28. Enescu, D.; Colella, P.; Russo, A. Thermal Assessment of Power Cables and Impacts on Cable Current Rating: An Overview. *Energies* **2020**, *13*, 5319. [CrossRef]
29. Cirino, A.W.; Paula, H.D.; Mesquita, R.C.; Saraiva, E. Cable parameter variation due to skin and proximity effects: Determination by means of finite element analysis. In Proceedings of the 2009 35th Annual Conference of IEEE Industrial Electronics, Porto, Portugal, 3–5 November 2009. [CrossRef]
30. Goga, V.; Paulech, J.; Váry, M. Cooling of electrical Cu conductor with PVC insulation - Analytical, numerical and fluid flow solution. *J. Electr. Eng.* **2013**, *64*, 13. [CrossRef]
31. Ferreira, J. Improved analytical modeling of conductive losses in magnetic components. *IEEE Trans. Power Electron.* **1994**, *9*, 127–131. [CrossRef]
32. Dowell, P. Effects of eddy currents in transformer windings. *Proc. Inst. Electr. Eng.* **1966**, *113*, 1387. [CrossRef]
33. *LV 112*; Electrical Cables for Motor Vehicles. ISO: Geneva, Switzerland, 2005.
34. *ISO 6722*; Road Vehicles—60 V and 600 V Single-Core Cables. ISO: Geneva, Switzerland, 2006.
35. Lammeraner, J.; Stafl, M. *Eddy Currents*; Iliffe Books: London, UK, 1966.
36. Tourkhani, F.; Viarouge, P. Accurate analytical model of winding losses in round Litz wire windings. *IEEE Trans. Magn.* **2001**, *37*, 914375. [CrossRef]
37. Cywiński, A.; Chwastek, K. A Multiphysics Analysis of Coupled Electromagnetic-Thermal Phenomena in Cable Lines. *Energies* **2021**, *14*, 8. [CrossRef]
38. Hameyer, K.; Driesen, J.; De Gersem, H.; Belmans, R. The classification of coupled field problems. *IEEE Trans. Magn.* **1999**, *35*, 1618–1621. [CrossRef]
39. Kondrath, N.; Kazimierczuk, M. Inductor winding loss owing to skin and proximity effects including harmonics in non-isolated pulse-width modulated dc–dc converters operating in continuous conduction mode. *IET Power Electron.* **2010**, *3*, 989. [CrossRef]

energies

MDPI

Article

Cloud-Based Deep Learning for Co-Estimation of Battery State of Charge and State of Health

Dapai Shi [1,†], Jingyuan Zhao [2,*,†], Zhenghong Wang [1], Heng Zhao [3], Chika Eze [4], Junbin Wang [5], Yubo Lian [5] and Andrew F. Burke [2,*]

1 Hubei Longzhong Laboratory, Hubei University of Arts and Science, Xiangyang 441053, China
2 Institute of Transportation Studies, University of California-Davis, Davis, CA 95616, USA
3 College of Big Data and Internet, Shenzhen Technology University, Shenzhen 518118, China
4 Department of Mechanical Engineering, University of California, Merced, CA 94720, USA
5 BYD Automotive Engineering Research Institute, Shenzhen 518118, China
* Correspondence: jyzhao@ucdavis.edu (J.Z.); afburke@ucdavis.edu (A.F.B.)
† These authors contributed equally to this work.

Abstract: Rechargeable lithium-ion batteries are currently the most viable option for energy storage systems in electric vehicle (EV) applications due to their high specific energy, falling costs, and acceptable cycle life. However, accurately predicting the parameters of complex, nonlinear battery systems remains challenging, given diverse aging mechanisms, cell-to-cell variations, and dynamic operating conditions. The states and parameters of batteries are becoming increasingly important in ubiquitous application scenarios, yet our ability to predict cell performance under realistic conditions remains limited. To address the challenge of modelling and predicting the evolution of multiphysics and multiscale battery systems, this study proposes a cloud-based AI-enhanced framework. The framework aims to achieve practical success in the co-estimation of the state of charge (SOC) and state of health (SOH) during the system's operational lifetime. Self-supervised transformer neural networks offer new opportunities to learn representations of observational data with multiple levels of abstraction and attention mechanisms. Coupling the cloud-edge computing framework with the versatility of deep learning can leverage the predictive ability of exploiting long-range spatio-temporal dependencies across multiple scales.

Keywords: lithium-ion battery; state of charge; state of health; deep learning; cloud; field application

Citation: Shi, D.; Zhao, J.; Wang, Z.; Zhao, H.; Eze, C.; Wang, J.; Lian, Y.; Burke, A.F. Cloud-Based Deep Learning for Co-Estimation of Battery State of Charge and State of Health. *Energies* **2023**, *16*, 3855. https://doi.org/10.3390/en16093855

Academic Editor: Simone Barcellona

Received: 11 April 2023
Revised: 22 April 2023
Accepted: 25 April 2023
Published: 30 April 2023

1. Introduction

With increased concerns about global warming, transportation electrification has recently emerged as an important step across the world. In electrified vehicles, rechargeable lithium-ion batteries are currently the most widely used systems for electrochemical energy storage and powering electric vehicles (EVs) due to their relatively high specific energy, acceptable cost and cycle life [1]. However, degradation and aging during the system's operational lifetime is still one of the most urgent and inevitable problems, especially under realistic conditions [2]. In field applications, such as an EV, an online battery management system (BMS) offers tools to monitor cell behavior under dynamic operating conditions. However, predicting real-life battery performance in field applications only using the online BMS is either difficult or impossible due to the limited data computing and storage ability of the onboard chips.

Over the past decade, scientists and researchers are increasingly storing and analyzing their big datasets by using remote 'cloud' computing servers [3]. On the cloud, researchers can interact with field data more flexibly and intelligently. Migrating observational data from custom servers to the cloud opened up a new world of opportunities to both assimilate the data sensibly and explore it in depth.

Several international companies have recognized this and have recently launched their cloud-based software, including Bosch [4], Panasonic [5] and Huawei [6]. Such public-cloud services are also termed software as a service (SaaS). The SaaS provided by Bosch—battery in the cloud—claimed that it is possible to improve the cycle life of batteries by 20% through the development of digital twins by using the big datasets from vehicle fleets. The universal battery management cloud (UBMC) service developed by Panasonic aims to identify the cell state and optimal battery operation. The SaaS launched by Huawei aims to provide a public cloud computing and storage service for EV companies. By learning from the historical battery data, the purely data-driven model embedded on its cloud monitoring system is applied to predict cell fault by discovering intricate structure in large EV-battery datasets. Beyond enterprise-level cloud services, a national-level big-data platform was built in 2017 in China, named the National Monitoring and Management Platform for New Energy Vehicles (NMMP-NEV) [7]. Up to now, the NMMP-NEV has provided remote fault diagnosis for more than six million EVs.

1.1. Literature Review

1.1.1. Modelling and Predicting Battery States

A battery is a sophisticated material system, with its functionality reliant on the transport of charge and ions through distinct phases and across interfaces, as well as both reversible and irreversible chemical reactions, among other material-dependent factors. The performance of a cell can be influenced by variations in components such as electrodes, electrolytes, interfaces, microstructures, current collectors, separators, binders, and cell or pack designs, as well as environmental factors and operating conditions. While significant progress has been made in first-principles, atomistic, and physics-based electrochemical modeling of battery systems, the absence of comprehensive predictive models remains a limiting factor for advancement. The battery management system (BMS) plays a pivotal role in maintaining the safe and reliable operation of battery systems for EV applications. Battery modelling is the core function of a BMS. Over the past few years, a variety of estimation techniques have been developed for the determination of the state of batteries in terms of two important parameters: SOC [8] and SOH [9]. In the literature, the most-studied methods in this regard for Li-ion batteries are equivalent circuit models (ECMs), physics-based models (PBMs), the observational filter model, and, more recently, data-driven, machine learning-based techniques. Each method has its own advantages and challenges. But there is always a trade-off between model accuracy and computational cost (Figure 1). For example, ECMs offer an effective tool to identity the cell states with low computational cost. Such a simple method has been widely used in onboard BMS for the last decade. However, it cannot provide accurate cell parameter values due to the simplification and assumptions in battery behaviors. Compared to ECMs, PBMs can approximate the physico-chemical processes that take place inside the cell during the system's operation, which provides accurate and physically consistent predictions. This method requires detailed information on cell specifications, including the materials and chemistry of the electrode, electrolyte, separator, current collectors, and so on. However, it is impossible to obtain the evaluation of these parameters during the operational lifetime under realistic conditions. In addition, battery problems in this case are governed by highly parameterized partial differential equations (PDEs). Solving the governing PDEs faces severe challenges and introduces multiple sources of uncertainty, especially in real-life physical problems with missing and noisy data and uncertain boundary conditions. Filter-based models are the most-commonly used methods for battery parameter estimation in the existing studies. Two issues constrain the wide application of filtering algorithms: (i) model parameters need to be updated and (ii) the algorithms may have poor generalization performance.

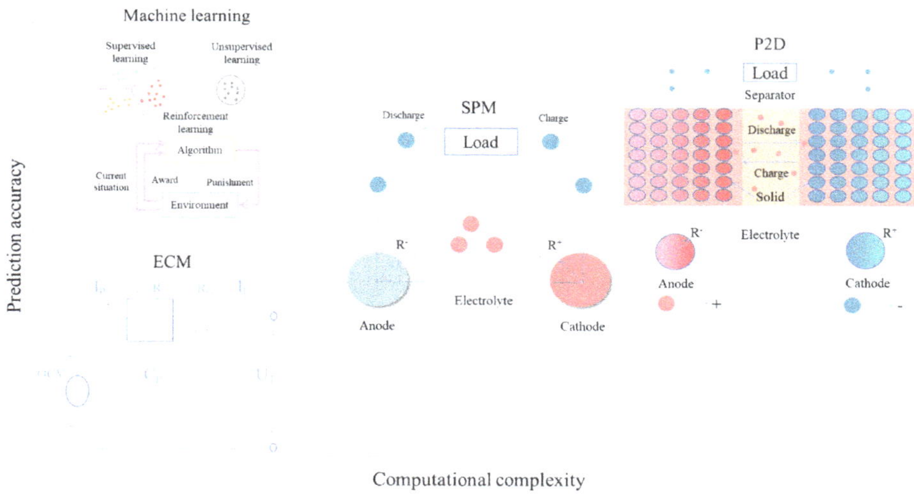

Figure 1. The trade-off between computational cost and model accuracy.

Conversely, the data-driven approach, especially machine-learning-based techniques, displays superior advantages in applications to materials and batteries, from the characterization of the material properties to the non-destructive evaluation of cell performance [10]. Machine learning allows computational models to discover intricate structure in the dataset and capture the statistics of the observational data [11].

The machine learning techniques used to predict the evolution of the battery can be classified into two main categories: traditional machine learning such as kernel-based approaches, and deep learning approaches such as deep neural networks. Conventional machine-learning techniques can be applied to process observational data in their raw form. The learning subsystem in widespread use in the machine learning community, deep or not, is supervised learning—that is, classification and regression. Such practical applications of machine learning use hand-engineered features or raw data for almost all recognition and predictive tasks. For example, extreme gradient boosting (XGBoost) was used to estimate the battery SOC of Li-ion batteries under dynamic loading conditions [12]. The XGboost technique offers a tool to improve the predictive performance by leveraging a set of weak learners and aggregating the outputs of each base model. The application of the multi-step forecasting strategy using XGBoost has been demonstrated to work in capturing, in real-time, the cell dynamics and predicting the terminal voltage and SOC under WLTP driving cycles. Moreover, gradient boosting also shows good performance in battery lifespan prediction [13]. Using the openly shared dataset provided by MIT/Toyota [14], a variety of features including voltage-related, capacity-related and temperature-related features were extracted and constructed for the gradient-boosting regression tree (GBRT) model. Gaussian process regression (GPR) is another common machine learning technique for battery SOC and SOH estimation. For example, battery SOC is identified by using electrochemical impedance spectroscopy (EIS) measurements based on GPR [15]. Through the feature selection from EIS data over frequencies of 1 mHz to 6 kHz, the GPR model can be trained to establish the mapping relationship between the selected features and the SOC under various temperatures. In another study, the GPR technique has been demonstrated as an effective tool to learn nonlinear battery systems and predict capacity fade in a variety of loading scenarios [16]. By introducing a Bayesian non-parametric transition, the model can incorporate estimates of uncertainty into predictions, allowing the determination of varying probabilities of the ranges of possible future health values across a long-term timescale.

Conventional machine learning offers a straightforward and effective tool for classification and regression tasks. However, constructing such a machine-learning system in general requires careful feature engineering and considerable domain-specific expertise to design a feature extractor that can transform the raw data (such as battery voltage, current, etc.) into suitable vector representations from which the learning algorithm could classify or predict patterns in the input.

In recent years, a new learning philosophy is the family of deep learning, which enables a machine to be fed with raw observations in mathematically useful latent spaces and to discover intricate structure in datasets automatically. One popular type of deep learning model is recurrent neural networks (RNNs) and their popular variants, including long-short term memory (LSTM) and gated recurrent units (GRU).

For example, a single hidden-layer GRU-RNN model was designed to estimate battery SOC by using the measured voltage and current [17]. In the proposed gradient method, the weight change direction takes a compromise of the gradient direction at current instant and at historical time to prevent the oscillation of the weight shift and to improve the training speed. Moreover, artificial noise was added to the observational data to improve the generalization and robustness of the neural networks. Recently, a hybrid neural network model was developed for SOC estimation of batteries at low temperatures by coupling a convolutional neural network (CNN) and GRU [18]. The CNN module was applied to learn the feature parameters of the inputs, while the bidirectional weighted GRU offers tools to improve the fitting performance of the network at low operating temperatures by tuning the weights.

In application to battery SOH estimation, a dynamic RNN model with good mapping ability was established for co-estimation of SOC and SOH for a lithium-ion battery [19]. The dynamic RNN model was suitable for estimating the nonlinear and dynamic cell behaviors. Meanwhile, self-adaptive weight particle swarm optimization was applied to improve the performance of the networks. Compared with the traditional gradient descent algorithm, particle swarm optimization offers an opportunity to improve the error convergence speed and avoid local optima. In a recent study, an encoder–decoder model based on the GRU was developed to be suitable for time series prediction of a Li-ion battery. The GRU-based encoder–decoder model has demonstrated its ability to predict the dynamic cell voltage response under complex current load profiles. In contrast to a conventional ECM model, the data-driven deep neural network does not require domain-specific knowledge and time-consuming tests under a well-controlled laboratorial environment.

Collectively, the results from these works demonstrate that RNNs and their variants are effective in modelling and predicting nonlinear battery systems [20]. However, they suffer from limitations due to the sequential processing and challenges related to back-propagation through time, particularly in the modelling of long-range connections across multiple timescales. These are manifested as training instabilities leading to vanishing and exploding back-propagated gradient problems [21]. The transformer model, primarily utilized for natural language processing, has recently achieved remarkable advancements in time series forecasting [22]. The transformer model allows for parallel processing, enabling efficient utilization of computing resources and faster training. Consequently, this methodology can be a promising option for battery state estimation. For example, one study proposed Dynaformer, a new deep learning architecture based on a transformer, which can predict the aging state and full voltage discharge curve for real batteries accurately, using only a limited number of voltage/current samples [23]. The study shows that the transformer-based model is effective for different current profiles and is robust to various degradation levels. Transformers tackle these obstacles by employing self-attention and positional encoding methods that simultaneously focus on and encode the order information while analyzing current data points within the sequence. These methods preserve the sequential information essential for learning while eliminating the traditional concept of recurrence. Transformers are capable of capturing such information through the utilization of multiple attention heads. In addition, the model bridge the gap between simulations and real data, enabling accurate planning and control over missions.

1.1.2. Cloud-Based Progress on Battery State Prediction

Engineers are increasingly exploring the option of outsourcing computing needs to the cloud, which has become popular among researchers dealing with large amounts of data, due to its rapid improvements [24]. For example, one study proposed a cloud-based approach to estimate battery life by analyzing charging cloud data, which includes capacity and internal resistance estimations [25]. The capacity estimation relied on the ampere-hour integral method, which was further improved using temperature data and optimized with the Kalman filter (KF). To increase the precision of the estimation results, the study also implemented fuzzy logic (FL) to manage observation noise. Finally, the battery life was predicted using the Arrhenius empirical model. Experimental results demonstrated that the proposed method exhibited a low error rate of less than 4% in estimating battery life. In another study, the authors proposed a cloud-assisted online battery management method based on AI and edge computing technologies for EVs [26]. A cloud-edge battery management system (CEBMS) was established to integrate cloud computation and big data resources into real-time vehicle battery management. The proposed method utilized a deep-learning-algorithm-based cloud data mining and battery modeling method to estimate the battery's voltage and energy state with high accuracy. The effectiveness of the proposed method was verified by experimental tests, demonstrating its potential for more effective battery use and management in EVs.

Solving real-life physical problems in practical applications can be a daunting task, especially when dealing with multiple sources of uncertainty and imperfect data, including missing or noisy data, and outliers. This study focuses on the potential of using an AI-powered cloud-based framework to predict a nonlinear multiphysics and multiscale electrochemical systems' evolution in real-world applications. The cloud-based, closed-loop framework utilizes machine learning models that can learn seamlessly from field battery data in EV applications [27]. The concept of establishing digital twins for battery systems is an innovative approach to generating longitudinal electronic health records in cyberspace. By creating a digital replica of the battery system and continually training it on a stream of field data, it allows for continual lifelong learning. This approach can lead to significant benefits such as improved robustness, higher accuracy and faster training times. The continual training on a stream of field data enables the digital twin to adapt to changing conditions and learn from real-world experiences. This leads to improved accuracy in predicting the behavior of the battery system, which can help optimize its performance and extend its lifespan. In summary, the establishment of digital twins for battery systems offers a powerful approach to achieving continual lifelong learning, improving the robustness and accuracy of the system and reducing downtime.

1.2. *Major Challenges Involved*

Onboard BMS has long been an important component for EVs in the monitoring and controlling of battery systems. Despite relentless progress, solving real-life battery problems with noisy data and uncertain boundary conditions through traditional approaches remains challenging. Modelling and predicting multiscale and multi-physics battery problems for EV applications require further developments. Challenges specific to battery SOC and SOH estimation will further stimulate the development of new methodologies and frameworks, and we identify four major issues why collaboration between onboard BMS and cloud BMS (Figure 2) is of great importance in achieving the task, as follows.

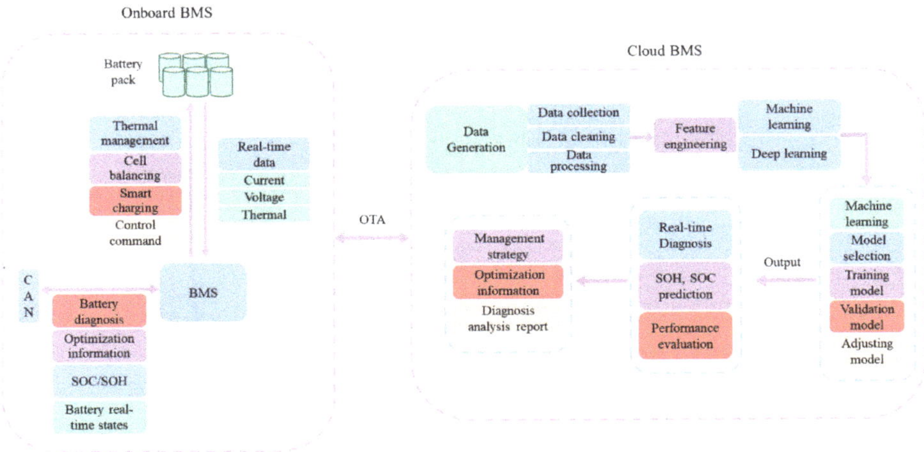

Figure 2. Synergy between onboard BMS and cloud BMS.

(1) Although data-driven machine learning techniques introduce considerable savings in computational cost compared with the traditional numerical methods (e.g., solving PDEs using finite elements), it still requires complex formulations and elaborate computer codes. Performing such tasks requires scheduling the training of computational algorithms in a more powerful computing environment. This is where cloud systems have come into play.

(2) Upon identifying cell conditions in real-world applications, there will be cell-to-cell, pack-to-pack and batch-to-batch variation, even with the most state-of-the-art manufacturing techniques. These cells would exhibit distinct states after long-term incubation. The specific approach to the predictive modelling of such battery systems significantly relies on the amount of data available and on the cell itself.

(3) In field applications such as EVs, the operation of the batteries depends not only on user driving patterns but also on environmental factors. Lab tests cannot incorporate diverse driving cycles and resting periods. Uncertainty arising from the randomness of high-dimensional parameter spaces make it difficult to perfectly match lab experiments to field applications.

(4) Last, but perhaps most important: even with open sharing of test data, reproducibility and generalization issues make it rather challenging to transfer academic progress to industry. However, the cloud-computing system provides a very flexible platform for analyzing, training and developing new frameworks and standardized benchmarks, which can be leveraged to improve our observational, empirical and physical understanding of real-life battery systems in a more intelligent manner.

1.3. Contributions of the Work

Despite relentless progress, predicting the dynamics of nonlinear battery systems by using traditional physical models inevitably faces severe challenges and introduces multiple sources of uncertainty. First-principle, phase-field, atomistic simulations may lead to insights into fundamental battery charging–discharging mechanisms, but they cannot truly predict cell performance for real-world applications [28]. We are well aware of the many benefits from cloud computing and storage. However, obviously, it is not enough to migrate data to a cloud platform; researchers need to be able to interact with the data more seamlessly and intelligently. Contributions of this study are as follows:

(a) Field data, which exhibits irregular loading conditions, dynamic operating scenarios, and path-dependent deterioration processes, is generated and uploaded to the cloud, reflecting real-world usage and making reliable predictions meaningful.

(b) A specialized attention-based transformer neural network model is designed to learn parameters in the high-dimensional stochastic thermodynamic and kinetic battery system. The proposed transformer model has the advantage of strong generalization and robustness in the small data regime.
(c) We examine the evolution of batteries using deep learning approaches in the time-resolved context and demonstrate how transformer neural networks, which automatically extract useful features, have the potential to overcome the limitations that have hindered the widespread adoption of data-driven machine learning-based techniques to date.
(d) The designed cloud-based data-driven framework provides a highly flexible digital solution for a wide range of diverse physical, chemical and electrochemical problems in a way that produces promising results for the target outputs.

2. Key Components of Cloud BMS

Machine learning has emerged as a promising technique, but training an intelligent machine requires plentiful, high-quality and relevant training data. Onboard BMS cannot competently perform this task due to the high computational complexity of the data-driven models [29]. To make the best use of such a flexible technique, a cloud-based BMS provides complementary skills and opportunities to improve our understanding and evaluate comprehensive battery behaviors. Sensor data can be transmitted continuously to the cloud, where machine learning models can learn seamlessly from labeled samples while exploiting the wealth of information in the observations (Figure 3). With the advancement of sensor networks, it is now feasible to monitor the battery system across multiple spatial and temporal scales.

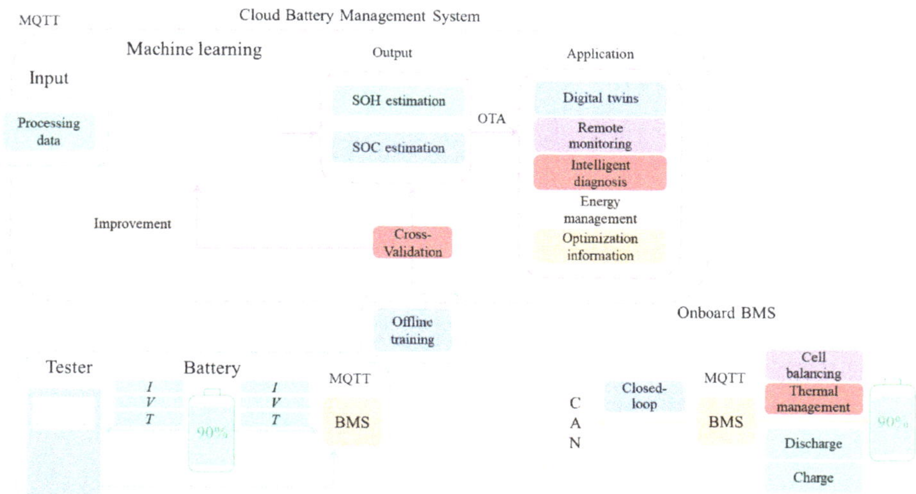

Figure 3. Cloud-based framework for commercial EV applications.

2.1. Physical Entity

In EV applications, lithium-ion batteries encounter complex operating conditions, including stochastic discharging processes for driving, dynamic charging processes for "refilling", and resting processes when parked. While most existing research in battery modeling has focused on either one cell or a specific, well-designed test, these efforts produce insights and improve our understanding of physical systems but cannot fully reflect the real-life situations with diverse aging mechanisms, significant cell-to-cell variability, and complex loading scenarios. The huge gap between lab tests and practical applications makes it challenging to transfer academic progress to engineering. However, by assimilating real sensor measurements to optimize computational models, a digital twin can be used to replicate the behavior of a physical entity in silico. Focusing on analyzing time-resolved battery data such as voltage, current and temperature can directly contribute to meeting certain goals. Ultimately, what matters is the predictive ability under realistic conditions. These three fundamental parameters are the only information that we can obtain from an operating a battery using the onboard BMS.

2.2. IoT

The widespread use of the internet of things (IoT) in end-use devices such as EVs enables a wealth of multi-fidelity observations to be explored across several spatial and temporal scales [30]. There is a growing realization that terminal devices embedded with electronics and connected to networks play a crucial role in monitoring the evolution of complex digital and physical systems. With the prospect of trillions of sensors in the coming decade, it will be possible to seamlessly incorporate multi-fidelity data streams from real-world cases into physical models. In electric vehicle (EV) applications, battery performance, states and mechanical properties can vary greatly with dynamic loading conditions such as charging–discharging current rate, operating voltage window, frequency of usage and temperature. This calls for sophisticated and continuous monitoring throughout the operational lifetime.

Sensor measurements of battery cells can be transmitted to IoT components by the onboard BMS using the Controller Area Network (CAN) protocol. A special IoT protocol, message queuing telemetry transport (MQTT), allows for dual-direction messaging between the device and cloud and requires minimal resources. A large amount of sequential data are generated and collected from both private and fleet vehicles, which can be easily scaled to connect with millions of IoT devices. Data stored in onboard memory can be seamlessly uploaded to the cloud using TCP/IP protocols. The IoT wireless system in modern cities provides infrastructure for real-time data transmission using IoT actuators and onboard sensors.

2.3. Cloud

Cloud storage and computing have been demonstrated as powerful tools for remote monitoring and diagnosis. For automotive industry uses, researchers and engineers can configure their cloud environment and infrastructure to suit their requirements. The cloud-based BMS can seamlessly learn the stream of time-series battery data and produce electronic health records in the cloud. The most popular programming languages for cloud development include Java and Go. In addition, PHP offers a simple, effective, and flexible tool for web developers to create dynamic interfaces and interact with data deluge. The servers in these systems should have high-performance CPUs, plenty of RAM, and fast storage such as solid-state drives (SSDs). Additionally, the storage arrays should have high capacity, high performance and redundancy features such as RAID or replication to ensure data availability and durability. Backup and recovery systems are also critical for protecting customer data in case of disasters or system failures, and they should have high capacity and reliability. Cloud-based digital twins have demonstrated practical value in closed-loop full-lifespan battery management, including material design, cell performance evaluation and system optimization [31].

2.4. Modelling

Despite the progress made on the electrochemical modeling of battery systems using first-principle, atomistic or physics-based methods, the lack of canonical predictive models that can associate cell properties and mechanisms underlying their behavior with cell states has been a bottleneck for widespread adoption. Mathematical and computational tools have been developing rapidly, yet the multiscale and multiphysics battery system behavior dominated by the spatial or temporal context underscores the need for a transformative approach. Machine learning technology is now a successful part of data-driven approaches, addressing a wide range of problems that have resisted the best efforts of the artificial intelligence (AI) community for many years [32]. The availability of shared data and open-source software, along with the ease of automation of materials tools, has brought machine learning into computational frameworks. Several software libraries, including TensorFlow, PyTorch and JAX, are contributing to the determination of cell performance by using various data modalities, such as time series, spectral data, lab tests, field data and more.

3. Methodologies

A learning algorithm that can seamlessly combine data and abstract mathematical operators plays a crucial role in discovering the representations needed for regression or classification. Deep learning techniques, in particular, naturally offer tools for extracting features and patterns from data automatically. To explore the observational data (which are uploaded to a private cloud system) that are characterized by multiple spatial and temporal coverages, a specialized self-attention transformer-based neural network model is designed in this study. Transformer-based deep learning (bidirectional encoder representations from transformers, known as BERT) has received a lot of attention since it was proposed in 2017 [33], particularly in natural language processing [34] and computer vision [35]. In comparison with recurrent neural networks (RNNs), transformer neural networks perform parallelization and solve the long-term dependencies problem and thus can process the observations much more quickly. Inspired by the successful operation, recently, various transformer-based models have been designed in the aspects of time-series prediction and analysis. The core idea of transformer networks is the self-attention mechanism, which belongs to a variant of the attention mechanism that can discover intricate structure in large time-series datasets and reduces dependence on the unimportant information across multiple timescales. In this study, we investigate the use of a transformer and design specialized network architectures that automatically satisfy the physical system for multivariate time series predictive tasks, as shown in Figure 4.

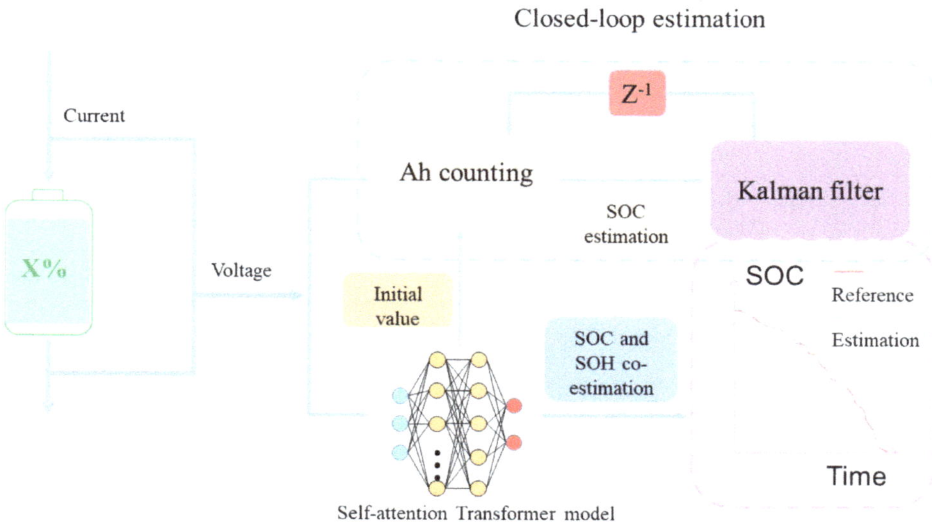

Figure 4. Specialized transformer architecture for the prediction of the battery system.

3.1. Transformer Neural Networks for Co-Estimation

In this study, the transformer deep learning model includes embedding, dual-encoder architecture and a gating mechanism. Time-series data are a collection of samples, observations and features recorded in a sequential manner over time. Self-supervised techniques are utilized to enhance transformers by providing them with the capability to learn, classify and forecast unlabeled data. The embedding vector replaces the original time-series data, i.e., the entire feature vector at a given time step. The specialized dual-encoder architecture is applied to battery prognostic and health management, which shows a higher test accuracy than a typical encoder–decoder architecture. A gating mechanism is used for coupling the predictive results of the two encoders. The success of a data-driven approach for predictive modeling of such real-life battery systems depends heavily on the amount of available data and the complexity of the model. Therefore, the model can be trained offline, and during the system's operational lifetime, online prediction only requires some sampled data points after data preprocessing (Figure 5).

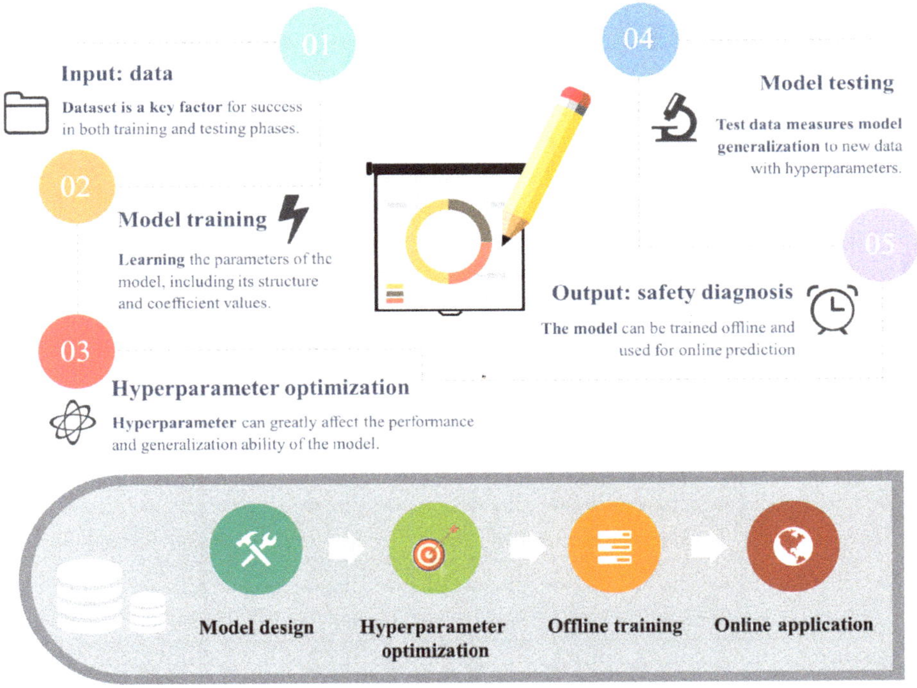

Figure 5. Architecture of the deep learning for co-estimation of battery states.

3.1.1. Data Processing

Data normalization has been demonstrated as a crucial step to a training process in a data-driven machine learning manner. A classical method of data normalization, named Z-score transformation, is used to normalize all the parameters of Li-ion batteries in the dataset into the vectors, characterized by the standard normal distribution with a mean of zero and variance of one. The Z-score method for each feature is calculated as:

$$z_{feature} = \frac{x_{obs.} - \mu}{\sigma} \quad (1)$$

where $x_{obs.}$ is the raw observational data, and μ and σ are the mean and standard deviation of the complete population, respectively.

The Transformer is an encoder-decoder structured sequence-to-sequence model, designed to accept a sequence of observational data as input. Transformers capture this information by employing multiple attention heads. To implement the transformer model efficiently, time-series data are separated into different segments based on the charging and discharging processes using an adaptive-length sliding window. Each time, the deep learning explores the long-range correlations across multiple timescales based on the sensor data inside the sliding window (Figure 6). It follows two rules: (a) the model outputs the SOC every 1 sample points (i.e., 10 s) when the direction of the current flow is constant. (b) the model re-starts the prediction process when the direction of the current flow changes. The sliding window offers a simple and effective tool to capture the structured relationships between the input and output under the charging and discharging processes.

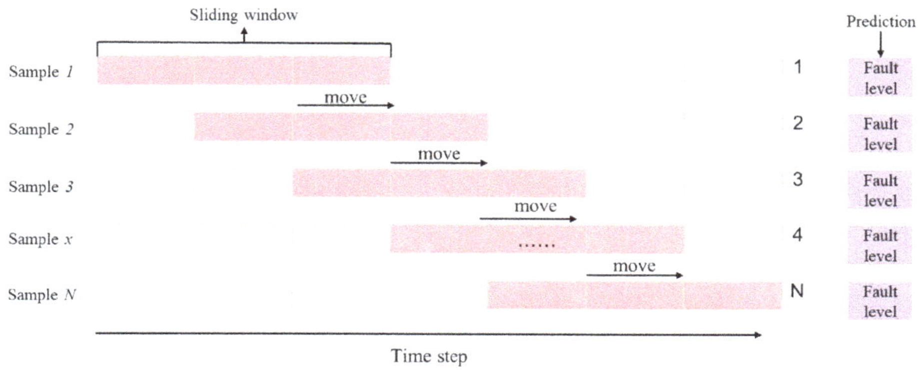

Figure 6. Sliding window mechanism of the transformer model.

3.1.2. Embedding

Unlike the LSTM or RNN models, The transformer model has no recurrence and no convolution. Instead, it models the sequence information using the encoding included in the input embeddings. The embedding of a typical BERT model includes token and position embeddings. By embedding time (seconds, minutes, hours, weeks, years, etc.) into the input, the model can effectively analyze time-series data while utilizing the computational advantages of modern hardware such as GPUs, TPUs and others. In some ways, our embedding strategy is analogous to BERT, but it has unique capabilities and merits for leveraging physical information. The token embedding of the original BERT is a discrete variable (word), while the observational data of our model is a time-series variable (cell parameters) with missing data and sensor noise. Moreover, fine-tuning ensures that the output embedding for each cell condition encodes contextual information that is more relevant to the multiscale and multiphysics battery system. The positional encoding applied to model the sequence information of the battery can be expressed as:

$$\mathrm{PE(t)_i} = \left\{ \begin{array}{c} \mathrm{P_i} = \sin\left(\frac{\mathrm{pos}}{10{,}000^{\frac{2i}{d}}}\right) \\ \mathrm{P_{i+1}} = \cos\left(\frac{\mathrm{pos}}{10{,}000^{\frac{2i}{d}}}\right) \end{array} \right\} \tag{2}$$

where pos is the position in the time-step of the input, and i is the dimension of the embedding vector. It allows the learning algorithm to easily learn to attend by relative positions [33].

3.1.3. Dual Encoder

In a typical transformer model, the encoder block consists of multi-head self-attention modules and position-wise feedforward neural networks. To meet the needs of the battery system, a dual-encoder architecture is designed to produce predictions that respect the physical invariants and principles. The transformer model has made breakthrough progress due to the self-attention mechanism, which offers an effective tool for the automatic extraction of abstract spatio-temporal features automatically. Such new paradigms of pretraining and fine-tuning enable large-scale scientific computations on long-range correlations across multiple timescales and thus enhance the generalization of neural network models. The multi-head attention mechanism allows the transformer model to extract information from different representation subspaces, which offers new opportunities for capturing the subtle differences between different battery cells within the pack.

In the self-attention module, multi-head self-attention sublayers simultaneously transform into query, key and value matrices. A sequence of vectors can be generated from the linear projections of the scaled dot-product attention:

$$\mathrm{Attention}(q_{h,} k_{h,} v_h) = \mathrm{softmax}(\frac{q_h k_h^T}{\sqrt{d_k}}) v_h \tag{3}$$

where $q_h \in R^{n*d_k}$, $k_h \in R^{m*d_k}$, $v_h \in R^{m*d_v}$ represents the query, key and value matrices, respectively; n and m denote the lengths of queries and keys/values, respectively; and d_k and d_v denote the dimensions of keys/queries and values, respectively. The multi-head attention mechanism with $h \in \{0, 1, \cdots, H\}$ different sets of learned projections can be expressed as:

$$\mathrm{Multi\text{-}head\_Attention}(q_{h,} k_{h,} v_h) = \mathrm{Concat}(\mathrm{head}_1, \mathrm{head}_h) \omega^o \tag{4}$$

where

$$\mathrm{head}_h = \mathrm{Attention}(q_h \omega_h^q, k_h \omega_h^k, v_h \omega_h^v) \tag{5}$$

3.2. Data Generation

Machine learning techniques encompass a collection of algorithms, techniques, normative structures and data that enable the derivation of a plausible model directly from observational data. The battery raw data under realistic conditions has been recorded and uploaded to a private cloud server, including cell voltage, current and temperature, which has been used to achieve a number of tasks for prognostics and health management, such as battery failure diagnosis [36] and battery SOH prediction [37]. In EV applications, multivariate time series represent the evolution of a group of variables: voltage, current and temperature over time. Table 1 lists the key cell specifications. The dataset is divided into training and test sets. The training set is utilized to learn the model for developing a base model and improve the accuracy and generalizability of predictions by fine-tuning the model under unseen battery charge-discharge protocols. The test set is used to quantitatively predict cell states: SOC and SOH.

Table 1. Cell chemistry and operating windows.

Parameter	Value
Cell type	Nickel Manganese Cobalt (NMC)
Nominal open circuit voltage	3.6 V
Nominal capacity	135 Ah
Operating voltage window	4.2 V to 2.5 V
Operating current density	Up to 1 C during charging vs. up to 6 C during discharging/driving
Operating temperature window	Less than 45 °C protected by thermal management

3.3. Evaluation Criteria

SOC and SOH are the two most important parameters in the prognostics and predictive health management, which are generally defined as:

$$\mathrm{SOC} = \frac{Ah_{current}}{Ah_{full}} \times 100\% \tag{6}$$

$$\mathrm{SOH} = \frac{Ah_{\mathrm{full\_pre.}}}{Ah_{\mathrm{full\_nom.}}} \times 100\% \tag{7}$$

where $Ah_{current}$ and Ah_{full} are the cell capacity in the present state and its full capacity, respectively, during the specific charging or discharging step, and $Ah_{\mathrm{full\_pre.}}$ and $Ah_{\mathrm{full\_nom.}}$ are the full capacity and nominal capacity, respectively.

The main output of the transformer model in this study is the prediction of SOC and SOH, which is compared with the observed values of the Li-ion cells. Three metrics are used to evaluate model performance, including root mean square error (RMSE), the mean absolute percentage error (MAPE) and the maximum absolute error (MAE). The inputs are the variables that follow a ground truth joint distribution. Specifically, RMSPE is defined as

$$\mathrm{RMSPE} = \sqrt{\frac{1}{n}\sum_{i=1}^{n} (\hat{y}_i - {y_i}^*)^2} \tag{8}$$

where $\hat{y}_i$ and are ${y_i}^*$ the observed and predicted value of the i-th sample in the observational data.

MAPE can be expressed as

$$\mathrm{MAPE} = \frac{1}{n}\sum_{i=1}^{n} \frac{|\hat{y}_i - {y_i}^*|}{y_i} \tag{9}$$

MAE can be given by:

$$\mathrm{MAE} = \max_{1 \leq i << n} |\hat{y}_i^* - y_i^*| \tag{10}$$

4. Performance of Cloud-Based BMS

4.1. SOC Estimation Results

The proposed transformer model was utilized to explore the intricate structures in battery time-series data and identify the representations necessary for predicting cell states. However, please note that solving real-life physical problems with missing, gappy or noisy boundary conditions requires pre-training of transformer models. In this study, tens of cells are randomly collected and used during their operational lifetime to pre-train the model initially. The observational data were fed into the transformer model, and its output was the SOC estimations corresponding to the sampling points (10 s sampling frequency using onboard sensor measurements). Due to the physico-chemical (thermodynamic and kinetic) principles, the model split the time-series data into several segments based on the charging and discharging processes. The model can thus discover intricate structures in two distinct operating conditions and then couple them together.

In the transformer model, multiple self-attention heads are introduced to operate on the same input in parallel. Each head uses distinct weight matrices to extract various levels of correlation between the input data. The transformer dual encoder offers predictive tools to extract vector representations of multivariate time-series, which can be considered as an autoregressive task of denoising the input [34]. The estimation results are independent in the charging and discharging conditions since the transformer model maps the input voltage and current sequences to SOC separately based on the direction of current flow. The initial SOC is calibrated at the time when the cell is fully charged or fully discharged during the system's operational lifetime, which depends on the usage behavior at an uncertain time. Once a precise value of the initial SOC is obtained, ampere-hour (Ah) counting can be directly introduced to provide the ground truth for those observations. Therefore, the model estimates the SOC of the cell from voltage, current and temperature data by coupling the transformer model and the Ah counting method. The transformer-based model is initially trained for Cell_1, and the SOC estimation result is shown in Figure 7. The data-driven model achieves a MAPE of 0.76% and an RMSPE of 0.68%, with a maximum absolute error of less than 2%.

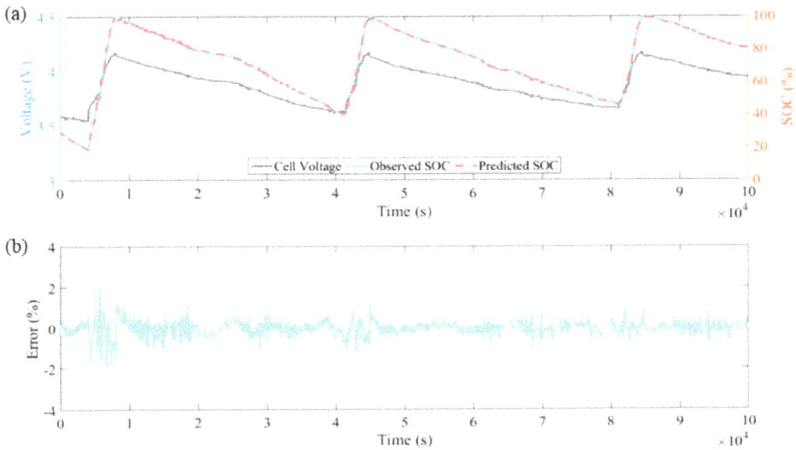

Figure 7. (**a**) SOC estimation results for Cell_1. (**b**) the prediction errors.

Subsequently, the self-attention transformer model is calibrated using another Cell_2, under totally different dynamic operating conditions. Regular calibration and maintenance of machine learning models require significant resources, including specialized personnel and technicians. Ensuring that the system remains accurate and effective necessitates constant attention and care, making it a crucial component of successful machine learning implementation. A good calibration process can be expressed as

$$P(\hat{y} = y | \hat{p} = p) = p \tag{11}$$

where the probability $p \in [0, 1]$ is over the joint distribution, and $\hat{y}$ and $\hat{p}$ are the predictions and the associated confidence (probability of correctness). Let NN be a neural network, and thus it can be given by $NN(x) = (\hat{y}, \hat{p})$. As shown in Figure 8, the developed model can accurately estimate the SOC for the NMC battery (Cell_2) over both the charging and discharging processes with a MAE of less than 2.5%, a MAPE of 0.96% and an RMSPE of 0.81%. The proposed transformer approaches, in particular, provide reliable SOC estimations during the plateau in charge–discharge profiles. While accurate SOC estimation through machine learning modeling is possible, it should also focus on accounting for SOC errors induced by aging, temperature and hysteresis. Despite these factors, data-driven estimation remains a reliable SOC reference for other methods.

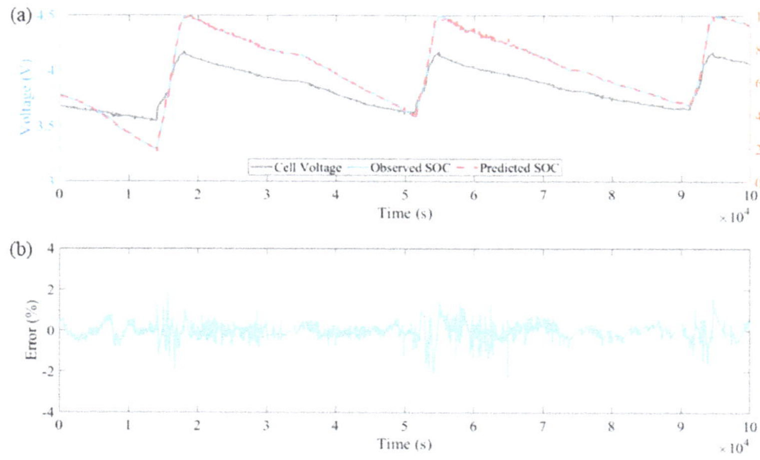

Figure 8. (**a**) SOC estimation results for Cell_2. (**b**) the prediction errors.

4.2. SOH Estimation Results

The definition, based on the Equation (12), is used to calculate the SOH in its current state. The estimation of the capacity can be expressed as:

$$\hat{C}_{mn} = transformer([x_{mn,t}, x_{mn,t+1}, \cdots, x_{mn,t+1+i}] | C_{mj}) \tag{12}$$

where $\hat{C}_{mn}$ is the estimated capacity of the m-th cell in the n-th cycle, $x_{mn,t}$ is the observational data in the n-th cycle, and C_{mj} is the observed value used as the ground truth. In field applications, the ground truth of the capacity cannot be obtained for every cycle. Therefore, $j << n$. The methodologies of the transformer model need to be revisited. For a complete explanation of the algorithm, refer to [22].

Herein, the loss value used to determine the hyperparameters of the self-attention transformer model can be given by:

$$\text{MSE} = \frac{1}{n}\sum_{i=1}^{n} (\hat{y}_i - y_i{}^*)^2 \tag{13}$$

Setting the hyperparameters for a transformer model can be a challenging task, which depends highly on the specific case, including the size and complexity of the training data and the available hardware. The model processes the encoder block's outputs for input into the linear layers. However, concatenation alone may yield poor prediction accuracy. Thus, a dense interpolation algorithm [38] with tunable hyperparameters is adopted to enhance performance. The validity of the trained transformer is demonstrated through the interpolated results in the time-space domain. Despite a decrease in accuracy with increasing feature differences between the test and training data, the proposed method still produces reasonable interpolation results. The trained transformer is then employed to reconstruct dense data with halved trace intervals for the field data. The reconstructed dense data exhibit greater spatial continuity, and the spatial aliasing effects disappear in the time domain. These reconstructed dense data hold the potential to enhance the accuracy of subsequent seismic data processing and inversion.

Hyperparameters are inherent in every machine learning system, and the fundamental objective of automated machine learning (AutoML) is to optimize performance by automatically setting these hyperparameters. Table 2 summarizes the hyperparameters used in this study.

Table 2. Hyperparameters of the transformer model.

Hyperparameter	Value/Method
Layers	10 layers with 8 heads per layer
Training	65 k gradient updates, 4096 sequence length, 64 batch size (262,144 tokens)
Learning Rate	Maximum of 1×10^{-3}, linear warmup of 500 steps
Dropout	0.2 rate
Optimizer	Adam with starting learning rate of 2
Implementation	Use TensorFlow and PyTorch for efficient implementation
Layer Normalization	Normalize input to each layer of the transformer
Weight Tying	Tie decoder and output layer weights
Label Smoothing	Apply label smoothing to target labels
Early Stopping	Stop training when validation loss stops improving

Figures 9 and 10 illustrates the performance of the transformer model in estimating the SOH of cell_1 and cell_2, which are used to train and test the model, respectively. In each group, we randomly selected 50 sampling points, and the results show that the proposed transformer-based model can achieve high predictive accuracy, with MAPE varying between ±2.5% within a 98% confidence interval during the system's operational lifetime. However, the model still needs further development in some field applications. Firstly, more efforts are required, such as hand-engineered feature extraction and establishing ground truth, to label the observational data for training. Secondly, more calibration work is needed to enhance the model's performance in short-length cycles, such as charging from 50% to 80% SOC. As the model lacks sufficient information to learn from, it may fail to provide accurate and physically consistent predictions for each field charging process, especially under random usage behaviors (e.g., charging for only a few minutes). This can lead to extrapolation or observational biases, which can negatively impact the model's performance.

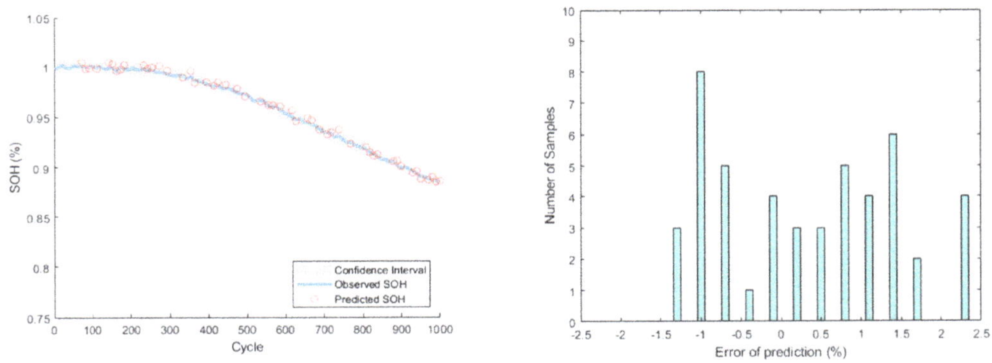

(a) (b)

Figure 9. The training set. (**a**) SOH estimation for Cell_1. (**b**) the prediction errors.

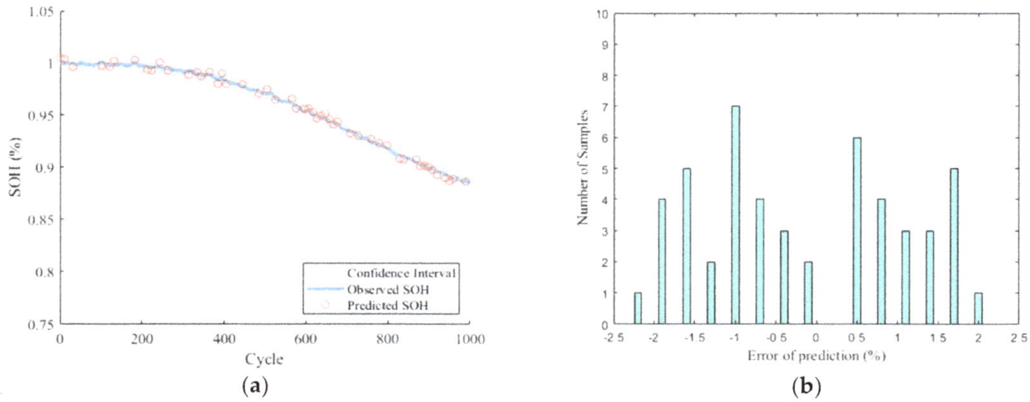

Figure 10. The test set. (**a**) SOH estimation for Cell_2. (**b**) the prediction errors.

5. Outlook

Developing cloud systems for battery and EV applications can pose challenges for many practitioners, but a user-friendly and accessible cloud development environment could help address some key issues. Observational data can be sparse and noisy, and may comprise vastly heterogeneous data modalities such as images, time series, lab tests, historical data, records, and more. The data for certain quantities of interest might not be readily available. To enhance the efficiency and accuracy of these systems, we propose five major recommendations:

(i) There is a significant opportunity for synergy between onboard-BMS and cloud-BMS technologies. Urgent and real-time tasks should be allocated to onboard BMS, while complex tasks that involve multiple scales and temporal dependencies should be distributed to cloud BMS.

(ii) Machine-learning models rely heavily on observational data, and new algorithms and mathematics are needed to yield accurate and robust methods that can handle high signal-to-noise ratios and outliers. These methods should also be able to generalize well beyond the training data. However, the model requires craftsmanship and elaborate implementations on different cell chemistries.

(iii) Battery behavior in EV applications is much more complex than in lab tests due to unprecedented spatial and temporal coverage. Working with noisy data and limited training sets and dealing with under-constrained battery problems with uncertain boundary conditions are major challenges that need to be addressed.

(iv) Developing deep learning architectures for modeling multiscale and multiphysics battery systems is currently done empirically, which is time-consuming. Training and optimizing deep neural networks can also be expensive. Emerging meta-learning techniques and transfer learning may offer promising directions to explore.

(v) Battery performance fluctuates unpredictably throughout its operational life. Precise forecasting and modeling of long-range spatio-temporal dependencies across cell, pack, and system levels are essential for efficient learning algorithms. A promising approach might involve hybrid modeling, combining physical process models with configurable, structured data-driven machine learning.

6. Conclusions

Field data have the potential to enhance the effectiveness of computational techniques developed for cloud-based battery management systems (BMS). In this study, we propose a cloud-based data-driven technique that utilizes state-of-the-art computational methods, specifically transformer neural networks, to accurately model cell behaviors for real-life electric vehicle (EV) applications. Our prediction model automatically extracts spatio-temporal features using an attention-based deep learning approach, without relying on data from experimental test cycles or prior knowledge of cell chemistry and degradation mechanisms. By combining IoT devices to generate field data and machine-learning modeling on the cloud, our work underscores the potential for understanding and forecasting complex physical systems such as lithium-ion batteries. Overall, modeling and estimation using cloud-based BMS can complement other approaches based on simplified battery models (such as equivalent circuit models), physical and semi-empirical models, and specialized diagnostics embedded in the onboard BMS.

Author Contributions: Methodology, Supervision, A.F.B.; Software, J.Z. and H.Z.; Resources, Project administration and Funding acquisition, Y.L.; data curation, J.W.; writing—original draft, D.S. and J.Z.; writing—review & editing, C.E.; visualization, J.Z. and Z.W. All authors have read and agreed to the published version of the manuscript.

Funding: This research was funded by [Independent Innovation Projects of the Hubei Longzhong Laboratory] grant number [2022ZZ-24], [Central Government to Guide Local Science and Technology Development fund Projects of Hubei Province] grant number [2022BGE267], [Basic Research Type of Science and Technology Planning Projects of Xiangyang City] grant number [2022ABH006759] and [Hubei Superior and Distinctive Discipline Group of "New Energy Vehicle and Smart Transportation"] grant number [XKTD072023].

Data Availability Statement: The data could not be shared due to confidentiality.

Acknowledgments: We thank [Independent Innovation Projects of the Hubei Longzhong Laboratory], [Central Government to Guide Local Science and Technology Development fund Projects of Hubei Province], [Basic Research Type of Science and Technology Planning Projects of Xiangyang City] and [Hubei Superior and Distinctive Discipline Group of "New Energy Vehicle and Smart Transportation"] for their financial support in this research.

Conflicts of Interest: The authors declare no conflict of interest.

References

1. Li, M.; Lu, J.; Chen, Z.; Amine, K. 30 years of lithium-ion batteries. *Adv. Mater.* **2018**, *30*, 1800561. [CrossRef] [PubMed]
2. Sulzer, V.; Mohtat, P.; Aitio, A.; Lee, S.; Yeh, Y.T.; Steinbacher, F.; Khan, M.U.; Lee, J.W.; Siegel, J.B.; Howey, D.A. The challenge and opportunity of battery lifetime prediction from field data. *Joule* **2021**, *5*, 1934–1955. [CrossRef]
3. Gibney, E. Europe sets its sights on the cloud: Three large labs hope to create a giant public–private computing network. *Nature* **2015**, *523*, 136–138. [CrossRef] [PubMed]
4. Bosch Mobility Solutions: Battery in the Cloud. Available online: https://www.bosch-mobility-solutions.com/en/solutions/software-and-services/battery-in-the-cloud/battery-in-the-cloud/ (accessed on 19 August 2022).
5. Panasonic Announces UBMC Service: A Cloud-Based Battery Management Service to Ascertain Battery State in Electric Mobility Vehicles. Available online: https://news.panasonic.com/global/press/data/2020/12/en201210-1/en201210-1.pdf (accessed on 19 August 2022).
6. HUAWEI: CloudLi. Available online: https://carrier.huawei.com/en/products/digital-power/telecom-energy/Central-Office-Power (accessed on 19 August 2022).
7. National Monitoring and Management Platform for NEVs. Available online: http://www.bitev.org.cn/a/48.html (accessed on 19 August 2022).
8. Zheng, Y.; Ouyang, M.; Han, X.; Lu, L.; Li, J. Investigating the error sources of the online state of charge estimation methods for lithium-ion batteries in electric vehicles. *J. Power Sources* **2018**, *377*, 161–188. [CrossRef]
9. Hu, X.; Xu, L.; Lin, X.; Pecht, M. Battery lifetime prognostics. *Joule* **2020**, *4*, 310–346. [CrossRef]
10. Berecibar, M. Machine-learning techniques used to accurately predict battery life. *Nature* **2019**, *568*, 325–326. [CrossRef] [PubMed]
11. Ng, M.F.; Zhao, J.; Yan, Q.; Conduit, G.J.; Seh, Z.W. Predicting the state of charge and health of batteries using data-driven machine learning. *Nat. Mach. Intell.* **2020**, *2*, 161–170. [CrossRef]
12. Dineva, A.; Csomós, B.; Sz, S.K.; Vajda, I. Investigation of the performance of direct forecasting strategy using machine learning in State-of-Charge prediction of Li-ion batteries exposed to dynamic loads. *J. Energy Storage* **2021**, *36*, 102351. [CrossRef]
13. Yang, F.; Wang, D.; Xu, F.; Huang, Z.; Tsui, K.L. Lifespan prediction of lithium-ion batteries based on various extracted features and gradient boosting regression tree model. *J. Power Sources* **2020**, *476*, 228654. [CrossRef]
14. Severson, K.A.; Attia, P.M.; Jin, N.; Perkins, N.; Jiang, B.; Yang, Z.; Chen, M.H.; Aykol, M.; Herring, P.K.; Braatz, R.D. Data-driven prediction of battery cycle life before capacity degradation. *Nat. Energy* **2019**, *4*, 383–391. [CrossRef]
15. Babaeiyazdi, I.; Rezaei-Zare, A.; Shokrzadeh, S. State of charge prediction of EV Li-ion batteries using EIS: A machine learning approach. *Energy* **2021**, *223*, 120116. [CrossRef]

16. Richardson, R.R.; Osborne, M.A.; Howey, D.A. Battery health prediction under generalized conditions using a Gaussian process transition model. *J. Energy Storage* **2019**, *23*, 320–328. [CrossRef]
17. Jiao, M.; Wang, D.; Qiu, J. A GRU-RNN based momentum optimized algorithm for SOC estimation. *J. Power Sources.* **2020**, *459*, 228051. [CrossRef]
18. Cui, Z.; Kang, L.; Li, L.; Wang, L.; Wang, K. A hybrid neural network model with improved input for state of charge estimation of lithium-ion battery at low temperatures. *Renew. Energy* **2022**, *198*, 1328–1340. [CrossRef]
19. Che, Y.; Liu, Y.; Cheng, Z. SOC and SOH identification method of li-ion battery based on SWPSO-DRNN. *IEEE J. Emerg. Sel. Top. Power Electron.* **2020**, *9*, 4050–4061. [CrossRef]
20. Schmitt, J.; Horstkötter, I.; Bäker, B. Electrical lithium-ion battery models based on recurrent neural networks: A holistic approach. *J. Energy Storage* **2023**, *58*, 106461. [CrossRef]
21. Sherstinsky, A. Fundamentals of recurrent neural network (rnn) and long short-term memory (lstm) network. *Phys. D Nonlinear Phenom.* **2020**, *404*, 132306. [CrossRef]
22. Ahmed, S.; Nielsen, I.E.; Tripathi, A.; Siddiqui, S.; Rasool, G.; Ramachandran, R.P. Transformers in time-series analysis: A tutorial. *arXiv* **2022**, arXiv:2205.01138.
23. Biggio, L.; Bendinelli, T.; Kulkarni, C.; Fink, O. Dynaformer: A Deep Learning Model for Ageing-aware Battery Discharge Prediction. *arXiv* **2022**, arXiv:2206.02555.
24. Drake, N. Cloud computing beckons scientists. *Nature* **2014**, *509*, 543–544. [CrossRef]
25. Li, K.; Zhou, P.; Lu, Y.; Han, X.; Li, X.; Zheng, Y. Battery life estimation based on cloud data for electric vehicles. *J. Power Sources* **2020**, *468*, 228192. [CrossRef]
26. Li, S.; He, H.; Wei, Z.; Zhao, P. Edge computing for vehicle battery management: Cloud-based online state estimation. *J. Energy Storage* **2022**, *55*, 105502. [CrossRef]
27. Zhao, J.; Nan, J.; Wang, J.; Ling, H.; Lian, Y.; Burke, A. Battery Diagnosis: A Lifelong Learning Framework for Electric Vehicles. In Proceedings of the 2022 IEEE Vehicle Power and Propulsion Conference (VPPC), Merced, CA, USA, 1–4 November 2022; IEEE: New York, NY, USA, 2022; pp. 1–6.
28. Aykol, M.; Herring, P.; Anapolsky, A. Machine learning for continuous innovation in battery technologies. *Nat. Rev. Mater.* **2020**, *5*, 725–727. [CrossRef]
29. Zhao, J.; Burke, A.F. Electric Vehicle Batteries: Status and Perspectives of Data-Driven Diagnosis and Prognosis. *Batteries* **2022**, *8*, 142. [CrossRef]
30. Mohammadi, F.; Rashidzadeh, R. An overview of IoT-enabled monitoring and control systems for electric vehicles. *IEEE Instru. Meas Mag.* **2021**, *24*, 91–97. [CrossRef]
31. Yang, S.; He, R.; Zhang, Z.; Cao, Y.; Gao, X.; Liu, X. CHAIN: Cyber hierarchy and interactional network enabling digital solution for battery full-lifespan management. *Matter* **2020**, *3*, 27–41. [CrossRef]
32. Lecun, Y.; Bengio, Y.; Hinton, G. Deep learning. *Nature* **2015**, *521*, 436–444. [CrossRef]
33. Vaswani, A.; Shazeer, N.; Parmar, N.; Uszkoreit, J.; Llion, J.; Gomez, A.N.; Kaiser, L.; Polosukhin, I. Attention is All you Need. *Adv. Neural. Inf. Process Syst.* **2017**, *30*, 3–5.
34. Wolf, T.; Debut, L.; Sanh, V.; Chaumond, J.; Delangue, C.; Moi, A.; Cistac, P.; Rault, T.; Louf, R.; Rush, A.M. Transformers: State-of-the-art natural language processing. In Proceedings of the 2020 Conference on Empirical Methods in Natural Language Processing: System Demonstrations, Online, 1 October 2020; EMNLP: New York, NY, USA, 2020; pp. 38–45.
35. Liu, Z.; Lin, Y.; Cao, Y.; Hu, H.; Wei, Y.; Zhang, Z.; Lin, S.; Guo, B. Swin transformer: Hierarchical vision transformer using shifted windows. In Proceedings of the IEEE/CVF International Conference on Computer Vision (ICCV), Montreal, QC, Canada, 10–17 October 2021; IEEE: New York, NY, USA, 2021; pp. 10012–10022.
36. Zhao, J.; Ling, H.; Wang, J.; Burke, A.F.; Lian, Y. Data-driven prediction of battery failure for electric vehicles. *Iscience* **2022**, *25*, 104172. [CrossRef]
37. Zhao, J.; Ling, H.; Liu, J.; Wang, J.; Burke, A.F.; Lian, Y. Machine learning for predicting battery capacity for electric vehicles. *eTransportation* **2023**, *15*, 100214. [CrossRef]
38. Wang, B.; Zhang, N.; Lu, W.; Wang, J. Deep-learning-based seismic data interpolation: A preliminary result. *Geophysics* **2019**, *84*, 11–20. [CrossRef]

energies

MDPI

Review

Cloud-Based Artificial Intelligence Framework for Battery Management System

Dapai Shi [1,2,†], Jingyuan Zhao [3,*,†], Chika Eze [4], Zhenghong Wang [2], Junbin Wang [5], Yubo Lian [5] and Andrew F. Burke [3,*]

1 Hubei Longzhong Laboratory, Hubei University of Arts and Science, Xiangyang 441000, China
2 Hubei Key Laboratory of Power System Design and Test for Electrical Vehicle, Hubei University of Arts and Science, Xiangyang 441053, China
3 Institute of Transportation Studies, University of California-Davis, Davis, CA 95616, USA
4 Department of Mechanical Engineering, University of California, Merced, CA 95343, USA
5 BYD Automotive Engineering Research Institute, Shenzhen 518118, China
* Correspondence: jyzhao@ucdavis.edu (J.Z.); afburke@ucdavis.edu (A.F.B.)
† These authors contributed equally to this work.

Abstract: As the popularity of electric vehicles (EVs) and smart grids continues to rise, so does the demand for batteries. Within the landscape of battery-powered energy storage systems, the battery management system (BMS) is crucial. It provides key functions such as battery state estimation (including state of charge, state of health, battery safety, and thermal management) as well as cell balancing. Its primary role is to ensure safe battery operation. However, due to the limited memory and computational capacity of onboard chips, achieving this goal is challenging, as both theory and practical evidence suggest. Given the immense amount of battery data produced over its operational life, the scientific community is increasingly turning to cloud computing for data storage and analysis. This cloud-based digital solution presents a more flexible and efficient alternative to traditional methods that often require significant hardware investments. The integration of machine learning is becoming an essential tool for extracting patterns and insights from vast amounts of observational data. As a result, the future points towards the development of a cloud-based artificial intelligence (AI)-enhanced BMS. This will notably improve the predictive and modeling capacity for long-range connections across various timescales, by combining the strength of physical process models with the versatility of machine learning techniques.

Keywords: lithium-ion battery; battery management system; machine learning; cloud; artificial intelligence; state of charge; state of health; safety; field; real-world application

Citation: Shi, D.; Zhao, J.; Eze, C.; Wang, Z.; Wang, J.; Lian, Y.; Burke, A.F. Cloud-Based Artificial Intelligence Framework for Battery Management System. *Energies* **2023**, *16*, 4403. https://doi.org/10.3390/en16114403

Academic Editor: Simone Barcellona

Received: 4 April 2023
Revised: 16 May 2023
Accepted: 26 May 2023
Published: 30 May 2023

1. Introduction

The transportation sector's shift towards electrification is crucial for reducing carbon emissions and improving air quality [1]. Improving battery performance will enhance the benefits of electrifying transportation. Lithium-ion batteries have undergone significant advancements over the past decade [2], but proper evaluation and management practices are still lacking [3]. The widespread adoption of battery-powered electric vehicles (EVs) has been hindered by numerous challenges, including range anxiety [4] and battery aging [5]. The implementation of onboard battery management systems (BMS) provides tools to address these issues by determining the state of charge (SOC) and state of health (SOH) of the battery as well as the thermal management and cell balancing during the system's operational lifetime [6–8]. Throughout the last decade, significant strides have been accomplished in reaching this objective via the evolution of sophisticated learning algorithms [9]. However, despite advances in multiphysics and multiscale battery modeling, seamless integration of academic progress into existing onboard BMS remains a challenge [10]. The computing ability of the onboard BMS is constrained by factors such as cost, power consumption, and size limitations [11].

The onboard BMS for EV applications requires compact and energy-efficient systems, limiting the processing power that can be incorporated. Furthermore, the high cost of advanced processors and components may be a significant hurdle, particularly in cost-sensitive automotive applications. Consequently, BMS is designed to execute essential tasks like battery cell monitoring and balancing, which do not demand extensive computing power. However, the accuracy of predicting battery characteristics under real-life operational conditions such as aging and dynamic environments is often limited. This is largely attributed to the calibration of the model under laboratory-controlled conditions, which may not accurately reflect the complex and varied conditions experienced in the field.

Recent developments in statistical modeling and machine learning present exciting opportunities for predicting cell behaviors by distilling key characteristics from an immense volume of multi-fidelity observational data [12,13]. Nonetheless, these advanced learning techniques often necessitate meticulous design and complex execution. Before implementation, it is essential to develop a comprehensive solution, and a cloud-based digital solution may be a viable option [14,15]. In recent years, general-purpose Central Processing Units (CPUs) that power cloud server farms have replaced specialized mainframe processors [16], providing researchers and start-up companies with access to public computing resources from commercial providers such as Amazon, Google, and Microsoft [17]. The EV and energy storage industries have also embraced this trend, with companies such as Bosch [18], Panasonic [19], and Huawei [20] launching cloud-based software, referred to as software as a service (SaaS). For instance, Bosch's 'battery in the cloud' SaaS offering, through leveraging vast data from vehicle fleets to create digital twins, promises to enhance battery life cycles by 20%. Meanwhile, Panasonic's Universal Battery Management Cloud (UBMC) service aspires to discern cell state and optimize battery operations. Huawei's SaaS, on the other hand, offers a public cloud computing and storage service tailored for EV companies. This service utilizes a purely data-driven model, embedded in its cloud monitoring system, aiming to predict cell faults by uncovering complex patterns within extensive EV battery datasets. On a broader scale, China has established the National Monitoring and Management Platform for New Energy Vehicles (NMMP-NEV) [21]. This expansive data platform provides remote fault diagnosis for over 6 million EVs.

In this review, we start by providing an overview of the functions and techniques utilized for onboard BMS, as discussed in Section 2. We then delve into the key technologies employed in cloud BMS in Section 3, followed by a comprehensive analysis of artificial intelligence (AI) and machine learning (ML) applications for battery state prediction in Section 4. Given the rapidly evolving nature of this field, we also offer insights into its current limitations and future directions.

2. Onboard BMS

For large-scale EV or grid-scale energy storage applications, BMS is a technology that monitors the performance of a battery system, which is typically composed of multiple battery cells arranged in a matrix configuration [22,23]. BMS ensures that the battery system can reliably work within a targeted range of voltage and current for a specific duration of time, even under varying load conditions. By monitoring the battery's system operations, BMS helps to keep operating conditions under control and stabilize employment. BMS can process and analyze data from various sensors and control algorithms in real-time and aims to improve performance and ensure safe operation by adjusting battery parameters [24]. BMS technology is essential for many applications, including EVs, renewable energy systems, and portable electronics, and is continually evolving to meet the demands of increasingly sophisticated battery systems. However, BMS systems typically have limited computing power and data storage capacity. The onboard BMS presently cannot be used as a specialized technology designed to optimize battery performance but rather a general-purpose computing system used to manage the battery system under a given program.

Estimation of SOC and SOH, thermal management, cell balancing, and so on are the main functions of the onboard BMS (Figure 1). An onboard BMS is a dedicated hardware and software system installed directly within the battery pack of an EV. It monitors and controls various parameters such as voltage, current, temperature, and SOC for individual cells or the entire battery pack. The primary objectives of an onboard BMS are to ensure safe and efficient operation, optimize battery performance, extend battery life, and prevent thermal runaway or other hazardous conditions. The onboard BMS communicates with other vehicle systems and provides real-time information to the driver or user.

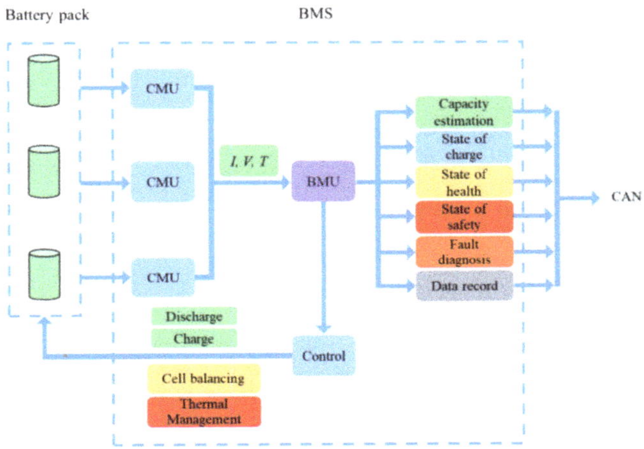

Figure 1. Onboard BMS for field applications (abbreviations: CMU, Communication Management Unit; BMU, Battery Management Unit).

2.1. SOC

A crucial function of an onboard BMS is to precisely ascertain the SOC. Essentially, SOC represents the comparison of the battery's current capacity to its fully charged capacity, serving as an equivalent to the stored charge measured in Coulombs. SOC in battery management is generally defined as:

$$\mathrm{SOC} = \frac{Ah_{\mathrm{cur.}}}{Ah_{\mathrm{ful.}}} \times 100\% \tag{1}$$

where $Ah_{\mathrm{cur.}}$ represents the battery's capacity in its present state, while $Ah_{\mathrm{ful.}}$ denotes the battery's capacity when fully charged. The Ampere-hour (Ah) counting [25] and open-circuit voltage (OCV) [26] are commonly used for onboard BMS due to their low computational complexity. However, it is susceptible to certain limitations that impact the accuracy of Ah counting, including erroneous SOC initialization, drifts caused by current sensor noise, and battery capacity variations. Furthermore, the OCV can only be accurately gauged when the battery is not in use, which hinders its ability to provide real-time SOC estimates during operation. In BMS for EVs, equivalent circuit models (ECMs) are chiefly used because of their lower computational demand and fewer input requirements than electrochemical models. Utilizing networks of resistors and capacitors, ECMs simulate cell behavior tied to diffusion and charge-transfer processes [27]. Hence, they serve as a pragmatic approach for real-time operation and management of onboard battery systems in EVs. Early and typical examples of ECMs are the Rint model, Randles model, Thevenin model, etc. Despite their computational efficiency, most equivalent circuit models (ECMs) have limited accuracy in predicting battery characteristics, particularly during complex loading conditions and cell aging. This limitation is due to the fact that model parameters are designed based on laboratory conditions and often lack the incorporation of multifrequency impedance measurements [28–30].

The SOC of a battery is a crucial parameter for field applications since it signifies the remaining energy capacity within the battery system. The necessity for precise and real-time monitoring of SOC is underlined by several reasons:

(a) Range estimation: SOC is a primary factor in determining the remaining driving range of an EV. By continuously monitoring the SOC, drivers can better plan their trips and avoid anxiety.
(b) Optimal battery performance: Maintaining the battery within an optimal SOC range helps preserve its health and prolong its life. Operating the battery at extreme SOC levels (either too high or too low) can accelerate battery degradation and reduce its overall lifespan.
(c) Charging management: Knowledge of the current SOC is crucial for optimizing charging strategies. It allows for better estimation of the required charging time and enables the use of smart charging algorithms that can balance the charging load on the grid and minimize charging costs.
(d) Energy management: SOC information is vital for the efficient management of energy consumption in EVs. The onboard energy management system uses SOC data to optimize power distribution between various vehicle systems, ensuring efficient use of energy and enhancing overall performance.
(e) Diagnostics and prognostics: Monitoring SOC over time, along with other battery parameters, can provide valuable insights into the battery's health and aid in the early detection of potential issues. This can help prevent unexpected battery failures and enable predictive maintenance, minimizing downtime and maintenance costs.

2.2. SOH

The SOH describes the capacity of a fully charged battery relative to its nominal capacity at the point of manufacture when it was brand new. Upon manufacturing, a battery's State of Health (SOH) starts at 100% and diminishes to 80% at its end of life (EOL). Within the battery manufacturing industry, EOL is typically characterized as the stage when the actual capacity at full charge dwindles to 80% of its initial nominal value. The count of charge/discharge cycles left until the battery attains its EOL is denoted as the battery's Remaining Useful Life (RUL). Consequently, SOH can be articulated as:

$$\mathrm{SOH} = \frac{Ah_{\mathrm{ful.}}}{Ah_{\mathrm{nom.}}} \times 100\% \tag{2}$$

where $Ah_{\mathrm{nom.}}$ represents the nominal capacity of the battery when it is brand new.

Battery degradation is a complex issue that involves numerous electrochemical reactions taking place in the anode, cathode, and electrolyte [31,32]. The operating conditions have a critical impact on the degradation process and ultimately impact the battery lifetime. Predicting the remaining battery lifespan with precision under a variety of operating conditions is of utmost importance to ensure reliable performance and timely maintenance, as well as for battery second-life applications [33]. Onboard SOH estimation is used to determine the health of a battery system during its operating lifetime. The battery capacity frequently serves as a health indicator, given its association with the energy storage potential of batteries and its immediate influence on the remaining operational duration and overall lifespan of the batteries. Computational tools have provided insights into fundamental battery physics, but despite the advances in first principles and atomistic calculations, they are unable to accurately predict battery performance under realistic conditions. As is the case for SOC estimation for online applications, the most commonly used onboard SOH estimation methods are ECMs with limited accuracy. Data-driven approaches can provide a better nonlinear fitting capability [34–37]. However, due to the computational complexity, it is challenging to make most existing advanced methods widespread and practical. This could potentially be attributed to the substantial computational resources required to accurately estimate the SOH of a battery, especially when it is exposed to various operating and environmental conditions during its lifetime. In addition, the need to continuously monitor and analyze battery performance can place a significant burden on the vehicle's onboard system and affect overall vehicle performance.

In field applications such as EVs, battery SOH provides an indication of the battery's overall condition and its remaining useful life. Precise and timely evaluation of SOH is crucial for various reasons:

(a) Battery life prediction: Monitoring SOH allows for better estimation of the battery's RUL, enabling vehicle owners and fleet managers to plan for battery replacements or upgrades, thus minimizing unexpected downtime and associated costs.
(b) Performance optimization: As a battery degrades, its capacity and power capabilities decrease, affecting the vehicle's range, acceleration, and overall performance. By keeping track of the battery's SOH, the energy management system can optimize the power distribution among various vehicle systems, ensuring consistent performance and preserving battery life.
(c) Safety assurance: A deteriorating battery may pose safety risks, such as an increased probability of thermal runaway events, which can lead to fires or explosions. Monitoring SOH can help identify potential safety hazards early, allowing for preventive measures to be taken in case of anomalous capacity degradation.
(d) Charging management: Knowledge of the battery's SOH is vital for adapting charging strategies that account for its current condition. As battery health declines, charging algorithms can be adjusted to minimize further degradation and maintain safe operation.
(e) Warranty management: SOH information can be used by manufacturers to manage warranty claims more effectively and ensure that battery performance remains within the specified warranty limits.
(f) Second-life applications: Accurate SOH assessment can facilitate the identification of batteries suitable for applications in their second life, like stationary energy storage systems, once their performance in EVs has degraded below acceptable levels.
(g) Residual value estimation: The SOH is a pivotal factor in establishing the residual value of an EV in the used vehicle market, as it directly impacts the battery's remaining useful life and the vehicle's overall performance.

2.3. Thermal Management

The thermal management system plays a crucial role in ensuring optimal performance and longevity of the battery system [38]. It aims to maintain an average temperature that balances performance and life, as determined by the battery manufacturer. The thermal management system should fulfill the requirements outlined by the vehicle manufacturer, such as compactness, affordability, reliability, easy installation, low energy consumption, accessibility for maintenance, compatibility with varying climate conditions, and provision for ventilation as needed [39]. Various methods can be employed for these processes, such as the utilization of air for temperature regulation and ventilation, liquid for thermal control, insulation for maintaining temperature levels, and phase change materials for thermal storage. Alternatively, a combination of these methods can be used. The approach can be either passive, where it relies on environmental conditions, or active, where an internal source is employed for heating or cooling purposes. The battery's electronic control unit manages the control strategy.

The battery's temperature directly affects its discharge power, energy, and charge acceptance during regenerative braking, which can impact the vehicle's fuel economy and driving experience. Moreover, temperature significantly determines the battery's lifespan [40]. Hence, batteries ought to function within a temperature spectrum that's optimal for their electrochemical processes, as specified by the manufacturer. However, this range may be narrower than the vehicle's specified operating range, as determined by the manufacturer. For instance, the optimal operating temperature window for a lithium-ion battery is typically between 20 °C and 40 °C. However, the operating temperature range for an EV battery system might be much wider, extending from as low as −20 °C to as high as 55 °C. Hence, thermal management is crucial for EV applications, and can be summarized for the following reasons:

(a) Performance optimization: Maintaining optimal temperature ranges for battery cells is essential for achieving peak performance levels, ensuring efficient energy utilization, and extending the driving range of electric vehicles.
(b) Safety: Effective thermal management helps prevent thermal runaway, which can lead to battery fires or explosions. By closely monitoring and regulating the temperature, potential hazards can be mitigated.
(c) Battery life extension: Prolonged exposure to extreme temperatures can degrade battery materials, leading to a reduction in overall battery life. Proper thermal management helps maintain the battery within its optimal operating temperature range, thus prolonging its lifespan.
(d) Charging efficiency: Effective thermal management enables faster and more efficient charging of batteries by minimizing temperature-related inefficiencies and maintaining safe charging conditions.
(e) Consistent performance: By maintaining consistent temperature conditions within the battery pack, thermal management systems ensure that the battery's performance remains stable and predictable, regardless of external environmental factors.

At present, EVs do not have temperature information for every cell within the battery pack due to practical constraints associated with the large number of required sensors. Adding more sensors and wiring can increase the battery pack's weight and complexity, leading to reduced vehicle efficiency and performance. Additionally, the cost of adding more sensors and wiring can be prohibitively expensive for mass-produced EVs that need to be cost-effective for consumers. Hence, EVs typically rely on strategically placed temperature sensors within the battery pack to provide an overall temperature reading, rather than individual readings for each cell. To enhance the batteries' safe operation, one possible way is to develop advanced data-driven learning algorithms that leverage time-resolved data (voltage and current). However, this comes at the cost of computing efficiency losses.

2.4. Cell Balancing

A crucial element of electric vehicles, the battery balancing system (BBS), is composed of two main components: the balancing circuitry and the control strategy governing the balancing process [41]. To transfer charge between cells and maintain balance, the balancing circuit can be either passive or active [42]. Passive balancing relies on resistance to convert excess energy in a high-charge cell into heat, which is then dissipated until the charge is equalized with a low-charge cell. Passive balancing is simple, cost-effective, and easy to implement, but it generates substantial heat and has low balancing efficiency. Unlike passive balancing, active balancing utilizes energy carriers to transfer energy from the cells with high SOC to the cells with low SOC inside the battery pack. While active balancing is the preferred choice for applications that operate at high temperatures and require rapid balancing, it does increase the complexity of the circuit.

The architectural composition of the BBS holds paramount importance, yet the balancing control strategy it employs shares an equally critical role in dictating the circuit's overall conversion efficiency and speed of balance, as indicated by references [43,44]. These balancing tactics can be classified according to the control variable they utilize, such as the state of charge (SOC), cell voltage, or capacity. Creating an optimized balancing control strategy tailor-made for a specific battery system and its corresponding application is vital to guaranteeing a balanced operation that is both efficient and effective.

The merits of utilizing cell balancing span several areas, including:

(a) Capacity utilization: Cell balancing ensures that all cells within a battery pack are utilized to their full capacity, maximizing the overall energy storage and extraction capability. This, in turn, optimizes the vehicle's driving range and performance.
(b) Lifespan extension: Imbalances in cell voltages can lead to some cells aging faster than others, ultimately reducing the overall battery pack's lifespan. Cell balancing helps equalize the charge and discharge cycles across all cells, promoting even wear and prolonging the battery pack's life.
(c) Safety enhancement: Unbalanced cells can cause overcharging or over discharging, which may lead to thermal runaway and other safety risks. Cell balancing prevents these issues by ensuring that all cells are charged and discharged within their safe operating limits.
(d) Performance consistency: Cell imbalances can result in inconsistent performance and reduced efficiency. By maintaining balanced cells, the battery pack can deliver predictable and stable performance, improving the overall driving experience.
(e) Reduced Maintenance: Employing cell balancing can minimize the frequency of maintenance checks and services. By ensuring uniformity in cell usage, the system reduces the possibility of individual cell failures and maintains the overall health of the battery pack. This, in turn, lowers maintenance costs and offers greater convenience to the user.

3. Key Components and Technologies of Cloud-BMS

Machine learning has emerged as a powerful instrument; however, it necessitates substantial quantities of high-quality and pertinent observational samples. The computational complexity associated with this requirement surpasses the capabilities of onboard Battery Management Systems (BMS). Cloud-based BMS (Figure 2) provides a brand-new digital solution, as it can process and analyze data in a more efficient and flexible manner. Sensor measurements can be uploaded to the cloud, enabling machine learning to continually learn from these data points while harnessing the vast wealth of information present in the samples. A cloud BMS enables remote monitoring, diagnostics, and even predictive maintenance, improving overall battery management and reducing the need for manual inspections or on-site intervention. The cloud BMS can also facilitate fleet management by aggregating data from multiple vehicles or energy storage systems, allowing operators to optimize energy consumption and plan maintenance schedules more efficiently. Additionally, a Cloud BMS can enable over-the-air (OTA) updates to the onboard BMS firmware and algorithms, further enhancing battery performance and extending its lifespan.

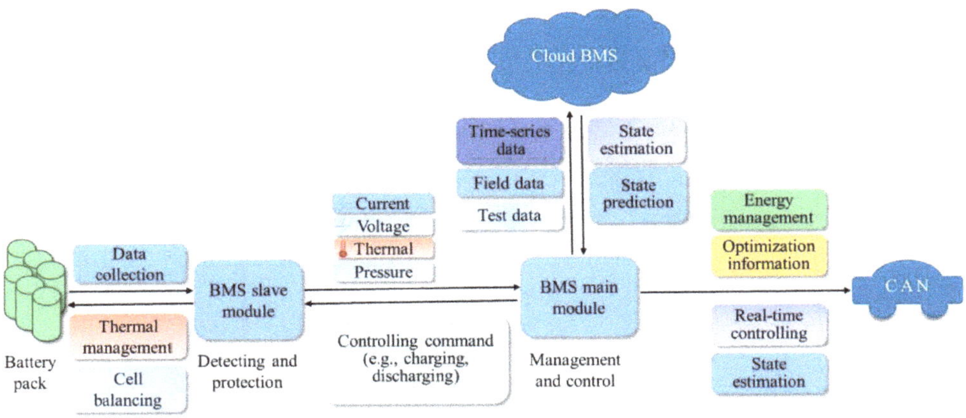

Figure 2. Cloud-based framework for battery management in EV applications.

3.1. IoT-Devices

Given the extensive embrace of Internet of Things (IoT) technology [45], end-use devices have gained the ability to collect and analyze vast amounts of data across various spatial and temporal scales. Equipped with electronics and network connectivity, these devices hold a key position in monitoring and management. As the number of sensors is expected to reach trillions in the near future, integrating data streams with diverse levels of fidelity into real-world applications and battery models becomes increasingly feasible.

The physical, chemical, and electrochemical performance of batteries can exhibit significant variations due to dynamic loading conditions such as current rate, operating voltage, temperature, and more. Consequently, continuous monitoring throughout the operational lifetime is of paramount importance [46]. The onboard Battery Management System (BMS) enables the transfer of sensor measurements from the battery cells to the IoT component, employing the Controller Area Network (CAN) protocol for communication. To optimize resource utilization while efficiently transmitting a substantial volume of sequential data generated by both private and fleet vehicles, the message queuing telemetry transport (MQTT) protocol [47] enables bidirectional communication between the device and the cloud. The infrastructure can effortlessly support millions of IoT devices, seamlessly accommodating their operations. Moreover, the data stored in the onboard memory can be efficiently transmitted to the cloud system using TCP/IP protocols, ensuring smooth and reliable upload processes. Modern cities' IoT systems provide infrastructure for remote data transmission through the use of IoT actuators and on-board sensors. For a more detailed explanation of the next-generation IoT, please refer to [48].

3.2. Cloud Server-Farm

A cloud server farm is a large-scale data center infrastructure that offers remote data storage and analysis capabilities, including real-time monitoring, early warning systems, and intelligent diagnosis over the internet. This beckons scientists as data sets continue to expand [49]. Cloud storage and computing has been widely recognized and acknowledged as a highly effective and flexible solution for remote monitoring, especially in the context of large-scale EV applications [50]. In this context, developers have the flexibility to seamlessly tailor their cloud to meet their specific needs and demands, thereby achieving maximum efficiency and convenience. The cloud based BMS has the capability to learn and analyze the continuous flow of the charging and discharging data of battery systems, enabling the generation of health information.

The cloud BMS can learn and analyze the continuous stream of time-series battery data and generate electronic health records, which provide insightful information about the battery's performance and health status. Java and Go are among the most commonly used programming languages for cloud development, providing developers with robust and efficient tools to create sophisticated and reliable cloud applications. Additionally, PHP offers a flexible and effective solution for web developers to design interactive interfaces and engage with the vast amount of data generated by the system [51].

In order to implement the battery-cloud system efficiently, it is essential for users to have some basic computing skills, but more importantly, it requires a deep understanding and knowledge of the learning task at hand, particularly in the context of complex, nonlinear multiphysics battery systems that exhibit gappy and noisy boundary conditions. Moreover, modeling of battery systems for field applications, such as prognostics and predictive health management (PHM), is often prohibitively expensive and requires complex formulations, new algorithms, and elaborate computer codes.

3.3. Machine Learning

In spite of the progress achieved in forecasting the dynamics of battery systems utilizing fundamental principles, atomic-level analysis, or methods rooted in physics, a notable obstacle persists due to the lack of all-encompassing prognostic models capable of establishing robust connections between cell properties, underlying mechanisms, and the states of the cell. The prognostication and modeling of battery systems' multi-dimensional behavior, influenced by various spatio-temporal factors, emphasize the necessity for a revolutionary approach. Deep learning has exhibited extraordinary advancements in addressing enduring quandaries faced by the artificial intelligence community [52]. The widespread availability of open-source software and the automation capabilities of material tools have seamlessly integrated machine learning into computational frameworks. Prominent software libraries like TensorFlow [53,54], PyTorch [55,56], and JAX [57] contribute significantly to the analysis of cell performance by harnessing diverse data modalities encompassing time series data, spectral data, laboratory tests, field data, and more.

In the realm of predictive modeling of battery systems, there has been a recent push towards synergistically integrating machine learning tools with cloud computing. In this context, researchers and engineers can access real-time data streams and perform real-time analysis and predictions of battery performance, which is pivotal when it comes to the design and optimization of battery systems. The integration of machine learning algorithms, cloud computing, and big data analysis has created a powerful ecosystem for the representation of multiscale and multiphysics battery systems. By incorporating actual sensor data to calibrate the models, a battery-powered digital twin strives to emulate the dynamics of the physical entity in a digital environment. Physics-informed learning is poised to emerge as a driving force in the transformative era of digital twins, thanks to its innate ability to seamlessly integrate physical models and data.

A recent illustration of this innovative learning approach is Physics-Informed Neural Networks (PINNs). The integration of data from measurements and partial differential equations (PDEs) is flawlessly accomplished by PINNs through the incorporation of these PDEs into the neural networks. This approach exhibits exceptional adaptability, allowing it to effectively handle a wide range of PDE types, including integer-order PDEs, fractional PDEs, and stochastic PDEs. To illustrate its effectiveness, the PINN model can be successfully employed to solve forward problems utilizing the viscous Burgers' equation, which can be represented as:

$$\frac{\partial u}{\partial t} + \rho \frac{\partial u}{\partial x} = \varphi \frac{\partial^2 u}{\partial x^2} \tag{3}$$

The physics-uninformed networks act as a surrogate for the PDE solution u(x, t), whereas the physics-informed networks characterize the PDE residual. The loss function encompasses both a supervised loss, incorporating data measurements of u obtained from initial and boundary conditions, and an unsupervised loss, which captures the PDE discrepancy:

$$\mathcal{L} = \omega_{\text{sample}} \mathcal{L}_{\text{sample}} + \omega_{\text{PDE}} \mathcal{L}_{\text{PDE}} \tag{4}$$

where

$$\mathcal{L}_{\text{sample}} = \frac{1}{N_{\text{sample}}} \sum_{i=1}^{N_{\text{sample}}} (u(x_i, t_i) - u_i)^2 \tag{5}$$

$$\mathcal{L}_{\text{PDE}} = \frac{1}{N_{\text{PDE}}} \sum_{j=1}^{N_{\text{PDE}}} \left(\frac{\partial u}{\partial t} + \rho \frac{\partial u}{\partial x} - \varphi \frac{\partial^2 u}{\partial x^2} \right) \Bigg| (x_j, t_j) \tag{6}$$

The two sets of points, $\{(x_i, t_i)\}$ and $\{(x_j, t_j)\}$, correspond to samples taken from initial and boundary locations and the complete domain, respectively. To effectively balance the relationship between the two loss terms, weight, ω_{sample} and ω_{PDE} are utilized. The neural network undergoes training using gradient-based optimizers like Adam to minimize the loss until it is below a predefined threshold ε. For a detailed discussion and introduction of PINN, one can refer to a comprehensive review [58].

4. AI Modelling for Battery State Prediction

Machine learning plays a vital role in AI modeling as it empowers systems to learn and enhance their performance through data-driven experiences, without the need for explicit programming. Machine learning algorithms analyze large datasets to identify patterns and relationships, which are then used to make predictions and decisions. Machine learning utilizes a customizable function with adjustable parameters to accurately predict battery behavior, often through the use of experimental training data [59]. This allows for generalization to other battery systems. A representative example includes transfer learning, which leverages data-driven mechanisms to govern personalized health status prediction tasks [60]. Machine learning has surfaced as a powerful tool for analyzing the ever-growing amount of time-series data, but it faces limitations when dealing with complex spatio-temporal systems. To overcome this, deep learning has attracted significant attention over the past few years, with its ability to automatically extract spatio-temporal features. Deep learning has made significant progress in addressing challenges that have previously proven difficult for the artificial intelligence community. It has proven to be particularly effective in identifying complex structures in large data sets for battery systems [61–64]. This would not only improve the accuracy of models for forecasting and long-range spatial connections across multiple timescales but also enable a deeper understanding of complex physical processes. In this section, we provide a concise overview of the recent advancements in battery state estimation achieved through diverse machine learning methods (Figure 3).

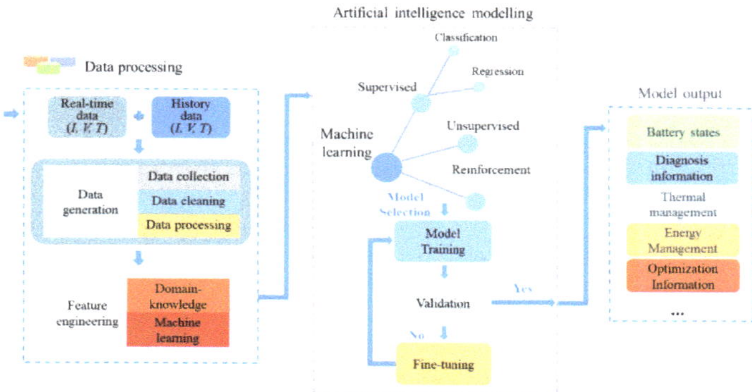

Figure 3. AI and machine learning for modelling and predicting battery states.

4.1. SOC

Machine learning (ML) methods have exhibited remarkable efficacy in accurately interpolating between data points, even for high-dimensional tasks. With the ability to learn complex patterns and relationships within data, ML models can accurately capture the underlying structure of the data, allowing for effective interpolation and prediction. For example, a gated recurrent unit (GRU)-based recurrent neural network (RNN) has shown good performance in estimating the battery SOC using data from varied loading patterns [65]. Despite the training process demanding several hours in a GPU environment, the testing phase demonstrated remarkably swift execution, even within a CPU environment. This underscores the efficiency and efficacy, in precisely estimating SOC, which serves as a crucial parameter for management and control in diverse applications. In another study, a stacked bidirectional long-short-term memory (LSTM) neural network was applied to estimate the cell SOC [66]. The study focuses on three main improvements: (1) the use of bidirectional LSTM to capture temporal dependencies in both forward and backward directions within time-series data; (2) the stacking of bidirectional LSTM layers to create a deep model with increased capacity to process nonlinear and dynamic LiB data; and (3) a detailed comparison and analysis of multiple parameters that affect the estimation performance of the proposed method. The results demonstrate the effectiveness of the approach and its potential to enhance SOC estimation. A single hidden layer GRU-RNN algorithm with momentum optimization for SOC estimation is proposed [67]. GRU is a streamlined variant of LSTM that integrates the forget and input gates into a singular update gate, resulting in reduced parameters and enhanced computational efficiency compared to LSTM. The algorithm employs the momentum gradient method, which balances the current gradient direction and historical gradient direction to prevent oscillations in the weight change and improve the speed of SOC estimation. The performance of the algorithm is evaluated under varying parameters, including β, noise variances σ, epochs, and the number of hidden layer neurons. The results of the study provide insights into the accuracy and efficiency of the GRU-RNN-based momentum algorithm in estimating the SOC of lithium batteries, demonstrating its potential as a promising approach for battery management and control in various applications. More recently, a combined SOC estimation method called gated recurrent unit adaptive Kalman filter (GRU-AKF) was proposed, which is both robust and efficient [68]. The method eliminates the requirement for developing a complex battery model by employing a GRU-RNN for initial SOC estimation and establishing a nonlinear relationship between observed data and SOC across the entire temperature range. Subsequently, the Adaptive Kalman Filter (AKF) is utilized to refine the SOC estimated by the GRU-RNN, resulting in the final estimated SOC. The proposed GRU-AKF exhibits enhanced adaptability to practical battery applications, facilitated by the improved adaptive approach. The design cost of the estimation method is reduced since the hyperparameters of the network do not need to be carefully designed as the output SOC is further processed by the AKF. AKF offers an effective tool for estimating the state of a dynamic system based on noisy measurements. The method is specifically designed to address the challenge of noisy data in dynamic systems, where conventional data-driven approaches may fall short in delivering accurate outcomes. The study's findings showcase the efficacy of the proposed method in accurately estimating SOC for batteries.

In addition to the RNN model, self-supervised transformer model is another deep learning method that has attracted a lot of attention for predicting cell SOC. For example, transformer-based SOC estimation was used to leverage self-supervised learning to achieve higher accuracy with limited data availability within a constrained timeframe [69]. The framework additionally integrates cutting-edge deep learning techniques, including the Ranger optimizer, time series data augmentation, and the Log-Cosh loss function, to enhance accuracy. The acquired parameters can be efficiently transferred to another cell by fine-tuning, even with limited data available within a short timeframe. Another study proposes a hybrid methodology for SOC estimation of batteries by employing a sliding window to pre-process data, using a Transformer network to capture the relationship between observational data and SOC, and feeding the result into an adaptive observer [70]. The effectiveness of the proposed method is validated across different temperatures using US06 data, demonstrating accurate SOC estimation with less than 1% Root Mean Square Error (RMSE) and maximum error in the majority of temperature scenarios. The proposed method surpasses LSTM-based approaches and exhibits the ability to provide reliable predictions even for temperatures not included in the training dataset.

4.2. SOH

In a recent research endeavor, a battery health and uncertainty management pipeline (BHUMP) is introduced as a machine learning-driven solution, showcasing its adaptability to various charging protocols and discharge current rates. Notably, BHUMP excels at making accurate predictions without the need for specific knowledge about battery design, chemistry, or operating temperature [71]. The study underscores the significance of incorporating machine learning techniques in conjunction with charge curve segments to effectively capture battery degradation within a limited timeframe. However, the authors stress that even if the algorithm produces low errors, it is crucial to perform uncertainty quantification tests to ensure its reliability before deploying it in real-world applications.

Differential approaches, namely incremental capacity and differential voltages are frequently employed to identify causes of deterioration in online applications. One research study, for example, combines the Support Vector Regression (SVR) algorithm with a multi-timescale parameter identification approach based on Extended Kalman Filter-Recursive Least Squares (EKF-RLS) and a known relationship model between representative RC

(Resistor-Capacitor) parameters and State of Health (SOH) [72]. The study's results showcase that the proposed method achieves Mean Absolute Error (MAE) and Root Mean Square Error (RMSE) values below 3% for SOH prediction, utilizing both static and dynamic observational data. This suggested technique demonstrates excellent capability in accurately estimating SOH in complex dynamic environments, offering high accuracy, robustness, and practicality.

In recent years, some large datasets relating to batteries during the daily operation of EVs have been collected and analyzed. For example, one study used 147 vehicle data points from two sources to verify a proposed method for estimating the capacity and internal resistance of EV batteries [73]. The results demonstrate that the estimation results converge to the true trend, with a maximum estimation error of less than 4% for the capacity of sampled real EVs. The proposed method can accurately estimate the battery capacity of EVs and enable life prediction using current cloud data. Another study proposed a SOH estimation method for EV batteries based on discrete incremental capacity analysis that is robust, compatible, computationally efficient, and memory-efficient [74]. The SOH of EVs does not decrease linearly with mileage but shows stagnation and fluctuations due to seasonal temperature variations, driving habits, and charging strategies.

To emphasize the importance of cloud-based AI modeling for battery BMS, ensemble machine learning offers opportunities to accurately predict SOH using only daily operating charging data (i.e., voltage, current, and temperature) [75]. A two-step approach is employed to reduce noise in battery data, while domain-specific features derived from IC (incremental capacity) and DV (differential voltage) analysis offer physically consistent representations of intricate battery degradation patterns. To enhance prediction accuracy and model generalization, a stacking technique is adopted, leveraging four base-level models (linear regression, random forest regression, gaussian process regression, and gradient boosting regression) along with a meta-learner. The proposed multi-model fusion method exhibits robustness, stability, and compatibility with diverse usage histories, making it a valuable tool for forecasting cell capacity and constructing battery pack trajectories. Furthermore, the study indicates that with the advancement of onboard computing capabilities, the proposed method can be seamlessly migrated from cloud-BMS to onboard-BMS by employing feature engineering techniques and constructing lookup tables. In summary, this study demonstrates the potential of integrating onboard observational samples with data-driven machine learning models to predict the dynamics of complex systems like lithium-ion batteries, even in the presence of missing/noisy data and uncertain boundary conditions.

Reinforcement learning, which combines machine learning principles with neuroscientific approaches, offers a normative framework for agents to learn policies and optimize their behavior in response to rewards received from interacting with the environment [76]. In battery prognostics and health management applications, such as optimizing fast-charging protocols, the BMS acts as the agent, making decisions (like determining the applied current) based on rewards for each possible action while interacting with the environment (the battery) [77]. A pseudo-two-dimensional electrochemical model, Doyle-Fuller-Newman [78], is employed to predict the evolution of multiphysics battery systems by capturing macro-scale physics, including lithium concentration in solids and electrolytes, solid electric potential, electrolyte electric potential, ionic current, molar ion fluxes, and cell temperature. The Deep Deterministic Policy Gradient (DDPG)-based reinforcement learning demonstrates a remarkable ability to handle continuous state and action spaces by updating the control policy in the actor-critic network architectures, thereby reducing the likelihood of safety hazards during fast-charging protocols.

4.3. Battery Safety and Thermal Management

In addition to SOC and SOH estimation, cloud-based BMS can also be tailored to a much more complex problem—that is, battery failure. Lithium-ion batteries are multiphysics and multiscale systems, and their safety and reliability are crucial due to their widespread adoption in various applications. However, given the intricate nature of battery behavior, accurately predicting failures remains a formidable challenge, given the lack of understanding of the underlying degradation mechanisms. In light of the ever-evolving cell and battery designs, the multitude of potential failure scenarios and associated risks make it impractical to comprehensively understand the origins and consequences of each through laboratory testing alone. While computational modeling can reduce the number of required experiments, its effective implementation can be limited by rigorous validation requirements and computational resources.

The establishment of a "safety envelope", defining the operational range in which individual cells can function safely, is essential for ensuring the overall safety of electric vehicle battery packs. However, the challenge lies in acquiring a substantial dataset of battery failure tests. In a recent study, researchers developed a highly accurate computational model for lithium-ion pouch cells, incorporating calibrated constitutive models for each material composing the cell [79]. To construct a data-driven safety envelope, supervised machine learning techniques were applied to a vast matrix of severe mechanical loading scenarios. This study demonstrates the synergistic combination of numerical data generation and machine learning modeling to forecast the safety of battery systems.

Emerging technologies are addressing previously challenging obstacles by providing accessible and effective solutions, highlighting the significance of cloud-based AI modeling in battery BMS. Machine learning approaches utilizing data-driven frameworks excel at accurately forecasting complex nonlinear systems. A specific research study [80] focuses on the development of a tightly integrated cloud-based machine learning system for predicting real-life EV battery failure. By leveraging graphite/NMC cells, a data-driven early-prediction model is created, enabling the generation of longitudinal electronic health records through digital twins. The proposed hybrid semi-supervised machine learning model combines observational, empirical, physical, and statistical insights, achieving a 7.7% test error utilizing field data. Cloud-based machine learning approaches exemplify the significance of adopting a multifaceted strategy for continuous lifelong learning. These approaches not only provide a novel means of forecasting battery failure but also underscore the value of incorporating diverse methods to enhance accuracy and robustness.

Thermal management is a critical aspect in the context of battery systems, and a specific study [81] conducted a comprehensive analysis of the performance of a liquid-cooled Battery Thermal Management System (BTMS). The study primarily concentrated on the analysis of experimental data pertaining to air conditioning and the exploration of design considerations for the liquid-cooled Battery Thermal Management System (BTMS). By integrating these thermal characteristics, a more accurate and efficient operation of the liquid cooled BTMS can be achieved, thus contributing to the overall improvement of the HPACS for EVs. This can be achieved by coupling the battery electrochemical model with the machine learning model of HPACS and optimizing the liquid cooled BTMS based on the automatic calibration model and battery electrochemical model, leading to more efficient system optimization. In another case study, a multiphysics approach was employed to demonstrate the temperature-position-dependent thermal conductivity of Heat Pipes (HPs) [82]. By leveraging the multiphysics nature of HPs, which provides variable thermal conductivity, valuable insights into heat pipe efficiency can be gained. Increasing the condensation surface area of the heat pipes enables a reduction in the size and number of heat pipes required for cooling applications. However, it is crucial to utilize advanced methods to analyze the complex equations, multiphysics phenomena, and boundary conditions associated with these systems. By employing such advanced techniques, a deeper understanding of thermal management can be achieved, leading to improved design and performance of battery systems.

Machine learning techniques, such as physics-informed machine learning [58] offer a promising direction to follow. Such learning approach blends mathematical models with noisy data, utilizing neural networks or other kernel-based regression networks. By incorporating physical invariants into specialized network architectures, this approach can improve accuracy, training speed, and generalization. Additionally, this technique can automate the satisfaction of certain physical invariants for more effective implementation.

5. Current Limitations

5.1. Multiscale and Multiphysics Problems

While physics-informed learning has achieved remarkable success in various applications, ongoing efforts are being made to address challenges that involve multiple scales and physics. It is recommended to initially study each physics in isolation before integrating them, as learning multiple physics concurrently can pose computational challenges. Additionally, it is important to utilize fine-scale simulation data selectively to gain a broader understanding of the physics at a coarser scale. The existing body of research primarily focuses on models that specialize in predicting the SOH and Remaining Useful Life (RUL) over multiple cycles, as well as the SOC within a single charge/discharge cycle. However, to achieve a more comprehensive understanding of battery performance, it is necessary to develop a model that can forecast the long-term SOH from any arbitrary point in the charge/discharge cycle. This can be accomplished through a hybrid approach that combines sophisticated models capable of accurately forecasting the SOC up to a specific point in the cycle, such as a fully charged state, with a SOH model that takes into account multiple cycles. By integrating both short-term and long-term dynamics models, a comprehensive model of battery development can be created, enabling more accurate and reliable predictions of battery performance.

5.2. Gap between Lab Tests and Field Conditions

High-throughput testing offers a valuable means to obtain large and reliable datasets for machine learning applications. Various electrochemical techniques, such as cyclic voltammetry, galvanostatic charge/discharge, and electrochemical impedance spectroscopy, enable precise and reliable measurements of batteries' lifetime, rate capability, capacity, and impedance. This comprehensive approach ensures that the acquired data reflects real-world conditions and provides a solid foundation for machine learning algorithms to analyze and extract valuable insights from the battery performance characteristics. Large amounts of meaningful data can be swiftly generated. Machine learning models can then be trained using this data, and the battery testing process can be expedited further by detecting poorly performing batteries based on their initial cycles.

One approach to reconciling standard laboratory tests with field data involves laboratory testing of batteries using representative loading patterns, based on characterized typical user driving patterns. This methodology offers the benefit of a controlled environment with high-precision equipment and frequent characterizations. However, it is crucial to supplement this approach with field data for multiple reasons. Firstly, laboratory experimentation is limited in scope and cannot encompass all necessary conditions over an extended duration, particularly as the number of aging parameters grows exponentially. In contrast, field data is readily accessible and covers the complete range of operating conditions, offering a relatively cost-effective option as the cells are already in practical use. Secondly, laboratory testing is artificial and may diverge from real-world usage. Constraints in time and equipment within the laboratory often result in extreme conditions and short resting periods between cycles, potentially leading to an underestimation of battery lifespan and an excessive design of battery packs. Thirdly, external factors encountered in real-world environments, such as seasonal temperature variations or mechanical vibrations that contribute to failure, are not accounted for in laboratory settings. Fourthly, the accumulation of additional data is always valuable in enhancing statistical confidence when constructing models for battery lifespan and performance, considering both inherent factors related to manufacturing variability and external factors associated with usage patterns. The standardization of methodologies for interpreting not only accelerated cycle aging data but also accelerated calendar aging and scenarios involving a combination of cycle and calendar aging is crucial to extending the applicability of models beyond laboratory-accelerated aging tests to real-world applications. Lastly, gaining a comprehensive understanding of the influence of cycles and calendar conditions on battery lifespan is imperative for both laboratory and field applications.

5.3. Data Generation and Model Training

Accurately predicting the state of batteries, both in real-time and offline, is critical to enhancing cycle life and ensuring safety by enabling informed engineering and adaptation to unfavorable conditions. However, due to the wide range of battery options and constantly evolving pack designs, it is challenging to predict cell behavior under various conditions. One promising approach to address this is to use finite-element model data to train machine learning algorithms that can predict cell performance when exposed to different loading conditions. While this approach offers numerous benefits, it also encounters a common obstacle faced by data-driven methods, which is the acquisition of trustworthy, abundant, high-quality, and pertinent experimental data. This requires overcoming obstacles such as experimental data collection and curation as well as ensuring that the data is representative of real-world usage scenarios. Overcoming these challenges and obtaining the necessary data will be crucial in the application of machine learning for battery prediction.

The battery dataset provided by the Prognostics Center of Excellence at NASA Ames is extensively utilized by researchers [83]. These datasets involve subjecting batteries to various operational profiles at different temperatures. Impedance measurements are recorded after each cycle. The NASA battery dataset is a valuable resource for researchers who are interested in studying battery performance, aging, and prognostics. Making open data and software available is a promising approach to enhancing the transferability of models and making them more useful for battery design. This involves the systematic generation of datasets and their release for reuse by other researchers.

Harnessing data-based methods for discerning materials, estimating the lifespan and efficiency of lithium-ion batteries, has proven promising, leading to a heightened interest in utilizing these methodologies to amplify the prognostic abilities related to cell behaviors. The recent developments in amassing and processing vast quantities of data have paved the way for real-time learning and forecasting of battery operations. For example, Severson and colleagues [84] recently furnished an openly accessible dataset packed with a comprehensive array of battery data. This dataset comprises 124 LFP-graphite cells, subjected to diverse quick-charging conditions, varying from 3.6 to 6 C, and evaluated within a temperature-controlled chamber at 30 °C, achieving up to 80% of their original capacities. The cells underwent one or two charging steps, such as 6 C charging from 0% to 40% State of Charge (SOC), followed by a 3 C charging process up to 80% SOC. Additionally, all cells were charged from 80% to 100% SOC using a 1 C Constant Current-Constant Voltage (CC-CV) phase to 3.6 V and depleted with a 4 C CC-CV phase down to 2.0 V, with the end current regulated at C/50. During the cyclic evaluation, both cell temperature and internal resistance were documented at 80% SOC. This dataset offers a valuable resource for those looking to delve into battery performance, especially pertaining to rapid charging conditions. The feature-based machine learning model adeptly utilized voltage and capacity data from the initial 100 cycles (roughly 10% of the overall lifespan) of equivalent commercial cells to build a straightforward regression model capable of forecasting the cycle lifespan with approximately 90% precision. Nevertheless, a pressing query persists: How might data-based methods be employed to anticipate cell behavior within ever-changing field uses? Additionally, can these strategies provide an efficient solution to comprehend and predict the reaction of emerging cell and pack configurations to authentic environmental conditions, thus enhancing the efficacy of battery systems?

In certain scenarios, gathering training data can be a costly and challenging task. In such cases, there is a growing need to develop high-performance learning models that can be trained using data from different domains that is more easily accessible. This technique is commonly known as transfer learning [85]. Transfer learning offers a practical solution to the challenge of obtaining sufficient training data and has become increasingly relevant in a wide range of fields. With the help of transfer learning, it is possible to achieve high-performing models without incurring excessive costs or resource allocation.

While the exploration of negative transfer remains somewhat sparse, infusing negative transfer methods into transfer learning frameworks could be a potent path for forthcoming research. One conceivable trajectory entails the creation of solutions that cater to multiple origin domains, potentially enhancing the filtering out of irrelevant data. A further promising field is the concept of optimal transfer, a process that aims to selectively convey certain information from a source domain to maximize the performance of the target learner. Although there is some intersection between negative and optimal transfer, optimal transfer concentrates on boosting the performance of the target, whereas negative transfer emphasizes the detrimental influence exerted by the source domain on the target learner.

6. Outlook

6.1. Cloud-End Collaboration

In BMS, a collaboration with cloud computing capitalizes on the substantial computational power and storage space offered by cloud servers, overcoming the constraints of traditional BMS and paving the way for the use of advanced algorithms such as deep learning and reinforcement learning. The BMS's 5G communication module is used to capture real-time battery data, which can then be employed to build battery models in the cloud. This allows for a two-way dynamic correlation between the digital twin model and the actual battery, enabling detailed and secure battery management throughout its lifespan through online learning and model updating. The data gathered from the batteries and their associated digital twin models throughout their full lifespan is used to construct an optimal performance improvement path via the application of smart OTA remote program update technology. In order to cater to the escalating needs of battery management, the immediate processing abilities of the embedded system are integrated with the high-level intelligence offered by the cloud platform. To enhance the efficiency of the system further, the notion of a collaborative management model that incorporates cloud, edge, and end is introduced.

6.2. Digital Twins

Leveraging a digital twin, a virtual counterpart of the physical object, may serve as a bridge linking laboratory experiments to real-world uses [86]. The digital twin, integrating sensor readings from real-world scenarios into computational models, can faithfully simulate the conductivity of lithium-ion batteries under a variety of operational states like random discharge, dynamic charge, and idle stages. Fundamental aspects like voltage, current, temperature, and so on can be derived from the onboard BMS and used to optimize the digital twin. The goal is to improve the predictive ability of the digital twin under realistic conditions, which can lead to a better understanding and evaluation of battery behavior. The use of IoT technology, such as the MQTT protocol, enables the collection of large amounts of sequential data from the ever-increasing running time of EVs. This data can be seamlessly transmitted to the cloud for analysis, improving our understanding of complex battery behavior under different operating conditions.

Despite the potential benefits of digital twins, several challenges need to be solved before their widespread implementation can become a reality. Firstly, acquiring accurate and comprehensive observational data can be difficult, as it may be scarce and noisy and can take various forms. Secondly, physics-based computational models can be complex to set up and calibrate, requiring significant effort in pre-processing and determining initial and boundary conditions, making their use in real-time applications impractical. Additionally, the physical models of complex systems are often only partially understood, with conservation laws that do not provide a complete system of equations without further assumptions. Physics-informed learning, however, offers a solution to these issues by seamlessly blending physical models with data and utilizing automatic differentiation to eliminate the need for mesh generation.

6.3. Data-Model Fusion

Simulations based on fundamental principles are commonly employed to probe material degradation, incorporating techniques like density functional theory and molecular dynamics. Additionally, the transformation of microstructures and formations, such as lithium dendrite creation and phase segregation of active electrode materials, is examined using physics-rooted models like the phase field method. Despite the phase field modeling not yet achieving complete cell simulation, the simulation results align well with experimental data. In order to carry out in situ computations, a machine learning model needs to be trained first, followed by the establishment of a database comprising previous outcomes from multiscale first principles and phase-field simulations. This model serves as a proxy for the simulations; if machine learning demonstrates significant uncertainty, an additional simulation is conducted, added to the database, and the machine learning model is retrained. This iterative process of active learning holds the potential to drastically reduce the number of simulations required to understand a system. Similar machine learning applications can be utilized in experimental design and to eliminate costly experiments. Further studies focusing on the mechanical properties of solid electrolytes and voltage demonstrate how machine learning can expedite simulations.

As the fusion of physics-based modeling and machine learning progresses, researchers are likely to frequently encounter situations where multiple models of the same phenomenon are developed using the same training data, or even data that is equally informative. Even though their predictions based on the training data are nearly identical, this could lead to differentially trained networks. To address this issue, the construction of machine learning-based transformations between theories, models of varying complexity, and predictive models that can be validated is crucial. This will ensure that a phenomenon retains a unique and clear physical interpretation, even when multiple models are used to describe it. The merging of data and models can yield enhanced representations of physical systems by capitalizing on the strengths of each information source.

6.4. Explainable AI

In several scientific disciplines, the prevailing trend is an overflow of observational data that often surpasses our ability to understand and analyze effectively. Despite machine learning (ML) methodologies showing substantial promise and early successes, they continue to face hurdles in deriving significant insights from the wealth of data at hand. Furthermore, a sole reliance on data-driven models can lead to accurate correlations with observed data, but such models might produce physically inconsistent or implausible predictions due to extrapolation or biases inherent in the data, potentially diminishing their generalizing capabilities. In many instances, AI systems fall short of offering clear explanations of their autonomous actions to human users. While some argue that the emphasis on explain ability is misguided and unnecessary for specific AI applications, it remains vital for a number of key applications where users need to understand, trust, and effectively manage their AI counterparts. Explainable AI (XAI) systems [87], striving to improve their understandability for human users by delivering explanations of their actions, hold promise for enriching materials science and battery modeling. They can contribute to a more thorough understanding of the underlying physics, more effective hypothesis testing, and a higher level of confidence in learning models. By granting researchers the ability to interpret and visualize decision-making processes in complex models, XAI can assist in identifying crucial features and parameters impacting material and cell characteristics. This understanding can further promote the creation of new materials with superior properties and deepen our comprehension of their behavior under varying conditions. Furthermore, the transparency and interpretability provided by XAI methods can foster trust in learning models, empowering researchers to make well-informed decisions and draw accurate conclusions. In designing more effective, user-friendly AI systems, certain basic principles and domain-specific knowledge must be taken into consideration. Specifically, an XAI system should be capable of articulating its capabilities and insights, justifying its past actions, outlining its current trajectory, and revealing crucial information that steers its decisions.

7. Conclusions

Artificial intelligence and machine learning methods are increasingly being utilized to reveal patterns and insights from the expanding volume of battery data. However, these approaches often require craftsmanship and intricate implementations, especially when system dynamics are predominantly influenced by spatio-temporal context. This is where cloud-based digital solutions come in. The cloud environment can be configured by us-ers/developers to meet their specific needs and requirements. Cloud-BMS opens up a new world for collecting observational data and assimilating it sensibly through the seamless integration of data and abstract mathematical operators. However, merely moving data to the cloud isn't enough. New physics-based learning algorithms and computational frameworks are vital in addressing the challenges faced by complex battery systems, especially in real-time EV scenarios. Integrating AI and machine learning into BMS could boost battery diagnosis and prognosis accuracy. Furthermore, integrating cloud-based frameworks into the BMS can improve battery monitoring and management efficiency and scalability. Advanced sensing and monitoring technologies, such as wireless sensor networks and IoT devices, could allow for real-time data collection and analysis, enhancing battery management precision. The fusion of data-driven and physics-based modeling through physics-informed machine learning techniques promises to further boost battery management performance. The potential to model long-range correlations across multiple time scales, simulate thermodynamics and kinetics, and explore the dynamics of nonlinear battery systems holds promise for accelerating technology transfer from academic progress to real-world applications.

Author Contributions: Conceptualization and methodology, J.Z.; software, J.Z. and J.W.; writing—original draft preparation, D.S. and J.Z.; writing—review and editing, J.Z., A.F.B. and C.E.; visualization, J.Z. and Z.W.; supervision, resources, project administration, A.F.B. and Y.L.; funding acquisition, D.S. and Y.L. All authors have read and agreed to the published version of the manuscript.

Funding: This study was financially supported by [Independent Innovation Projects of the Hubei Longzhong Laboratory] grant number [2022ZZ-24], [Central Government to Guide Local Science and Technology Development fund Projects of Hubei Province] grant number [2022BGE267], [Basic Research Type of Science and Technology Planning Projects of Xiangyang City] grant number [2022ABH006759] and [Hubei Superior and Distinctive Discipline Group of "New Energy Vehicle and Smart Transportation"] grant number [XKTD072023].

Data Availability Statement: The data could not be shared due to confidentiality.

Conflicts of Interest: The authors declare no conflict of interest.

References

1. Liang, X.; Zhang, S.; Wu, Y.; Xing, J.; He, X.; Zhang, K.M.; Wang, S.; Hao, J. Air quality and health benefits from fleet electrification in China. *Nat. Sustain.* **2019**, *2*, 962–971. [CrossRef]
2. Li, M.; Lu, J.; Chen, Z.; Amine, K. 30 years of lithium-ion batteries. *Adv. Mater.* **2018**, *30*, 1800561. [CrossRef] [PubMed]
3. Meng, H.; Li, Y.F. A review on prognostics and health management (PHM) methods of lithium-ion batteries. *Renew. Sustain. Energy Rev.* **2019**, *116*, 109405. [CrossRef]
4. Neubauer, J.; Wood, E. The impact of range anxiety and home, workplace, and public charging infrastructure on simulated battery electric vehicle lifetime utility. *J. Power Source* **2014**, *257*, 12–20. [CrossRef]
5. Xiong, R.; Pan, Y.; Shen, W.; Li, H.; Sun, F. Lithium-ion battery aging mechanisms and diagnosis method for automotive applications: Recent advances and perspectives. *Renew. Sustain. Energy Rev.* **2020**, *131*, 110048. [CrossRef]
6. Gabbar, H.A.; Othman, A.M.; Abdussami, M.R. Review of battery management systems (BMS) development and industrial standards. *Technologies* **2021**, *9*, 28. [CrossRef]
7. Vichard, L.; Ravey, A.; Venet, P.; Harel, F.; Pelissier, S.; Hissel, D. A method to estimate battery SOH indicators based on vehicle operating data only. *Energy* **2021**, *225*, 120235. [CrossRef]

8. Li, D.; Wang, L. Onboard health estimation approach with segment warping and trajectory self-learning for swappable lithium battery. *J. Energy Storage* **2022**, *55*, 105749. [CrossRef]
9. Zhao, J.; Burke, A.F. Electric Vehicle Batteries: Status and Perspectives of Data-Driven Diagnosis and Prognosis. *Batteries* **2022**, *8*, 142. [CrossRef]
10. Sulzer, V.; Mohtat, P.; Aitio, A.; Lee, S.; Yeh, Y.T.; Steinbacher, F.; Khan, M.U.; Lee, J.W.; Siegel, J.B.; Howey, D.A. The challenge and opportunity of battery lifetime prediction from field data. *Joule* **2021**, *5*, 1934–1955. [CrossRef]
11. Peng, J.; Meng, J.; Chen, D.; Liu, H.; Hao, S.; Sui, X.; Du, X. A Review of Lithium-Ion Battery Capacity Estimation Methods for Onboard Battery Management Systems: Recent Progress and Perspectives. *Batteries* **2022**, *8*, 229. [CrossRef]
12. Berecibar, M. Machine-learning techniques used to accurately predict battery life. *Nature* **2019**, *568*, 325–326. [CrossRef]
13. Ng, M.F.; Zhao, J.; Yan, Q.; Conduit, G.J.; Seh, Z.W. Predicting the state of charge and health of batteries using data-driven machine learning. *Nat. Mach. Intell.* **2020**, *2*, 161–170. [CrossRef]
14. Zhao, J.; Nan, J.; Wang, J.; Ling, H.; Lian, Y.; Burke, A.F. Battery Diagnosis: A Lifelong Learning Framework for Electric Vehicles. In Proceedings of the 2022 IEEE Vehicle Power and Propulsion Conference (VPPC), Merced, CA, USA, 1–4 November 2022; IEEE: New York, NY, USA, 2022; pp. 1–6.
15. Zhao, J.; Burke, A.F. Battery prognostics and health management for electric vehicles under industry 4.0. *J. Energy Chem.* **2023**. [CrossRef]
16. Gibney, E. Europe sets its sights on the cloud: Three large labs hope to create a giant public-private computing network. *Nature* **2015**, *523*, 136–138. [CrossRef]
17. Drake, N. How to catch a cloud. *Nature* **2015**, *522*, 115–116. [CrossRef]
18. Bosch Mobility Solutions: Battery in the Cloud. Available online: https://www.bosch-mobility-solutions.com/en/solutions/software-and-services/battery-in-the-cloud/battery-in-the-cloud/ (accessed on 19 August 2022).
19. Panasonic Announces UBMC Service: A Cloud-Based Battery Management Service to Ascertain Battery State in Electric Mobility Vehicles. Available online: https://news.panasonic.com/global/press/data/2020/12/en201210-1/en201210-1.pdf (accessed on 19 August 2022).
20. HUAWEI: CloudLi. Available online: https://carrier.huawei.com/en/products/digital-power/telecom-energy/Central-Office-Power (accessed on 19 August 2022).
21. National Monitoring and Management Platform for NEVs. Available online: http://www.bitev.org.cn/a/48.html (accessed on 19 August 2022).
22. Xing, Y.; Ma, E.W.; Tsui, K.L.; Pecht, M. Battery management systems in electric and hybrid vehicles. *Energies* **2011**, *4*, 1840–1857. [CrossRef]
23. Liu, K.; Li, K.; Peng, Q.; Zhang, C. A brief review on key technologies in the battery management system of electric vehicles. *Front. Mech. Eng.* **2019**, *14*, 47–64. [CrossRef]
24. Xiong, R.; Li, L.; Tian, J. Towards a smarter battery management system: A critical review on battery state of health monitoring methods. *J. Power Source* **2018**, *405*, 18–29. [CrossRef]
25. Yang, N.; Zhang, X.; Li, G. State of charge estimation for pulse discharge of a LiFePO4 battery by a revised Ah counting. *Electrochim. Acta* **2015**, *151*, 63–71. [CrossRef]
26. Lee, S.; Kim, J.; Lee, J.; Cho, B.H. State-of-charge and capacity estimation of lithium-ion battery using a new open-circuit voltage versus state-of-charge. *J. Power Source* **2008**, *185*, 1367–1373. [CrossRef]
27. Tran, M.K.; Mathew, M.; Janhunen, S.; Panchal, S.; Raahemifar, K.; Fraser, R.; Fowler, M. A comprehensive equivalent circuit model for lithium-ion batteries, incorporating the effects of state of health, state of charge, and temperature on model parameters. *J. Energy Storage* **2021**, *43*, 103252. [CrossRef]
28. Tran, M.K.; DaCosta, A.; Mevawalla, A.; Panchal, S.; Fowler, M. Comparative study of equivalent circuit models performance in four common lithium-ion batteries: LFP, NMC, LMO, NCA. *Batteries* **2021**, *7*, 51. [CrossRef]
29. Wei, Y.; Wang, S.; Han, X.; Lu, L.; Li, W.; Zhang, F.; Ouyang, M. Toward more realistic microgrid optimization: Experiment and high-efficient model of Li-ion battery degradation under dynamic conditions. *eTransportation* **2022**, *14*, 100200. [CrossRef]
30. Wei, Y.; Han, T.; Wang, S.; Qin, Y.; Lu, L.; Han, X.; Ouyang, M. An efficient data-driven optimal sizing framework for photovoltaics-battery-based electric vehicle charging microgrid. *J. Energy Storage* **2022**, *55*, 105670. [CrossRef]
31. Wei, Y.; Yao, Y.; Pang, K.; Xu, C.; Han, X.; Lu, L.; Li, Y.; Qin, Y.; Zheng, Y.; Wang, H.; et al. A Comprehensive Study of Degradation Characteristics and Mechanisms of Commercial Li (NiMnCo) O2 EV Batteries under Vehicle-To-Grid (V2G) Services. *Batteries* **2022**, *8*, 188. [CrossRef]
32. Han, X.; Lu, L.; Zheng, Y.; Feng, X.; Li, Z.; Li, J.; Ouyang, M. A review on the key issues of the lithiumion battery degradation among the whole life cycle. *eTransportation* **2019**, *1*, 100005. [CrossRef]
33. Hu, X.; Xu, L.; Lin, X.; Pecht, M. Battery lifetime prognostics. *Joule* **2020**, *4*, 310–346. [CrossRef]
34. Deng, Z.; Hu, X.; Li, P.; Lin, X.; Bian, X. Data-driven battery state of health estimation based on random partial charging data. *IEEE Trans. Power Electron.* **2021**, *37*, 5021–5031. [CrossRef]
35. Khaleghi, S.; Hosen, M.S.; Karimi, D.; Behi, H.; Beheshti, S.H.; Van Mierlo, J.; Berecibar, M. Developing an online data-driven approach for prognostics and health management of lithium-ion batteries. *Appl. Energy* **2022**, *308*, 118348. [CrossRef]
36. Gou, B.; Xu, Y.; Feng, X. An ensemble learning-based data-driven method for online state-of-health estimation of lithium-ion batteries. *IEEE Trans. Transp. Electrif.* **2020**, *7*, 422–436. [CrossRef]

37. Li, R.; Hong, J.; Zhang, H.; Chen, X. Data-driven battery state of health estimation based on interval capacity for real-world electric vehicles. *Energy* **2022**, *257*, 124771. [CrossRef]
38. Wu, W.; Wang, S.; Wu, W.; Chen, K.; Hong, S.; Lai, Y. A critical review of battery thermal performance and liquid based battery thermal management. *Energy Convers. Manag.* **2019**, *182*, 262–281. [CrossRef]
39. Kim, J.; Oh, J.; Lee, H. Review on battery thermal management system for electric vehicles. *Appl. Therm. Eng.* **2019**, *149*, 192–212. [CrossRef]
40. Wang, X.; Wei, X.; Dai, H. Estimation of state of health of lithium-ion batteries based on charge transfer resistance considering different temperature and state of charge. *J. Energy Storage* **2019**, *21*, 618–631. [CrossRef]
41. Barsukov, Y. *Battery Cell Balancing: What to Balance and How*; Texas Instruments: Dallas, TX, USA, 2009.
42. Daowd, M.; Omar, N.; Van Den Bossche, P.; Van Mierlo, J. Passive and active battery balancing comparison based on MATLAB simulation. In Proceedings of the 2011 IEEE Vehicle Power and Propulsion Conference, Chicago, IL, USA, 6–9 September 2011; IEEE: New York, NY, USA, 2011; pp. 1–7.
43. Turksoy, A.; Teke, A.; Alkaya, A. A comprehensive overview of the dc-dc converter-based battery charge balancing methods in electric vehicles. *Renew. Sustain. Energy Rev.* **2020**, *133*, 110274. [CrossRef]
44. Moral, C.G.; Laborda, D.F.; Alonso, L.S.; Guerrero, J.M.; Fernandez, D.; Pereda, C.R.; Reigosa, D.D. Battery internal resistance estimation using a battery balancing system based on switched capacitors. *IEEE Trans. Ind. Appl.* **2020**, *56*, 5363–5374. [CrossRef]
45. Madakam, S.; Lake, V.; Lake, V.; Lake, V. Internet of Things (IoT): A literature review. *J. Comput. Commun.* **2015**, *3*, 164. [CrossRef]
46. Mahadik, Y.; Thakre, M.; Kamble, S. Electric Vehicle Monitoring System Based on Internet of Things (IoT) Technologies. In *IoT and Analytics for Sensor Networks: Proceedings of ICWSNUCA 2021*; Springer: Singapore, 2022; pp. 311–322.
47. Jun, H.J.; Yang, H.S. Performance of the XMPP and the MQTT protocols on IEC 61850-based micro grid communication architecture. *Energies* **2021**, *14*, 5024. [CrossRef]
48. Zikria, Y.B.; Ali, R.; Afzal, M.K.; Kim, S.W. Next-generation internet of things (iot): Opportunities, challenges, and solutions. *Sensors* **2021**, *21*, 1174. [CrossRef]
49. Drake, N. Cloud computing beckons scientists. *Nature* **2014**, *509*, 543–544. [CrossRef] [PubMed]
50. Zhang, Y.; Liu, H.; Zhang, Z.; Luo, Y.; Guo, Q.; Liao, S. Cloud computing-based real-time global optimization of battery aging and energy consumption for plug-in hybrid electric vehicles. *J. Power Source* **2020**, *479*, 229069. [CrossRef]
51. Laaziri, M.; Benmoussa, K.; Khoulji, S.; Larbi, K.M.; Yamami, E.A. A comparative study of laravel and symfony PHP frameworks. *Int. J. Electr. Comput. Eng.* **2019**, *9*, 704. [CrossRef]
52. Lecun, Y.; Bengio, Y.; Hinton, G. Deep learning. *Nature* **2015**, *521*, 436–444. [CrossRef] [PubMed]
53. Han, D.; sang Uk, K.; Kim, S.; Lim, C.; Kin, J. Tensorflow-based Long-Short term Memory Network for SOH Prediction of a Lithium-ion Cell. In Proceedings of the 2019 IEEE Transportation Electrification Conference and Expo, Asia-Pacific (ITEC Asia-Pacific), Seogwipo, Republic of Korea, 8–10 May 2019; IEEE: New York, NY, USA, 2019; pp. 1–5.
54. Ulleri, P.; Mukund, S.; Sudarshan, K.; Zenith, K.; NB, S.S. Development of contactless employee management system with mask detection and body temperature measurement using TensorFlow. In Proceedings of the 2021 Sixth International Conference on Wireless Communications, Signal Processing and Networking (WiSPNET), Chennai, India, 25–27 March 2021; IEEE: New York, NY, USA, 2021; pp. 235–240.
55. Xu, Q.; Wu, M.; Khoo, E.; Chen, Z.; Li, X. A Hybrid Ensemble Deep Learning Approach for Early Prediction of Battery Remaining Useful Life. *IEEE CAA J. Autom. Sin.* **2023**, *10*, 177–187. [CrossRef]
56. Xu, W.; Yan, C. Prediction of Lithium-ion Battery Remaining Useful Life Based on Long Short Term Memory. In Proceedings of the 2022 IEEE International Conference on Advances in Electrical Engineering and Computer Applications (AEECA), Dalian, China, 20–21 August 2022; IEEE: New York, NY, USA, 2022; pp. 942–948.
57. Sulzer, V.; Marquis, S.G.; Timms, R.; Robinson, M.; Chapman, S.J. Python battery mathematical modelling (PyBaMM). *J. Open Res. Softw.* **2021**, *9*, 14. [CrossRef]
58. Karniadakis, G.E.; Kevrekidis, I.G.; Lu, L.; Perdikaris, P.; Wang, S.; Yang, L. Physics-informed machine learning. *Nat. Rev. Phys.* **2021**, *3*, 422–440. [CrossRef]
59. Aykol, M.; Herring, P.; Anapolsky, A. Machine learning for continuous innovation in battery technologies. *Nat. Rev. Mater.* **2020**, *5*, 725–727. [CrossRef]
60. Ma, G.; Xu, S.; Jiang, B.; Cheng, C.; Yang, X.; Shen, Y.; Yang, T.; Huang, Y.; Han, D.; Yuan, Y. Real-time personalized health status prediction of lithium-ion batteries using deep transfer learning. *Energy Environ. Sci.* **2022**, *15*, 4083–4094. [CrossRef]
61. Samanta, A.; Chowdhuri, S.; Williamson, S.S. Machine learning-based data-driven fault detection/diagnosis of lithium-ion battery: A critical review. *Electronics* **2021**, *10*, 1309. [CrossRef]
62. Sharma, P.; Bora, B.J. A Review of Modern Machine Learning Techniques in the Prediction of Remaining Useful Life of Lithium-Ion Batteries. *Batteries* **2022**, *9*, 13. [CrossRef]
63. Shi, D.; Zhao, J.; Wang, Z.; Zhao, H.; Eze, C.; Wang, J.; Lian, Y.; Burke, A.F. Cloud-Based Deep Learning for Co-Estimation of Battery State of Charge and State of Health. *Energies* **2023**, *16*, 3855. [CrossRef]
64. Chandran, V.; Patil, C.K.; Karthick, A.; Ganeshaperumal, D.; Rahim, R.; Ghosh, A. State of charge estimation of lithium-ion battery for electric vehicles using machine learning algorithms. *World Electr. Veh. J.* **2021**, *12*, 38. [CrossRef]
65. Yang, F.; Li, W.; Li, C.; Miao, Q. State-of-charge estimation of lithium-ion batteries based on gated recurrent neural network. *Energy* **2019**, *175*, 66–75. [CrossRef]

66. Bian, C.; He, H.; Yang, S. Stacked bidirectional long short-term memory networks for state-of-charge estimation of lithium-ion batteries. *Energy* **2020**, *191*, 116538. [CrossRef]
67. Jiao, M.; Wang, D.; Qiu, J. A GRU-RNN based momentum optimized algorithm for SOC estimation. *J. Power Source* **2020**, *459*, 228051. [CrossRef]
68. Chen, J.; Zhang, Y.; Li, W.; Cheng, W.; Zhu, Q. State of charge estimation for lithium-ion batteries using gated recurrent unit recurrent neural network and adaptive Kalman filter. *J. Energy Storage* **2022**, *55*, 105396. [CrossRef]
69. Hannan, M.A.; How, D.N.; Lipu, M.H.; Mansor, M.; Ker, P.J.; Dong, Z.Y.; Sahari, K.S.M.; Tiong, S.K.M.; Blaabjerg, F. Deep learning approach towards accurate state of charge estimation for lithium-ion batteries using self-supervised transformer model. *Sci. Rep.* **2021**, *11*, 19541. [CrossRef]
70. Shen, H.; Zhou, X.; Wang, Z.; Wang, J. State of charge estimation for lithium-ion battery using Transformer with immersion and invariance adaptive observer. *J. Energy Storage* **2022**, *45*, 103768. [CrossRef]
71. Roman, D.; Saxena, S.; Robu, V.; Pecht, M.; Flynn, D. Machine learning pipeline for battery state-of-health estimation. *Nat. Mach. Intell.* **2021**, *3*, 447–456. [CrossRef]
72. Tan, X.; Zhan, D.; Lyu, P.; Rao, J.; Fan, Y. Online state-of-health estimation of lithium-ion battery based on dynamic parameter identification at multi timescale and support vector regression. *J. Power Source* **2021**, *484*, 229233. [CrossRef]
73. Li, K.; Zhou, P.; Lu, Y.; Han, X.; Li, X.; Zheng, Y. Battery life estimation based on cloud data for electric vehicles. *J. Power Source* **2020**, *468*, 228192. [CrossRef]
74. Hong, J.; Wang, Z.; Chen, W.; Wang, L.Y.; Qu, C. Online joint-prediction of multi-forward-step battery SOC using LSTM neural networks and multiple linear regression for real-world electric vehicles. *J. Energy Storage* **2020**, *30*, 101459. [CrossRef]
75. Zhao, J.; Ling, H.; Liu, J.; Wang, J.; Burke, A.F.; Lian, Y. Machine learning for predicting battery capacity for electric vehicles. *eTransportation* **2023**, *15*, 100214. [CrossRef]
76. Sutton, R.S.; Barto, A.G. *Reinforcement Learning*, 2nd ed.; MIT Press Academic: Cambridge, MA, USA, 2018; pp. 443–480.
77. Park, S.; Pozzi, A.; Whitmeyer, M.; Perez, H.; Kandel, A.; Kim, G.; Choi, Y.; Joe, W.T.; Raimondo, D.M.; Moura, S. A Deep Reinforcement Learning Framework for Fast Charging of Li-ion Batteries. *IEEE Trans. Transp. Electrif.* **2022**, *8*, 2770–2784. [CrossRef]
78. Drummond, R.; Bizeray, A.M.; Howey, D.A.; Duncan, S.R. A feedback interpretation of the Doyle–Fuller–Newman lithium-ion battery model. *IEEE Trans. Control Syst. Technol.* **2019**, *24*, 1284–1295.
79. Li, W.; Zhu, J.; Xia, Y.: Gorji, M.B.; Wierzbicki, T. Data-driven safety envelope of lithium-ion batteries for electric vehicles. *Joule* **2019**, *3*, 2703–2715. [CrossRef]
80. Zhao, J.; Ling, H.; Wang, J.; Burke, A.F.; Lian, Y. Data-driven prediction of battery failure for electric vehicles. *Iscience* **2022**, *25*, 104172. [CrossRef]
81. Tang, X.; Guo, Q.; Li, M.; Wei, C.; Pan, Z.; Wang, Y. Performance analysis on liquid-cooled battery thermal management for electric vehicles based on machine learning. *J. Power Source* **2021**, *494*, 229727. [CrossRef]
82. Boonma, K.; Mesgarpour, M.; NajmAbad, J.M.; Alizadeh, R.; Mahian, O.; Dalkılıç, A.S.; Wongwises, S. Prediction of battery thermal behaviour in the presence of a constructal theory-based heat pipe (CBHP): A multiphysics model and pattern-based machine learning approach. *J. Energy Storage* **2022**, *48*, 103963. [CrossRef]
83. "Battery Data Set", NASA Prognostics Data Repository, NASA Ames Research Center, Moffett Field, CA. Available online: https://www.nasa.gov/content/prognostics-center-of-excellence-data-set-repository (accessed on 19 August 2022).
84. Severson, K.A.; Attia, P.M.; Jin, N.; Perkins, N.; Jiang, B.; Yang, Z.; Braatz, R.D. Data-driven prediction of battery cycle life before capacity degradation. *Nat. Energy* **2019**, *4*, 383–391. [CrossRef]
85. Weiss, K.; Khoshgoftaar, T.M.; Wang, D. A survey of transfer learning. *J. Big Data* **2016**, *3*, 9. [CrossRef]
86. Tao, F.; Qi, Q. Make more digital twins. *Nature* **2019**, *573*, 490–491. [CrossRef]
87. Carvalho, D.V.; Pereira, E.M.; Cardoso, J.S. Machine learning interpretability: A survey on methods and metrics. *Electronics* **2019**, *8*, 832. [CrossRef]

energies

MDPI

Article

Comparing Machine Learning Strategies for SoH Estimation of Lithium-Ion Batteries Using a Feature-Based Approach †

Iacopo Marri, Emil Petkovski, Loredana Cristaldi * and Marco Faifer

Department of Electronics, Information and Bioengineering, Politecnico di Milano, 20133 Milan, Italy; iacopomarri@gmail.com (I.M.); emil.petkovski@polimi.it (E.P.); marco.faifer@polimi.it (M.F.)
* Correspondence: loredana.cristaldi@polimi.it

† This paper is an extended version of our paper published in the 18th IMEKO TC10 Conference on Measurement for Diagnostic, Optimization and Control to Support Sustainability and Resilience 2022, pp. 109–113.

Abstract: Lithium-ion batteries play a vital role in many systems and applications, making them the most commonly used battery energy storage systems. Optimizing their usage requires accurate state-of-health (SoH) estimation, which provides insight into the performance level of the battery and improves the precision of other diagnostic measures, such as state of charge. In this paper, the classical machine learning (ML) strategies of multiple linear and polynomial regression, support vector regression (SVR), and random forest are compared for the task of battery SoH estimation. These ML strategies were selected because they represent a good compromise between light computational effort, applicability, and accuracy of results. The best results were produced using SVR, followed closely by multiple linear regression. This paper also discusses the feature selection process based on the partial charging time between different voltage intervals and shows the linear dependence of these features with capacity reduction. The feature selection, parameter tuning, and performance evaluation of all models were completed using a dataset from the Prognostics Center of Excellence at NASA, considering three batteries in the dataset.

Keywords: lithium-ion battery; machine learning; SoH; battery degradation; prognostics

Citation: Marri, I.; Petkovski, E.; Cristaldi, L.; Faifer, M. Comparing Machine Learning Strategies for SoH Estimation of Lithium-Ion Batteries Using a Feature-Based Approach. *Energies* **2023**, *16*, 4423. https://doi.org/10.3390/en16114423

Academic Editor: Djaffar Ould-Abdeslam

Received: 20 April 2023
Revised: 16 May 2023
Accepted: 19 May 2023
Published: 30 May 2023

1. Introduction

One of the primary challenges of modern life is global warming caused by the emission of greenhouse gases from burning fossil fuels, as well as the urgency to diminish reliance on non-renewable resources. Renewable energy generation has become a top priority for governments all around the world. Focusing on photovoltaics (PVs) and wind, which are generally considered non-dispatchable and only partially participate in maintaining grid stability [1], makes battery and other energy storage systems essential. Out of the various battery technologies currently in use, lithium-ion batteries have become the preferred choice, owing to their high power and energy density as well as long service life [2]. For these reasons, exclusively Li-ion batteries are used in electric vehicles [3], where maximizing energy density and minimizing the weight of the battery pack is crucial. Continuous research and investments are focused on this technology to improve its performance, robustness, and stability.

A battery management system (BMS) is commonly used to ensure safe and efficient operation of the battery pack by controlling the charge and discharge processes of the cells and providing cell balancing. To achieve this task, the BMS must accurately estimate crucial battery parameters, such as state of charge (SoC), state of health (SoH), and the remaining useful life (RUL) [4]. The SoC is related to the available capacity of the batteries. By knowing this factor, the BMS prevents overcharging or discharging of batteries. The SoH provides information about the aging status of the battery and is indicated by the rise of internal resistance or capacity decrease. On the other hand, the goal of the RUL prediction is to understand how long a battery will continue to operate before it fails or has unacceptable performance. Battery degradation is a highly variable process, depending on cell chemistry, the BMS, ambient conditions, and use patterns. For this reason, a considerable amount of model-based [5–9] and data-driven battery aging methods used for the SoH and end-of-life predictions of batteries can be found in the literature.

Recently, data-driven methods have gained popularity due to the availability of vast amounts of data, gathered through sensors and other monitoring devices, and advancements in the field of machine learning (ML). They do not necessarily rely on prior knowledge of the particular battery cell and are less expensive to develop compared to the model-based ones. Data-driven methods use large data sets to identify patterns and relationships that may not be easily discernible using traditional analytical techniques. Many different features, otherwise known as health indicators (His), have been used to build various ML strategies. Apart from capacity and resistance changes, these His are based on voltage charge–discharge limits, the amount of current, battery temperature [10–14], and incremental capacitance analysis (ICA) [15–18], as well as features derived from statistical analysis of the other health indicators [19].

In the literature, many machine learning techniques have been studied and used to perform SoH estimation [20,21] apply regression to model battery aging behavior and compare the RUL-prediction capabilities of two fitting functions, while in [22], a combination of an exponential function and regression analysis is used. The authors of [23] discuss a strategy based on support vector regression (SVR) and ICA curves obtained from partial charging data. Similarly, [14] uses partial charging segments of voltage under constant current charging and a support vector machine model. Another SVR strategy is presented in [24], based on curves of battery voltage as a function of charging capacity (V-Q). Finally, in [25], a solution is proposed based on the random forest algorithm. In [13], a gaussian process regression (GPR) model is used with four specific inputs extracted from the charging curves, and a grey relational analysis method is applied to analyze the relationship between features and SoH. The authors of [26] apply GPR to discover the relationship between capacity, storage temperature, and SoC of lithium-ion batteries. By optimizing the feature selection process with an automatic-relevance-determination (ARD) structure, they provide predictions for the calendar aging of batteries tested under different conditions. GPR combined with electrochemical impedance spectroscopy is used in [27], adopting many wave shapes to obtain an estimation of the capacity of the batteries. Novel health indicators related to the lithium diffusion coefficient are provided and validated.

In [28], a capacity-estimation method based on back-propagation neural networks (NN) and partial charging voltage segments, corresponding to 10–50% SOC, has been developed. Another solution based on recurrent neural networks (RNNSs) is proposed in [29], while in [30], an echo state network (ESN) has been used together with a model-based approach to predict the SoH evolution curve of the tested batteries, starting from cycles 80, 100, or 120. From the generated curves, predictions are made for the RUL. Due to the problem of vanishing or exploding gradient, traditional RNNs are not capable of dealing with long sequences in practice. The emergence of long short-term memory (LSTM) has provided a solution to this problem [31], and [32] utilized LSTM to build a RUL model of the lithium-ion battery. In [33], another method is proposed based on LSTM NNs and signal processing methods for SoH monitoring and RUL prediction of lithium-ion batteries.

In [34], the authors proposed an approach for SoH estimation based on SVR and a feature extraction procedure. In this paper, SVR is compared to other ML approaches, including multiple linear and polynomial regression and random forest. These classical ML strategies have been chosen because they offer a good compromise between light computational effort, applicability, and accuracy of results, while also providing higher model interpretability than complex NNs. The performances of all strategies are compared using a dataset from the Prognostics Center of Excellence at NASA, considering three batteries of the dataset. This work differentiates itself from the other aforecited papers, including the ones employing the same NASA dataset [10–13,18,28,30,33], based on the specific ML strategies implemented, the features used, and their feature numbers. Discussion is provided on the feature selection process based on partial charging times between different voltage limits, as well as the parameter tuning process of the different strategies. Finally, this research had the goal of minimizing the necessary number of features, considering models based on one-to-four features, and achieving optimal results with only two features for all considered ML strategies.

2. NASA Dataset

The NASA Ames Prognostics Center of Excellence (PCoE) released a data repository composed of six datasets of aged Li-ion batteries [35]. However, only the first of these datasets is suitable for prognostic degradation prediction, according to their guidelines. In this work, batteries 5, 6, and 7 were considered, which were tested until failure. The charging process follows the constant-current (CC) and constant-voltage (CV) protocol. More specifically, the cells are charged with a current of 1.5 A until the upper voltage limit of 4.2 V is met, after which CV charging proceeds until the current drops below 20 mA. The discharge phase is carried out at 2.7 V, 2.5 V, and 2.2 V, depending on the battery. Cycles are grouped into charge, discharge, or impedance cycles. For every cycle of every cell, various quantities are measured, including current, time, temperature, voltage, and discharge capacity. To control the environmental temperature, the tests were carried out in a climatic chamber.

3. Considered Machine Learning Strategies

3.1. Multiple Linear Regression and Stepwise Regression

Multiple linear regression (MLR) is a statistical approach for modeling the relationship between a target variable (y) and two or more available descriptor variables (x_i), otherwise called features, using a linear equation. Regression models are usually fitted using the least-squares approach, which minimizes the sum of the squared differences between the predicted and actual values of the target variable. However, fitting based on other criteria can be performed, such as least absolute deviations or minimization of a penalized version of the least-squares function, as in the case of ridge and lasso regression. MLR is a powerful tool for analyzing complex relationships between variables, but it assumes that the relationships are linear. When this is not the case, better results could be obtained using polynomial regression, which is a statistical technique that models the relationship between x_i and y as an n-th degree polynomial, thus fitting a nonlinear relationship. Polynomial regression utilizing multiple features can have many potential terms resulting from the features raised to a certain power or their combination.

Stepwise regression can be used to automatically identify the most important terms. It involves iteratively adding or removing different terms according to a stopping criterion, which can be based on the p-value, Akaike information criterion (AIC), Bayesian information criterion (BIC), value of the coefficient of determination (R^2), or adjusted R^2. The most popular stepwise methods are forward selection (FS), backward elimination (BE), and bidirectional elimination. In FS, the model starts with no terms and iteratively adds them until a stopping criterion is met. In BE, the model starts with all combinations and iteratively removes terms until a stopping criterion is met. For both the BE and the FS methods, the decision regarding a term is final and is not reconsidered. This is not the case with bidirectional elimination, which is a combination of forward and backward stepwise regression and starts with no terms. If the adjusted R^2 is considered as a stopping criterion, this method will first add the terms that produce the largest increase in the adjusted R^2 value. Eventually, the removal of terms can also occur if this results in maximum increases of the adjusted R^2.

In this work, models based on MLR, as well as second- and third-degree polynomial terms, have been constructed using bidirectional-elimination stepwise regression. The generated models based on stepwise regression were limited to second- and third-degree polynomial terms, including combinational terms. The adjusted R^2 was used as the stopping criterion. In all cases, fitting was performed by using the least-squares method.

3.2. Support Vector Regression

The support vector machine (SVM), in ML, is a well-known supervised learning model, used mainly for binary classification tasks. It has been extensively applied in predictive and diagnostic tasks, such as in [36], where a partial-discharge-curve approach is combined with the least-squares SVM to estimate the state of health (SoH) of Li-ion batteries. Similarly, in [37], the SVM is utilized on an electric-vehicle (EV) battery-usage-profile dataset generated by simulations to determine the SoH. The SVM searches for the optimal hyperplane that maximizes the distance from each training point, making it not only effective in classifying points but also in finding the most robust hyperplane. When the points are not linearly separable and a higher-dimensional feature space is needed, the kernel trick is used.

SVR is a version of the SVM adapted to perform regression tasks. SVR fits the error of its predictions within the limit ϵ while minimizing the loss function in Equation (1), which is called the L2 loss:

$$\begin{cases} \min \frac{1}{2}\beta'\beta \\ |Y_n - (X_n'\beta + b)| \le \epsilon \forall n \end{cases} \tag{1}$$

where

β', β—values that weight arrays, normal and transposed

Y_n—target values

X_n'—transposed descriptor array

b—bias

ϵ—maximum allowed error.

The ϵ constraint is then relaxed, introducing the slack variables and applying what is called the soft margin approach.

$$\begin{cases} \min \frac{1}{2}\beta'\beta + C\sum\limits_{n=1}^{N} (\xi_n + \xi_n^*) \\ Y_n - (X_n'\beta + b) \le \epsilon + \xi_n \forall n \\ (X_n'\beta + b) - Y_n \le \epsilon + \xi_n^* \forall n \end{cases} \tag{2}$$

where

ξ_n, ξ_n^*—slack variables for positive and negative error

C—weight associated with slack variables.

The prediction is expressed as a function of the training samples in Equation (3), in particular of those data points with either α_i or α_i^* different from 0, which are called support vectors.

$$\begin{cases} \beta = \sum\limits_{n=1}^{N} (\alpha_n - \alpha_n^*) X_n \\ f(x) = \sum\limits_{n=1}^{N} (\alpha_n - \alpha_n^*)(X_n'X) + b \end{cases} \tag{3}$$

In this paper, the SVR hyperparameters have been initially tuned with the MATLAB built-in function for SVR models, using the Bayesian optimization algorithm, and run for 500 iterations to define a good starting point for the hyperparameters. The tunable hyperparameters are as listed:

Box constraint: Coefficient C that weights the slack variables in Equation (1) and helps regulate overfitting.

Epsilon (ε): The value that defines the radius of the epsilon tube where the algorithm tries to contain the points or, in other words, the maximum error allowed.

Kernel scale: The value that rescales the predictors. Each value in the predictors is divided by the kernel scale value.

Kernel function: The value used to compute the similarity between data points in a higher-dimensional feature space.

Additional tuning of the hyperparameters was carried out during the validation process. The final values of the hyperparameters are shown in Table 1. The linear kernel function was selected because the features are quite proportional to the target value to estimate and working in a higher-dimensional space was unnecessary. In fact, different kernel functions led to lower validation accuracy.

Table 1. Hyperparameter values tuned for the implemented SVR model.

	Hyperparameter	Value
1	Box constraint	0.1989
2	Kernel scale	11.55
3	Epsilon	0.030
4	Kernel function	Linear

3.3. Random Forest

A random forest (RF) is an ensemble learning method that puts together many decision trees (DT) as weak learners and is one of the best-known and most used algorithms for supervised learning tasks. In [38], a RF is used to perform an incremental capacity analysis to estimate the capacity of lithium batteries by only feeding raw measurements of new data to the model.

A decision tree is a non-parametric algorithm that develops a tree by splitting the dataset over the values of its features and associates different subsets of the dataset to different nodes of the tree. First, the entire dataset is paired with the root of the tree. Next, the dataset is split into two parts according to a decision made over some of the features, and each part is associated with a new node child of the root, forming the second level of the tree. This behavior is recursively iterated until subsets of the dataset contain only one value or a stop criterion is met, with the final subsets representing the leaves of the tree. Each splitting is made over the value of typically one feature, and the choice for the optimal split is made by finding the feature and its splitting value that optimize a given metric. *MSE* metric minimization was used in this work:

$$\begin{cases} MSE(S) = \frac{1}{N}\sum_{i=1}^{N}(y_i - y)^2 \\ splitMSE(F,V) = \frac{N_L}{N}MSE(S_L) + \frac{N_R}{N}MSE(S_R) \\ (F^*,V^*) = argmin(splitMSE(F,V)) \end{cases} \quad (4)$$

where y is the mean of the target values in the set S, y_i is the i-th target value, and N is the number of samples in the set. $SplitMSE$, S_L, and S_R are the weighted error, left, and right subsets, respectively, generated by splitting S over the feature F at value V, while N_L and N_R are the numbers of data points, respectively, in the left and right subsets. F^* and S^* are the optimal feature-value pair to split the set. Other metrics, such as Gini impurity or information gain, can be used.

However, decision trees are considered weak learners and strongly tend to overfit. A random forest is an ensemble algorithm whose mechanism consists of combining multiple decision trees with a bagging technique to provide higher accuracy and robustness than a single tree, reducing overfitting. Bagging is, in fact, known for reducing the variance of the model (as opposed to boosting, which reduces bias) by training each tree (or learner in general) on a randomly selected subset of the training data with replacement (bootstrapping), hence introducing diversity in the training data. What diversifies the random forest from the standard tree bagging ensemble is the use of subsets of randomly selected features for each tree in the forest, which helps to reduce correlation between each learner, thus reducing overfitting. In this work, one-third of the total features were randomly used to train each single decision tree.

4. Feature Selection

As aforementioned, the considered approaches were applied by considering a specific feature of the batteries. In most battery applications, the charging stage is conducted in a more repeatable way. While different chargers can be used, which will result in different charging profiles, many charging cycles will be the same or very similar. On the other hand, the battery discharge cycles vary greatly depending on the application and use patterns. Even though the charging phase is more similar between different cycles, complete charging cycles are by no means guaranteed. For this reason, a small portion of the charging curve of voltage was used to extract useful information. More specifically, the extracted feature is the partial charging time (PCT) necessary for the battery to charge by some small voltage range.

In Figure 1, the battery voltage versus time during charging for different cycles is represented. Unsurprisingly, the charging time decreases as the battery ages and the global capacity decreases. In fact, the charging time is halved near the final cycles compared to the initial ones. It is further noted that the beginning of the charging process is characterized by a high derivative and is therefore difficult to appreciate the time differences between different cycles. On the other hand, the middle part extends for a longer period of time and is more suitable for PCT feature extraction. This is why, in this work, the lower voltage limit of 3.7 V was set for the feature extraction process.

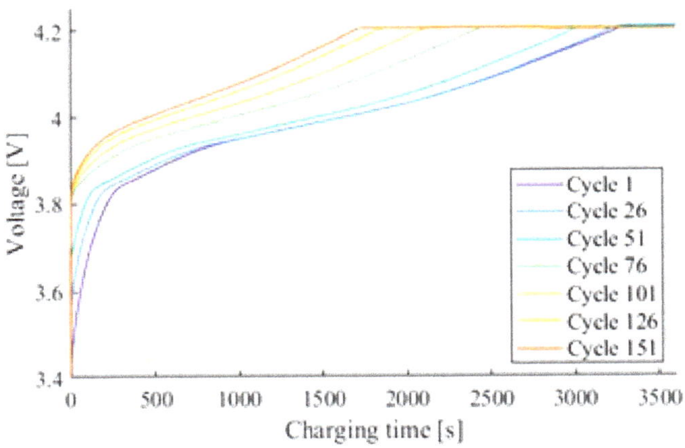

Figure 1. Charging curves at different cycles for battery 5.

5. Results and Discussion

The initial choice of the voltage range and limits was made empirically by computing four features over the limits of 3.7–4.1 volts with a voltage range of 0.1 V. More specifically, the first feature represents the evolution of the charging time between 3.7 V and 3.8 V over the number of cycles, the second feature uses the range of 3.8 V to 3.9 V, etc. In Figure 2, the value of the considered features as a function of the number of cycles is plotted. The first PCT feature computed for the lowest voltage values, from 3.7 V to 3.8 V, appeared to be an almost flat curve, containing no variance and thus very little information regarding the data. Conversely, the features computed from 3.8 V to 4.1 V have a higher variance and hence are more descriptive of the aging phenomena.

To find the optimal features and model parameters, from the voltage limits of 3.7–4.1 V, many feature sets were created. These sets differ from each other depending on the number of features, the upper and lower voltage limits used, and the voltage range. For each feature set, the models obtained using the different ML strategies are compared.

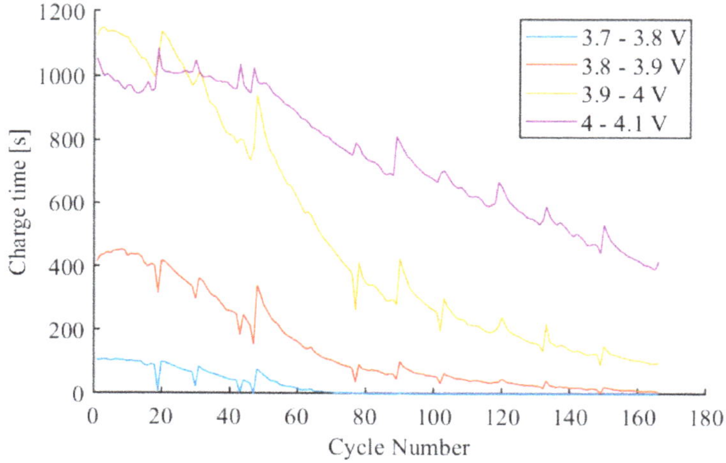

Figure 2. PCT values calculated considering the voltage limits of 3.7 V to 4.1 V, with a voltage range of 0.1 V, as a function of number of cycles, for battery 5.

The fitting accuracy of the various models was assessed through the value of the coefficient of determination (R^2). It is a measure used in statistics, indicating how much a hypothesis describes the variance of the data. In other words, it is a measure of how well a model can fit the data. R^2 is described as

$$\begin{cases} R^2 = 1 - \frac{SS_r}{SS_t} \\ SS_r = \sum_i (y_i - f_i)^2 \\ SS_t = \sum_i (y_i - \overline{y})^2 \end{cases} \quad (5)$$

where

SS_r—residual sum of squares
SS_t—total sum of squares
y_i—target value
f_i—estimated value
$\overline{y}$—mean of the target values.

A three-fold cross-validation (CV) procedure was applied to the three batteries of the dataset to find the best features and ML strategies. This means the SoH evolution of each battery was estimated based on the data of the other two batteries. The results are shown in Tables 2 and 3 for the voltage ranges of 0.1 V and 0.05 V, respectively. Initially, a smaller voltage range of 0.025 V and a larger voltage range of 0.2 V were also considered. However, the smaller voltage range resulted in features with low variability for most voltage limits and produced inferior results compared to the ones presented in Tables 2 and 3. The larger range of 0.2 V and higher ranges did not improve the SoH-estimation capability of the models. Since minimizing the voltage range was one of the objectives to ensure that the features would be available, even in the case of partial charging cycles, the ranges of 0.05 V and 0.1 V were regarded as optimal, and the higher voltage ranges were not further analyzed or presented.

Table 2. Three-fold CV results for a voltage range of 0.1 V.

Feature Set	Voltage Range	Number of Features	Mean Validation R^2				
			Linear Regression	Second-Degree Polynomial Regression	Third-Degree Polynomial Regression	SVR	Random Forest
A1	3.7–3.8 V	1	0.538	0.595	0.631	0.613	0.660
A2	3.8–3.9 V	1	0.918	0.916	0.921	0.939	0.902
A3	3.9–4 V	1	0.947	0.927	0.930	0.963	0.904
A4	4–4.1 V	1	0.554	0.535	/	0.759	0.743
A5	3.7–3.9 V	2	0.901	0.897	0.894	0.917	0.838
A6	3.8–4 V	2	0.945	0.939	0.946	0.971	0.909
A7	3.9–4.1 V	2	0.942	0.835	0.652	0.961	0.877

Table 3. Three-fold CV results for a voltage range of 0.05 V.

Feature Set	Voltage Range	Number of Features	Mean Validation R^2				
			Linear Regression	Second-Degree Polynomial Regression	Third-Degree Polynomial Regression	SVR	Random Forest
B1	3.8–3.85 V	1	0.781	0.896	0.897	0.810	0.878
B2	3.85–3.9 V	1	0.939	0.900	0.947	0.947	0.908
B3	3.9–3.95 V	1	0.937	0.918	0.916	0.949	0.896
B4	3.95–4 V	1	0.895	0.900	0.880	0.938	0.898
B5	3.8–3.9 V	2	0.931	0.909	0.928	0.941	0.901
B6	3.85–3.95 V	2	0.934	0.928	0.936	0.947	0.909
B7	3.9–4 V	2	0.950	0.912	0.922	0.968	0.903
B8	3.75–3.9 V	3	0.915	0.885	0.893	0.935	0.883
B9	3.8–3.95 V	3	0.899	0.911	0.895	0.948	0.905
B10	3.85–4 V	3	0.943	0.938	0.896	0.964	0.910
B11	3.9–4.05 V	3	0.939	0.756	0.884	0.962	0.898
B12	3.8–4 V	4	0.936	0.922	0.885	0.966	0.907
B13	3.85–4.05 V	4	0.931	0.864	/	0.958	0.911
B14	3.9–4.1 V	4	0.934	0.775	/	0.972	0.892

Table 2 shows the feature sets of partial charging times obtained for a voltage range of 0.1 volts. The first four single feature sets (A1–A4) explore the whole voltage range of 3.7 to 4.1 volts. Unsurprisingly, they show that all ML strategies perform better when the voltage limits of 3.8–3.9 V (A2) or 3.9–4 V (A3) are used as a feature. More specifically, the best results are obtained for the voltage limits 3.9–4 V when a single feature is used. Additionally, Table 2 shows that if the feature set is built from two features based on the limits of 3.8–3.9 V and 3.9–4 V (A6), there is only a marginal improvement in the R^2 value. In any case, the best results for single and double feature sets are A3 and A6.

Table 3 presents the feature sets obtained for a voltage range of 0.05 volts. In this case, feature sets consisting of one to four features were constructed. For example, B1 is a feature set of a single feature, which is the PCT between the voltage limits of 3.8 V to 3.85 V, while B6 consists of two features, which are the PCTs between the limits of 3.8 to 3.85 V and 3.85 to 3.9 V. The best results, per number of features, are B3, B7 B10, and B14. Using a single feature, even for the voltage range of 0.05 V, is sufficient if the voltage limits are between 3.85 and 4 volts. There is marginal improvement when two features are used; however, a further increase in the number of features does not lead to any meaningful increase of R^2. Considering the models of both tables, it can be noted that SVR delivers slightly better results than the other considered ML strategies for all feature sets. Still, using MLR also leads to satisfactory results. Furthermore, when comparing the three strategies based on regression, no significant improvement in the R^2 value is observed when increasing the polynomial order using stepwise regression. That means the PCT features and capacity reduction, as functions of the number of cycles, have a strong linear dependence. Hence, high model complexity will not result in an improvement of the results if the correct voltage range of 3.8–4 V has been selected. Actually, a drop in the mean validation R^2 value can even be observed in some cases due to overfitting the training data of the higher-complexity models. This is especially apparent in models built from a higher number of features (A7 and B12 to B14). However, some improvement when increasing the polynomial order can also be observed for the voltage range of 3.7–3.8 V, which has low variance. Finally, the models based on RF demonstrated worse performance than those of MLR and SVR.

The models based on feature sets A3, A6, B3, B7, B10, and B14 all represent satisfactory performance. Having the goal of minimizing the number of features and the voltage range, the authors consider the models based on feature set B7 as the overall best. The plots for the capacity estimation of all the batteries using MLR, SVR, and RF are plotted in Figures 3–5, respectively.

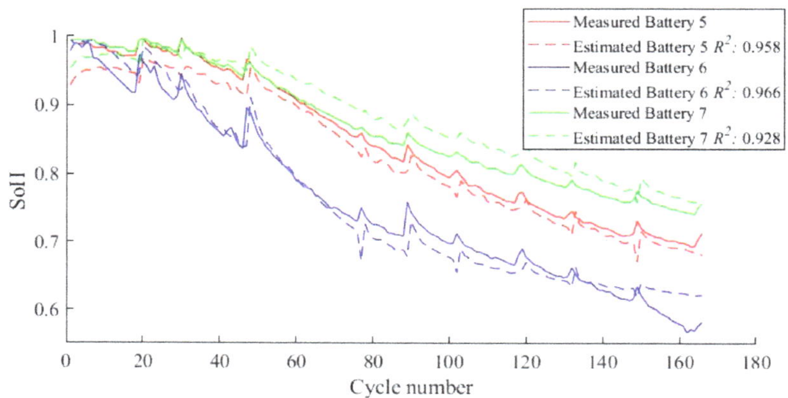

Figure 3. SoH estimation of each battery achieved using a model based on MLR and trained on the other two batteries.

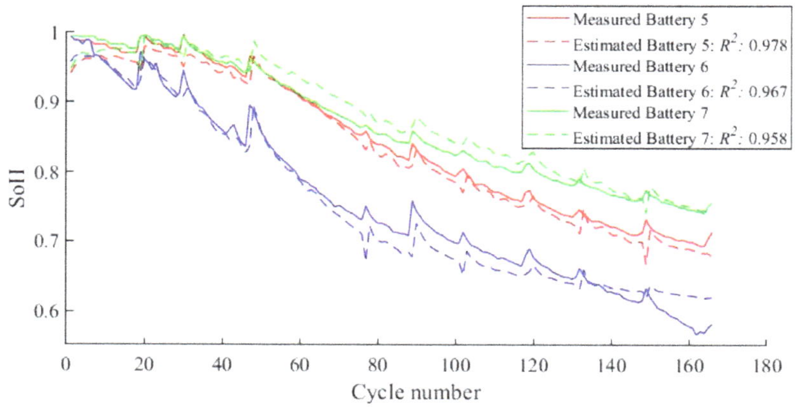

Figure 4. SoH estimation of each battery achieved using a model based on SVR and trained on the other two batteries.

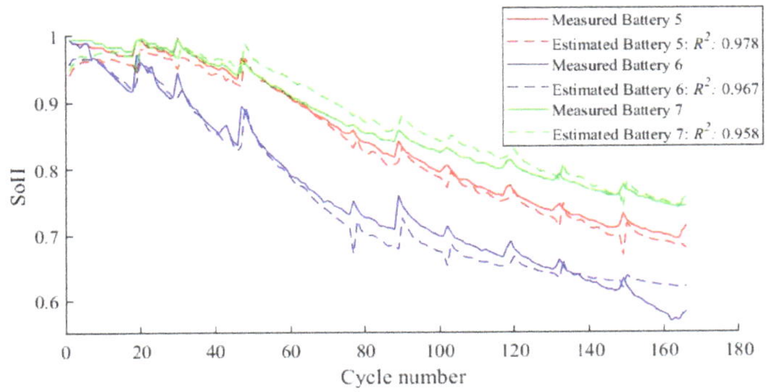

Figure 5. SoH estimation of each battery achieved using a model based on RF and trained on the other two batteries.

All three figures display the previously mentioned three-fold CV. For example, the SoH estimation of battery 5 was done with a model trained using the data of the chosen feature set of batteries 6 and 7. The full lines represent the measured SoH for the batteries, while the dashed lines represent the estimated SoH over the number of cycles. Figures 3 and 4 show that MLR and SVR accurately model the SoH of the batteries, even registering the peaks in the SoH function that are due to the rest time of the battery. Likewise, the RF is able to model batteries 5 and 7 with similar success, but the same cannot be said about battery 6, as is evident in Figure 5. After the SoH of battery 6 falls to around 0.7, the estimation begins to diverge from the measurement because batteries 5 and 7, which were used for training, do not contain data with SoH lower than 0.7.

The random forest and decision trees are indeed well known for their inability to extrapolate, that is, make estimations for predictor values lying outside of the range of the observed data. From Figure 5, it is clear that the SoH value of battery 6 from cycle 90 onwards is lower than that of any other cycle of the training batteries; hence, the decision trees will not be able to correctly estimate that target value. Furthermore, Figure 6 shows that also the feature value for battery 6 is lower than that of the other batteries. Consequently, the branches of the decision trees built on batteries 5 and 7 will "explore" the features in a range that does not include the values of battery 6 predictors after cycle 90. Hence, after this cycle number, all the decision trees of the random forest will infer the lowest observed SoH value for battery 6, which will be around 0.7 because the training data is composed of batteries 5 and 7. This is the reason for the observed flat line output. It is important to specify that this result does not imply that the RF is not a suitable solution for the general problem of battery prognostic because this precise case is strictly related to the dataset distribution and data scarcity.

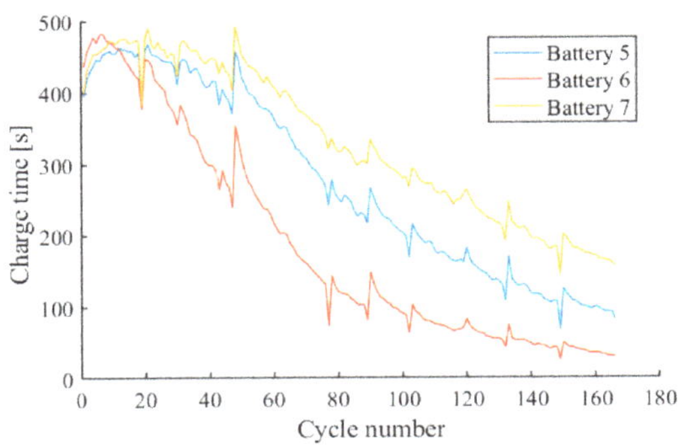

Figure 6. PCT for voltage range 3.9–3.95 V for all three batteries.

6. Conclusions

Accurate SoH estimation is essential for the safe and reliable operation of lithium-ion batteries. This paper compares SoH-estimation models based on the classical ML strategies of MLR, polynomial regression, SVR, and RF, which offer good trade-offs between applicability, light computation effort, and accuracy of results. Discussion is provided on the feature selection process and optimal number of features.

The partial charging time proved to be a good indicator of battery aging as long as the proper voltage limits were selected, and the partial charging phase was equal at every cycle. To find the optimal features, 21 feature sets were built considering different voltage limits and the two voltage ranges of 0.1 and 0.05 V. The best results were obtained when considering the voltage limits of 3.8 to 4 volts for both ranges of 0.1 V and 0.05 V. The quality of the features degrades significantly for a minimum voltage of less than 3.7 V due to small variance. Results showed that models based on one or two features are optimal.

Furthermore, the PCT feature demonstrated a linear dependence with capacity reduction as a function of number of cycles. Consequently, MLR produced very accurate results, and the use of polynomial regression was not justified. The overall best performance for all feature sets was achieved using SVR, especially when slightly lower voltage limits were considered. Finally, the RF had the worst performance when facing the limited dataset.

Author Contributions: Conceptualization, E.P., I.M. and L.C.; methodology, E.P. and I.M.; software, I.M. and E.P.; validation, I.M. and M.F.; formal analysis, E.P.; investigation, I.M.; resources, I.M.; data curation, L.C.; writing—original draft preparation, E.P. and I.M.; writing—review and editing, E.P. and I.M.; visualization, I.M.; supervision, L.C., M.F. and E.P.; project administration, L.C. and M.F.; funding acquisition, L.C. and M.F. All authors have read and agreed to the published version of the manuscript.

Funding: This research received no external funding.

Data Availability Statement: Not applicable.

Conflicts of Interest: The authors declare no conflict of interest.

References

1. Cristaldi, L.; Faifer, M.; Laurano, C.; Ottoboni, R.; Petkovski, E.; Toscani, S. Power Generation Control Algorithm for the Participation of Photovoltaic Panels in Network Stability. *IEEE Trans. Instrum. Meas.* **2023**, *72*, 1–9. [CrossRef]
2. Pelletier, S.; Jabali, O.; Laporte, G.; Veneroni, M. Battery Degradation and Behaviour for Electric Vehicles: Review and Numerical Analyses of Several Models. *Transp. Res. Part B Methodol.* **2017**, *103*, 158–187. [CrossRef]
3. Werling, T.; Geuting, P.; Höschele, P.; Ellersdorfer, C.; Sinz, W. Investigation of the Electro-Mechanical Behavior of Automotive High Voltage Busbars under Combined Electrical Load with Varying Indenter Geometry and Environmental Conditions. *J. Energy Storage* **2020**, *32*, 101861. [CrossRef]
4. Campestrini, C.; Horsche, M.F.; Zilberman, I.; Heil, T.; Zimmermann, T.; Jossen, A. Validation and Benchmark Methods for Battery Management System Functionalities: State of Charge Estimation Algorithms. *J. Energy Storage* **2016**, *7*, 38–51. [CrossRef]
5. Tanim, T.R.; Rahn, C.D.; Wang, C.Y. A Temperature Dependent, Single Particle, Lithium Ion Cell Model Including Electrolyte Diffusion. *J. Dyn. Syst. Meas. Control Trans. ASME* **2015**, *137*, 011005. [CrossRef]
6. Li, X.; Fan, G.; Rizzoni, G.; Canova, M.; Zhu, C.; Wei, G. A Simplified Multi-Particle Model for Lithium Ion Batteries via a Predictor-Corrector Strategy and Quasi-Linearization. *Energy* **2016**, *116*, 154–169. [CrossRef]
7. Petit, M.; Prada, E.; Sauvant-Moynot, V. Development of an Empirical Aging Model for Li-Ion Batteries and Application to Assess the Impact of Vehicle-to-Grid Strategies on Battery Lifetime. *Appl. Energy* **2016**, *172*, 398–407. [CrossRef]
8. Barcellona, S.; Colnago, S.; Dotelli, G.; Latorrata, S.; Piegari, L. Aging Effect on the Variation of Li-Ion Battery Resistance as Function of Temperature and State of Charge. *J. Energy Storage* **2022**, *50*, 104658. [CrossRef]
9. Xu, B.; Oudalov, A.; Ulbig, A.; Andersson, G.; Kirschen, D.S. Modeling of Lithium-Ion Battery Degradation for Cell Life Assessment. *IEEE Trans. Smart Grid* **2018**, *9*, 1131–1140. [CrossRef]
10. Cui, Z.; Wang, C.; Gao, X.; Tian, S. State of Health Estimation for Lithium-Ion Battery Based on the Coupling-Loop Nonlinear Autoregressive with Exogenous Inputs Neural Network. *Electrochim. Acta* **2021**, *393*, 139047. [CrossRef]
11. Liu, D.; Zhou, J.; Liao, H.; Peng, Y.; Peng, X. A Health Indicator Extraction and Optimization Framework for Lithium-Ion Battery Degradation Modeling and Prognostics. *IEEE Trans. Syst. Man. Cybern. Syst.* **2015**, *45*, 915–928. [CrossRef]

12. Cao, M.; Zhang, T.; Wang, J.; Liu, Y. A Deep Belief Network Approach to Remaining Capacity Estimation for Lithium-Ion Batteries Based on Charging Process Features. *J. Energy Storage* **2022**, *48*, 103825. [CrossRef]
13. Yang, D.; Zhang, X.; Pan, R.; Wang, Y.; Chen, Z. A Novel Gaussian Process Regression Model for State-of-Health Estimation of Lithium-Ion Battery Using Charging Curve. *J. Power Sources* **2018**, *384*, 387–395. [CrossRef]
14. Feng, X.; Weng, C.; He, X.; Han, X.; Lu, L.; Ren, D.; Ouyang, M. Online State-of-Health Estimation for Li-Ion Battery Using Partial Charging Segment Based on Support Vector Machine. *IEEE Trans. Veh. Technol.* **2019**, *68*, 8583–8592. [CrossRef]
15. Tian, J.; Xiong, R.; Yu, Q. Fractional-Order Model-Based Incremental Capacity Analysis for Degradation State Recognition of Lithium-Ion Batteries. *IEEE Trans. Ind. Electron.* **2019**, *66*, 1576–1584. [CrossRef]
16. Ansean, D.; Garcia, V.M.; Gonzalez, M.; Blanco-Viejo, C.; Viera, J.C.; Pulido, Y.F.; Sanchez, L. Lithium-Ion Battery Degradation Indicators Via Incremental Capacity Analysis. *IEEE Trans. Ind. Appl.* **2019**, *55*, 2992–3002. [CrossRef]
17. Stroe, D.I.; Schaltz, E. SOH Estimation of LMO/NMC-Based Electric Vehicle Lithium-Ion Batteries Using the Incremental Capacity Analysis Technique. In Proceedings of the 2018 IEEE Energy Conversion Congress and Exposition (ECCE), Portland, OR, USA, 23–27 September 2018; pp. 2720–2725. [CrossRef]
18. He, J.; Wei, Z.; Bian, X.; Yan, F. State-of-Health Estimation of Lithium-Ion Batteries Using Incremental Capacity Analysis Based on Voltage-Capacity Model. *IEEE Trans. Transp. Electrif.* **2020**, *6*, 417–426. [CrossRef]
19. Severson, K.A.; Attia, P.M.; Jin, N.; Perkins, N.; Jiang, B.; Yang, Z.; Chen, M.H.; Aykol, M.; Herring, P.K.; Fraggedakis, D.; et al. Data-Driven Prediction of Battery Cycle Life before Capacity Degradation. *Nat. Energy* **2019**, *4*, 383–391. [CrossRef]
20. Barcellona, S.; Cristaldi, L.; Faifer, M.; Petkovski, E.; Piegari, L.; Toscani, S. State of Health Prediction of Lithium-Ion Batteries. In Proceedings of the 2021 IEEE International Workshop on Metrology for Industry 4.0 & IoT (MetroInd4.0&IoT), Virtual Event, 7–9 June 2021; pp. 12–17.
21. Lashgari, F.; Petkovski, E.; Cristaldi, L. State of Health Analysis for Lithium-Ion Batteries Considering Temperature Effect. In Proceedings of the 2022 IEEE International Workshop on Metrology for Extended Reality, Artificial Intelligence and Neural Engineering, MetroXRAINE 2022—Proceedings, Rome, Italy, 26–28 October 2022; pp. 40–45. [CrossRef]
22. Chen, C.; Pecht, M. Prognostics of Lithium-Ion Batteries Using Model-Based and Data-Driven Methods. In Proceedings of the IEEE 2012 Prognostics and System Health Management Conference, PHM-2012, Beijing, China, 23–25 May 2012; pp. 12–17. [CrossRef]
23. Weng, C.; Cui, Y.; Sun, J.; Peng, H. On-Board State of Health Monitoring of Lithium-Ion Batteries Using Incremental Capacity Analysis with Support Vector Regression. *J. Power Sources* **2013**, *235*, 36–44. [CrossRef]
24. Weng, C.; Sun, J.; Peng, H. Model Parametrization and Adaptation Based on the Invariance of Support Vectors with Applications to Battery State-of-Health Monitoring. *IEEE Trans. Veh. Technol.* **2015**, *64*, 3908–3917. [CrossRef]
25. Chen, Z.; Sun, M.; Shu, X.; Shen, J.; Xiao, R. On-Board State of Health Estimation for Lithium-Ion Batteries Based on Random Forest. In Proceedings of the 2018 IEEE International Conference on Industrial Technology (ICIT), Lyon, France, 19–22 February 2018; pp. 1754–1759.
26. Liu, K.; Li, Y.; Hu, X.; Lucu, M.; Widanage, W.D. Gaussian Process Regression with Automatic Relevance Determination Kernel for Calendar Aging Prediction of Lithium-Ion Batteries. *IEEE Trans. Industr. Inform.* **2020**, *16*, 3767–3777. [CrossRef]
27. Su, X.; Sun, B.; Wang, J.; Zhang, W.; Ma, S.; He, X.; Ruan, H. Fast Capacity Estimation for Lithium-Ion Battery Based on Online Identification of Low-Frequency Electrochemical Impedance Spectroscopy and Gaussian Process Regression. *Appl. Energy* **2022**, *322*, 119516. [CrossRef]
28. Tian, Y.; Dong, Q.; Tian, J.; Li, X.; Kukkapalli, V.K.; Kim, S.; Thomas, S.A. Capacity Estimation of Lithium-Ion Batteries Based on Multiple Small Voltage Sections and BP Neural Networks. *Energies* **2023**, *16*, 674. [CrossRef]
29. Eddahech, A.; Briat, O.; Bertrand, N.; Delétage, J.Y.; Vinassa, J.M. Behavior and State-of-Health Monitoring of Li-Ion Batteries Using Impedance Spectroscopy and Recurrent Neural Networks. *Int. J. Electr. Power Energy Syst.* **2012**, *42*, 487–494. [CrossRef]
30. Catelani, M.; Ciani, L.; Fantacci, R.; Patrizi, G.; Picano, B. Remaining Useful Life Estimation for Prognostics of Lithium-Ion Batteries Based on Recurrent Neural Network. *IEEE Trans. Instrum. Meas.* **2021**, *70*, 3524611. [CrossRef]
31. Zhang, Y.; Xiong, R.; He, H.; Pecht, M.G. Long Short-Term Memory Recurrent Neural Network for Remaining Useful Life Prediction of Lithium-Ion Batteries. *IEEE Trans. Veh. Technol.* **2018**, *67*, 5695–5705. [CrossRef]
32. Marri, I.; Petkovski, E.; Cristaldi, L.; Faifer, M. Battery Remaining Useful Life Prediction Supported by Long Short-Term Memory Neural Network. In Proceedings of the IEEE International Instrumentation and Measurement Technology Conference (I2MTC), Kuala Lumpur, Malaysia, 22–25 May 2023; pp. 1–6.
33. Qu, J.; Liu, F.; Ma, Y.; Fan, J. A Neural-Network-Based Method for RUL Prediction and SOH Monitoring of Lithium-Ion Battery. *IEEE Access* **2019**, *7*, 87178–87191. [CrossRef]
34. Marri, I.; Petkovski, E.; Cristaldi, L.; Faifer, M. Lithium-Ion Batteries Soh Estimation, Based on Support-Vector Regression and a Feature-Based Approach. In Proceedings of the 18th IMEKO TC10 Conference on Measurement for Diagnostic, Optimisation and Control to Support Sustainability and Resilience, Warsaw, Poland, 26–27 September 2022; pp. 109–113.
35. Saha, B.; Goebel, K. Nasa Ames Prognostic Data Repository. NASA Ames Moffet Field, CA, USA. 2007. Available online: https://www.nasa.gov/content/prognostics-center-of-excellence-data-set-repository (accessed on 15 January 2022).
36. Chen, Z.; Xia, X.; Sun, M.; Shen, J.; Xiao, R. State of Health Estimation of Lithium-Ion Batteries Based on Fixed Size LS-SVM. In Proceedings of the 2018 IEEE Vehicle Power and Propulsion Conference (VPPC), Chicago, IL, USA, 27–30 August 2018; pp. 1–6.

37. Klass, V.; Behm, M.; Lindbergh, G. A Support Vector Machine-Based State-of-Health Estimation Method for Lithium-Ion Batteries under Electric Vehicle Operation. *J. Power Sources* **2014**, *270*, 262–272. [CrossRef]
38. Li, Y.; Zou, C.; Berecibar, M.; Nanini-Maury, E.; Chan, J.C.W.; van den Bossche, P.; Van Mierlo, J.; Omar, N. Random Forest Regression for Online Capacity Estimation of Lithium-Ion Batteries. *Appl. Energy* **2018**, *232*, 197–210. [CrossRef]

energies

MDPI

Article

Calendar Aging Effect on the Open Circuit Voltage of Lithium-Ion Battery

Simone Barcellona *, Lorenzo Codecasa, Silvia Colnago and Luigi Piegari

Department of Electronics, Information and Bioengineering, Politecnico di Milano, 20133 Milan, Italy; lorenzo.codecasa@polimi.it (L.C.); silvia.colnago@polimi.it (S.C.); luigi.piegari@polimi.it (L.P.)
* Correspondence: simone.barcellona@polimi.it

Abstract: In recent years, lithium-ion batteries (LiBs) have gained a lot of importance due to the increasing use of renewable energy sources and electric vehicles. To ensure that batteries work properly and limit their degradation, the battery management system needs accurate battery models capable of precisely predicting their parameters. Among them, the state of charge (SOC) estimation is one of the most important, as it enables the prediction of the battery's available energy and prevents it from operating beyond its safety limits. A common method for SOC estimation involves utilizing the relationship between the state of charge and the open circuit voltage (OCV). On the other hand, the latter changes with battery aging. In a previous work, the authors studied a simple function to model the OCV curve, which was expressed as a function of the absolute state of discharge, q, instead of SOC. They also analyzed how the parameters of such a curve changed with the cycle aging. In the present work, a similar analysis was carried out considering the calendar aging effect. Three different LiB cells were stored at three different SOC levels (low, medium, and high levels) for around 1000 days, and an analysis of the change in the OCV-q curve model parameters with the calendar aging was performed.

Keywords: lithium-ion batteries; calendar aging; OCV curve; state of charge estimation

Citation: Barcellona, S.; Codecasa, L.; Colnago, S.; Piegari, L. Calendar Aging Effect on the Open Circuit Voltage of Lithium-Ion Battery. *Energies* **2023**, *16*, 4869. https://doi.org/10.3390/en16134869

Academic Editor: Djaffar Ould-Abdeslam

Received: 15 May 2023
Revised: 17 June 2023
Accepted: 19 June 2023
Published: 22 June 2023

1. Introduction

The recent energy crisis and the issue of climate change are spurring governments to incentivize the production of energy through renewable energy sources (RESs) and electric mobility to reduce gas emissions and reliance on fossil fuels. Unfortunately, the power produced by RESs is intermittent and unpredictable, which means that they may not provide energy when needed or, conversely, produce it when not required. Therefore, to make the most of them, it is convenient to install energy storage systems. Additionally, energy storage systems are also necessary for electric mobility, preferably with high energy density, high efficiency, and a long lifecycle. In this scenario, lithium-ion batteries (LiBs) are currently the preferred technology due to their high energy density, relatively high power density, and low self-discharge, making them suitable for both mobile and stationary applications [1].

LiBs are subjected to various degradation mechanisms that limit their lifespan and degrade their performance. These degradation mechanisms are caused by storage and operating conditions. The former is known as calendar aging and mainly depends on the temperature and state of charge (SOC) as a function of the time for which the battery has been stored under those conditions. The latter, commonly referred to as cycle aging, mainly depends on the temperature, current rate, SOC, and charging and discharging cut-off voltages as a function of the total charge exchanged with the battery [2,3]. To reduce the degradation and extend the lifespan of batteries, it is essential to control battery operations properly and store them under optimal conditions. The former is ensured by the battery management system (BMS), which predicts battery parameters such as the

SOC, temperature, and state of health (SOH), using appropriate models and controlling the battery accordingly. Therefore, it is crucial for the aforementioned battery models to be as accurate as possible [4].

The estimation of the SOC is of paramount importance as it allows the determination of the remaining available battery capacity and ensures that the battery operates within its safety limits. In fact, properly controlling the SOC of the battery helps avoid working in an operating area where its degradation increases. Many different SOC estimation methods can be found in the literature. In recent years, some filter-based methods (Kalman filter [5], unscented Kalman filter [6], etc.) together with some observer methods (sliding mode observer [7], nonlinear observer [8], etc.) and intelligent algorithms (fuzzy logic [9], neural networks [10], etc.) have been proposed. However, traditional methods are still preferred in BMSs [11]. Among them, Coulomb counting remains the most widely used method due to its simplicity and execution speed [12]. Unfortunately, this method has some drawbacks, including heavy reliance on initial SOC values, the need for highly accurate current sensors, and susceptibility to cumulative errors. Other traditional methods are the model-based approaches which are quite accurate and simple [13]. On the other hand, they require a good knowledge of the open circuit voltage (OCV) curve [14], which is typically obtained through lengthy tests involving full charge or discharge of the battery at very low current rates or through the hybrid pulse power characterization method [15]. Furthermore, the model-based approaches are generally used to correct the Coulomb counting estimation method [11].

Different models can be found in the literature that are able to describe the dependence of the OCV curve as a function of the SOC. They can be mainly divided into table-based models and analytical models. In the former, pairs of OCV and SOC values are stored in a table, and interpolation is performed between the stored values [16,17]. The main advantage of this method is its low computational requirements, but it may require significant memory if high precision is desired. The analytical models, instead, use mathematical functions to describe the OCV curve and can be further classified into linear regression models and nonlinear regression models. Linear regression models consist of a sum of products between parameters and linear or nonlinear functions of SOC, while nonlinear regression models involve a general function of SOC and parameters. In linear regression models, the parameters can be determined using linear regression methods such as the linear least square method. In contrast, nonlinear regression methods like the nonlinear least square method are required for parameter estimation in nonlinear regression models. In some cases, it is also possible to linearize the nonlinear regression model and employ a linear regression method.

Among the mathematical expressions used in the OCV-SOC curve modeling, polynomial functions are the most commonly employed. These polynomials can range from the second-degree [18] to the twelfth-degree [19,20], depending on the desired accuracy and complexity. Lower-degree polynomials are simpler but offer limited precision and can accurately represent only a small portion of the OCV-SOC curve. Conversely, the higher-degree polynomials can provide excellent precision and fit the entire OCV curve well. However, they require many parameters to be fitted and may exhibit incorrect trends outside the range or between the experimental points. Other analytical models, which are possible to find in the literature, are based on logarithmic functions, also called Nernst models [15,21,22] or exponential functions [23]. Logarithmic functions offer a good accuracy with only three parameters to be fitted but cannot be defined for an SOC equal to 0 or 1. Moreover, there are a lot of different combinations of the aforementioned functions [24–28], some of which yield higher accuracy than others. Among them, the model proposed in [23] demonstrates high accuracy and low complexity. The latter is composed of two exponentials and a quadratic term, with a total of five parameters.

The OCV-SOC curve changes with the battery temperature and aging. Therefore, accurately estimating this curve and understanding its dependence on these factors is crucial for building a reliable battery model and, consequently, a reliable BMS. In [29], the authors corrected the SOC estimation with the value of the actual battery capacity and stated that the OCV curves were the same with this correction, despite different aging conditions. A similar procedure was employed in [30], where the SOC was defined as a function of the SOH and then used inside the OCV-SOC relation. The changes in the OCV-SOC curve and the consequent variations in the incremental capacity of a LiB were analyzed in [25], in which an extended Kalman filter was used for the parameter estimation. In [31], a correction of the OCV-SOC curve based on the SOH and temperature was proposed, while the authors in [32] used a convolutional neural network to estimate the electrode aging parameters, which were then used for the OCV-SOC curve estimation.

In all cases, there seems to be a lack of understanding regarding how the parameters of an OCV-SOC curve change as a function of aging. In a previous work [33], the authors used a double exponential function to model the discharge OCV curve as a function of the absolute state of discharge, q, instead of the SOC, and investigated the dependency of its parameters on cycle aging at a fixed temperature.

In light of the above, the focus of the present paper was to examine the variation of the discharge OCV-q curve as a function of the calendar aging at a fixed temperature. The same procedure and OCV-q curve model adopted in [33] to analyze the cycle aging of a LiB were applied to three different LiB cells of the same type. The latter were stored at three different SOC levels (low, medium, and high levels) for a period of almost three years under the same temperature conditions. Therefore, the parameter variation of the OCV-q curve model was studied for different calendar aging levels (i.e., different storage times) and three different SOC levels, developing a calendar aging model. Finally, the proposed model was validated through a wide campaign of experimental tests.

2. Battery Model

To develop an aging model that considers how the discharge OCV-q curve changes as a function of calendar aging, we needed to start from an equivalent electric circuit battery model. This allowed us to model the discharge OCV-q curve and develop a test procedure to extract it. In the literature, it is possible to retrieve many equivalent electric circuit models able to predict both the static and dynamic behavior of batteries. They can be very simple, as the ones reported in [34–36], or much more complex, as the ones reported in [37–39]. The choice of the appropriate equivalent electric circuit model depends on the desired accuracy and the specific aspects to be described for the application.

For the scope of the present paper, the simplest model, shown in Figure 1, was suitable for characterizing the discharge OCV-q curves. This model consists of a voltage source, E, as a function of the absolute state of discharge, q, which models the OCV of the battery, and a series resistor R_{in}, which represents the total internal resistance of the battery. This resistance is related to the ohmic resistance of the electrode and electrolyte, charge transfer chemical reaction resistance, solid electrolyte interface (SEI) resistance, and diffusion resistance.

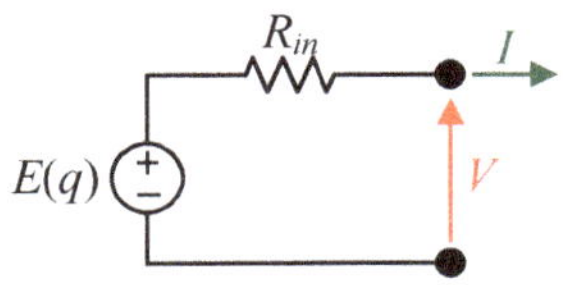

Figure 1. Equivalent electric circuit model.

According to this model, it is possible to express the battery terminal voltage as follows:

$$V = E - R_{in} \cdot I \quad (1)$$

where I is the battery current.

The analytical expression chosen to model the discharge OCV-q curve was the same as that reported in [33], which, in that case, was used to model the changes in the discharge

OCV-q curve as a function of the cycle aging. This analytical expression consists of two exponential terms and a constant one:

$$E(q) = a \cdot e^{b \cdot q} + c \cdot e^{d \cdot q} + f \tag{2}$$

where a, b, c, d, and f are five parameters that, in our case, are functions of the calendar aging, e is the Euler's number, while q is the absolute state of discharge in Ah and defined as follows:

$$q = \frac{1}{3600} \int_0^t I \cdot d\tau + q(0) \tag{3}$$

where $q(0)$ is the initial state of discharge. In this way, when q is null, the battery is fully charged, i.e., SOC = 100%; conversely, when q equates the actual capacity of the battery, C_a, the latter is fully discharged, i.e., SOC = 0%. Therefore, the SOC can be expressed as a function of the absolute state of discharge, as follows:

$$SOC = \left(1 - \frac{q}{C_a}\right) \cdot 100. \tag{4}$$

To quantify calendar aging, the calendar time t_{aging} expressed in days was used. Thus, the five parameters of (2) could be expressed as a function of this calendar time.

In this work, the procedure to obtain the discharge OCV-q curve was performed by discharging the battery at the nominal current rate (1C) to speed up the tests, as performed in [33]. Therefore, it was essential to correct the discharge voltage curve by eliminating the voltage drop over the total internal resistance of the battery. Thus, it was necessary to estimate the value of such an internal resistance. For the sake of simplicity, as assumed in [33], the battery's internal resistance was considered to be quite constant as a function of the absolute state of discharge. Moreover, assuming the battery was fully charged, when the discharge at constant current started, different phenomena, with different time constants, led to an electrical transient. To obtain the total battery's internal resistance, the entire electric transient had to be extinguished. To do this, we had to consider a proper time interval related to the largest time constant τ, which corresponds to the diffusion process of the lithium/lithium ions into the electrodes and electrolyte. Through the procedure reported in [26,27], it was possible to identify that time interval, considered to be five times the largest time constant τ, in the first part of the discharge voltage curve (Figure 2). Therefore, the resulting voltage variation related to that time interval was estimated. By calculating the ratio between that voltage variation and the current step, the battery's internal resistance was calculated, and the discharge voltage curve was corrected, obtaining the discharge OCV-q curve. Finally, the part of the OCV-q curve corresponding to the considered electric transient was eliminated to avoid errors in the following analysis.

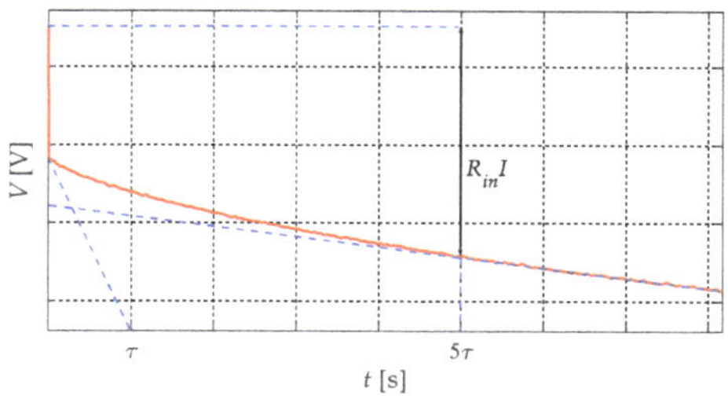

Figure 2. Electric transient at the beginning of the battery discharge.

3. Experimental Setup and Test Procedure

In the present work, three $LiCoO_2$ 8773160K pouch cells of the same batch, manufactured by General Electronics Battery Co. Ltd. (Shenzhen, China), were employed for the experimental tests. The main specifications of this type of cell are reported in Table 1.

Table 1. Battery cell specifications.

Parameter	Value
Nominal capacity	10 Ah
Maximum voltage	4.2 V
Discharge cut-off voltage	2.75 V
Maximum continuous discharge current	100 A (10C)

The three battery cells were used to test how the discharge OCV-q curve changed as a function of the calendar aging under three different storage conditions related to low SOC (about 7%), medium SOC (about 50%), and high SOC (about 93%).

3.1. Experimental Setup

Figure 3 shows the experimental setup that was composed of a potentiostat (SP-150) connected to a 100 A booster (VMP3B-100). Both instruments are manufactured by Biologic Science Instrument, and they are connected to a PC via an ethernet cable. The PC controlled them through the EC-Lab software. The 100 A booster was connected to the battery cell under test through a power cable.

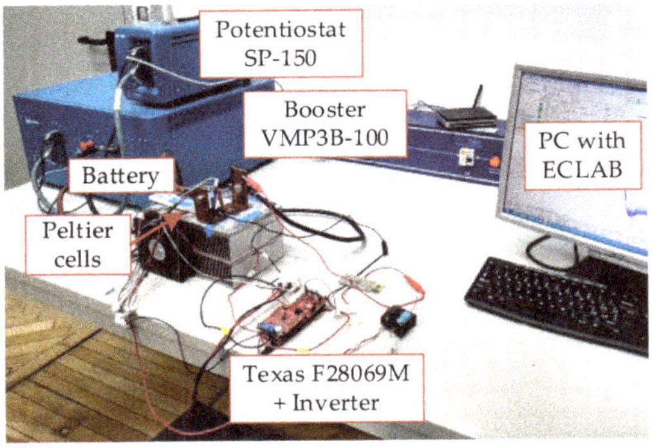

Figure 3. Experimental setup.

The battery cell arrangement consisted of a heatsink, two fans, three Peltier cells, a dc voltage source, a temperature probe, and a Texas Instrument DRV8303 inverter controlled by an F28069M controller board. The three Peltier cells were put between the battery cell under test and the heatsink to maintain the battery temperature as uniform and constant as possible at 25 °C during the discharge voltage curve measurements. This was carried out to avoid changes on the OCV-q curve due to temperature variation. To do this, they were connected in series and powered by the inverter, which was, in turn, controlled by the F28069M with a PI controller. Finally, the inverter was supplied using a dc voltage source. Figure 4 shows the schematics of the temperature control.

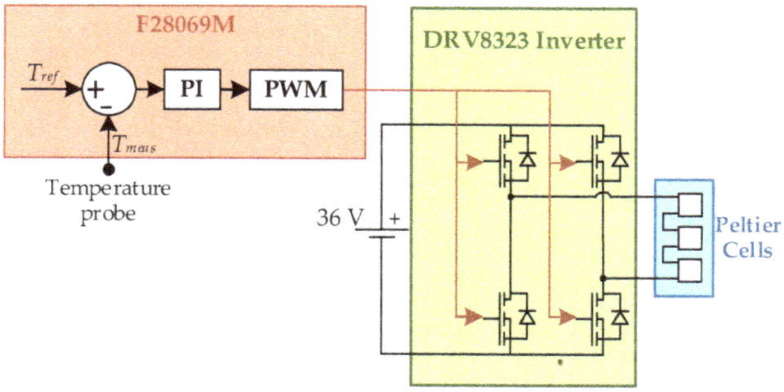

Figure 4. Temperature control scheme.

3.2. Test Procedure

The test procedure used for the analysis consisted of two main phases: the discharge voltage curve measurement phase and the calendar aging phase. The former phase was conducted at the beginning of the test and after each calendar aging phase for all three battery cells. In this way, it was possible to assess how the discharge OCV-q curve changed as a function of the calendar aging at different SOC levels.

3.2.1. Voltage Discharge Curve Measurement Phase

This phase consisted of fully charging and discharging the battery cell under test at the nominal current (1C) using the well-known constant current–constant voltage (CC-CV) protocol for the charging phase and the constant current protocol for the discharging phase. In particular, the battery cell was charged with a constant current of 10 A (1C) until the battery terminal voltage reached 4.2 V. After that, this voltage was applied until the battery current dropped by 100 mA (0.01C). At that stage, the battery cell could be considered charged at 100% of SOC. After completing the previous step, the battery cell was discharged at a constant current of 10 A (1C) until the battery terminal voltage reached 2.75 V. This entire procedure was repeated seven times for each of the three battery cells. In fact, for the first use of the batteries and after each calendar aging phase, the battery cells needed to be activated.

Finally, after each voltage discharge curve measurement phase, the three battery cells were brought to about 7%, 50%, and 93% of SOC for the low, medium, and high SOC tests, respectively. In particular, for the low and medium SOC tests, the cells were initially fully discharged with the CC-CV protocol, and then a charge current of 10 A (1C) was applied to move 0.7 Ah (7% of the nominal capacity) and 5 Ah (50% of the nominal capacity), respectively. On the other hand, for the high SOC test, the battery cell was first fully charged with the CC-CV protocol, and then a discharge current of 10 A (1C) was applied to move 0.7 Ah (7% of the nominal capacity). It is worth noting that since, during the aging, the actual capacity of the battery cells decreases, the moved charge (0.7 Ah and 5 Ah) corresponds to different percentages of SOC. In any case, these differences are minimal (about 5%) and do not change the SOC ranges at which the three cells were stored.

3.2.2. Calendar Aging Phase

This phase consisted of storing the three battery cells in a cabinet at a temperature between 20 °C and 30 °C for a certain time interval. The duration of this time interval was not constant throughout the entire analysis, but it varied according to the availability of the experimental setup.

4. Results and Model Validation

According to the test procedure reported in the previous section, a total of twelve calendar aging phases were evaluated at specific time intervals: 0th, 35th, 70th, 99th, 136th,

205th, 273rd, 344th, 437th, 682nd, 743rd, and 997th day. The entire duration of the testing period was about 1000 days. Among these twelve tests, a subset of four (at the 0th, 205th, 437th, and 997th day) was selected and used to tune the proposed aging model. As stated before, for each calendar aging phase and battery cell under test, seven voltage discharge curves were performed to activate the battery cells and stabilize their capacities. The latter were estimated in Ah at the rated nominal current of 10 A (1C) by integrating the current over the entire discharging time. Figure 5a shows an example of the battery capacity trend, as a function of the seven voltage discharge curve measurements, for the high SOC test battery cell on the 0th day. From this figure, it is possible to recognize that the values of the battery capacity become quite stable after at least five full charge–discharge cycles. Therefore, the last two voltage discharge curves were averaged and used for the analysis. Figure 5b illustrates the related temperature profile during the entire discharge curve measurement. Finally, according to the procedure proposed in [26,27], the time interval related to the lithium-ion diffusion was estimated to be approximately 50 s for all discharge curves. Therefore, for each of them, the total battery's internal resistance was evaluated along with the related voltage drops, and the initial section of the curve was eliminated. In this way, the OCV-q curves for each calendar aging phase and SOC level were obtained.

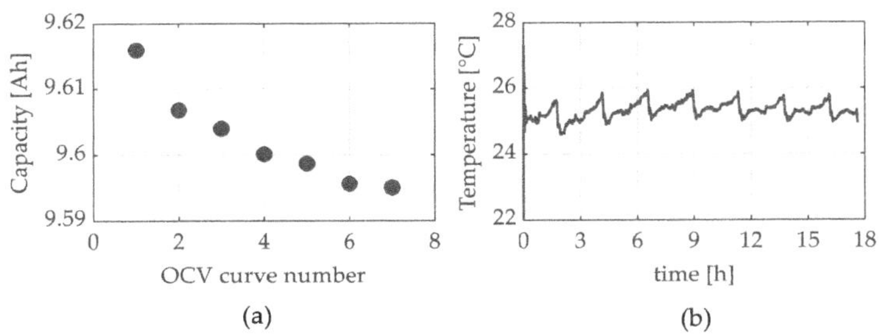

Figure 5. (**a**) battery capacity; (**b**) battery temperature.

4.1. Model Characterization

In the discharge voltage curve measurements, each battery cell under test was initially fully charged up to 4.2 V using the CC-CV protocol and then fully discharged with the CC protocol to obtain the OCV-q curves. Consequently, we can assume that all the OCV-q curves started from the maximum battery voltage of 4.2 V, and the parameter f of (2) can be rearranged as follows:

$$f = 4.2 - a - c. \tag{5}$$

Therefore, the total number of parameters to be tuned in (2) reduces to just four. At this stage, the OCV-q curves were fitted using the proposed OCV-q analytical expression through the nonlinear least square regression method, and the coefficients of determination (R^2) were evaluated.

Figure 6 illustrates the experimental OCV-q curves along with the corresponding fitting functions for the four tests chosen for the characterization of the model at each SOC level, and Figure 7 shows the related R^2. From these figures, it is possible to recognize a very good agreement between the experimental and modeled data for the low, medium, and high SOC tests. Moreover, the R^2 is greater than 0.995 for all tests.

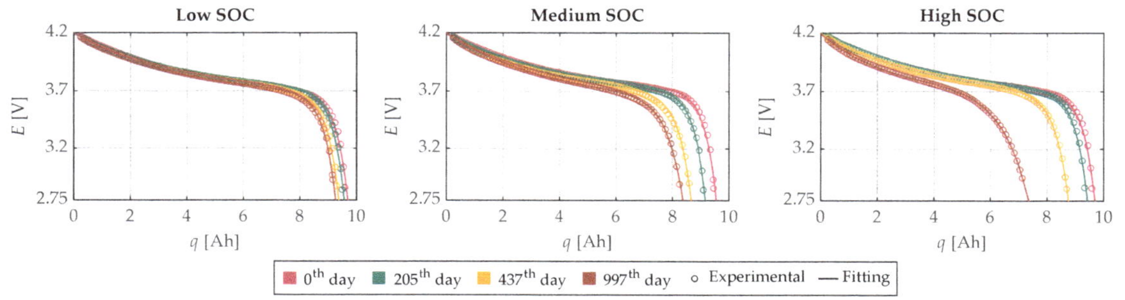

Figure 6. Experimental OCV-q curves and their fitting functions obtained through the nonlinear least square method for the characterization subset.

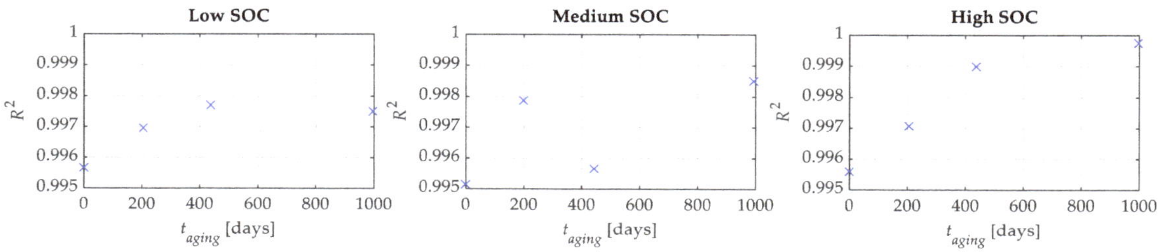

Figure 7. Coefficients of determinations R^2 for the characterization subset.

Furthermore, it is worth noting that the capacity fade experienced by the battery cells becomes more pronounced as the storage SOC level increases. In fact, for the battery cell stored at a low SOC, the capacity fade is about 5%, while for the medium SOC, it is about 13%, and for the high SOC, it is about 25%. This can be attributed to a significant potential disequilibrium at the electrode/electrolyte interface resulting from a high SOC. Therefore, secondary chemical reactions such as corrosion, electrolyte decomposition, and SEI decomposition occur, leading to the loss of lithium inventory and active material, thus resulting in faster capacity fade [40,41].

Figure 8 shows the behavior of the parameters a, b, c, and d of (2) as a function of the calendar time for the three SOC levels. From this figure, it is possible to note that, similar to the results obtained in [33] for cycle aging, the parameters a, b, and d do not present a well-defined trend as a function of the calendar time. Therefore, following the analysis performed in [33], we opted to consider, for each SOC level, the values of the parameters b and d constant at their mean values obtained previously and reported in Table 2. In this way, the function of the discharge OCV-q curve becomes linear in the parameters a and c; thus, the linear least square regression method could be applied.

Figure 9 illustrates the comparison between the experimental OCV-q curves and the corresponding fitting functions for the four tests chosen for the characterization of the model and for each SOC level. Additionally, Figure 10 shows the related R^2 values obtained by maintaining parameters b and d constant. From these figures, it is possible to recognize a good agreement between the experimental and modeled data for the low and medium SOC tests, while the agreement for the high SOC tests is a little worse. In any case, the R^2 is greater than 0.985 for all the tests.

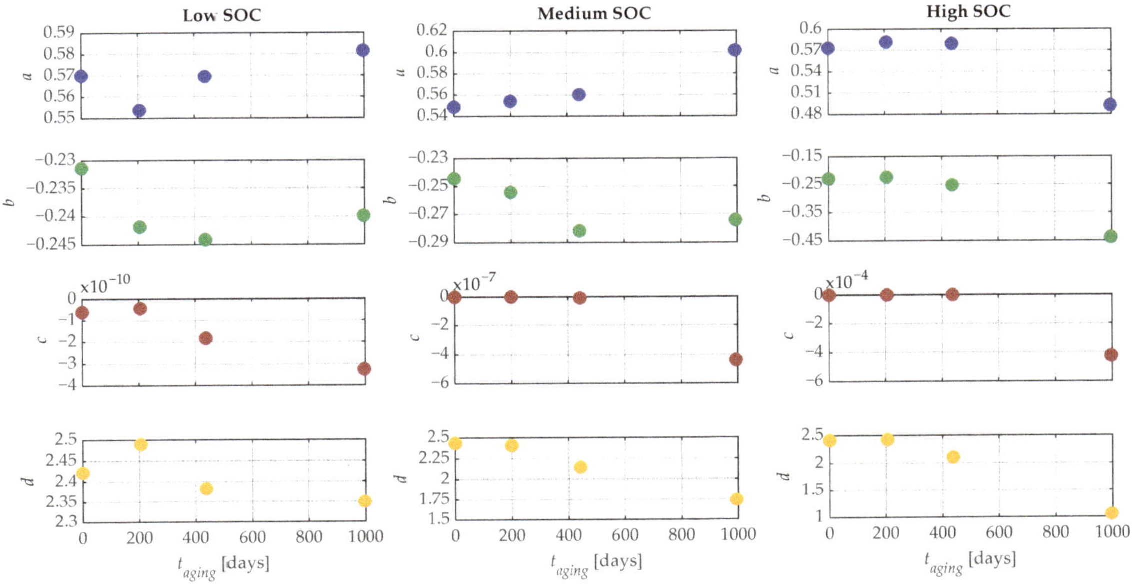

Figure 8. Parameters a, b, c, and d as a function of the calendar time for the characterization subset.

Table 2. Mean values of parameters b and d.

SOC Level	b [Ah^{-1}]	d [Ah^{-1}]
Low SOC	−0.2393	2.411
Medium SOC	−0.2635	2.183
High SOC	−0.2856	2.000

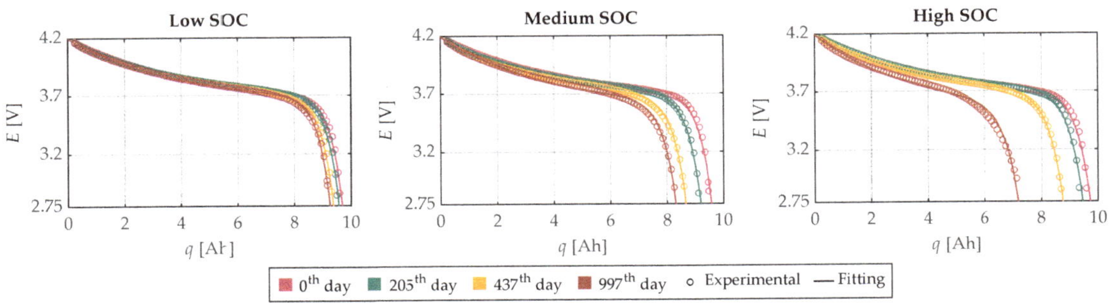

Figure 9. Experimental OCV-q curves and their fitting functions obtained by fixing the parameters b and d and through the linear least square method for the characterization subset.

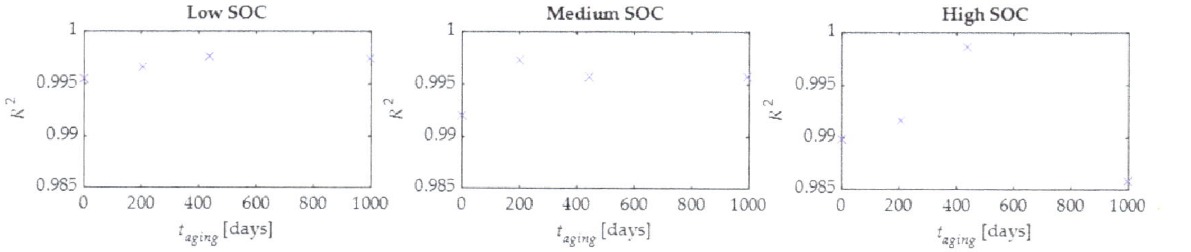

Figure 10. Coefficients of determination R^2 obtained by fixing the parameters b and d for the characterization subset.

In all cases, the new values of parameters a and c exhibited a well-defined behavior, enabling them to be fitted as a function of the calendar time. For the parameter a, a linear least square regression was performed using the following fitting function:

$$a = \alpha_a \cdot t_{aging} + \beta_a \tag{6}$$

where α_a and β_a are the coefficients of the fitting function of the parameter a. For the parameter c, a nonlinear least square regression was performed instead, using the following expression:

$$c = \alpha_c \cdot t_{aging}^{\delta_c} + \beta_c. \tag{7}$$

where α_c, β_c, and δ_c are the coefficients of the fitting function of the parameter c. The choice to use the expression (7) for parameter c was made to find an analytical expression that was suitable for all three SOC levels. Finally, Figure 11 shows the behavior of parameters a and c of (2) as a function of the calendar time for the three SOC levels, along with their related fitting functions. Table 3 reports the coefficients of (6) and (7). In this way, an aging model with only two parameters that depend on calendar aging was derived.

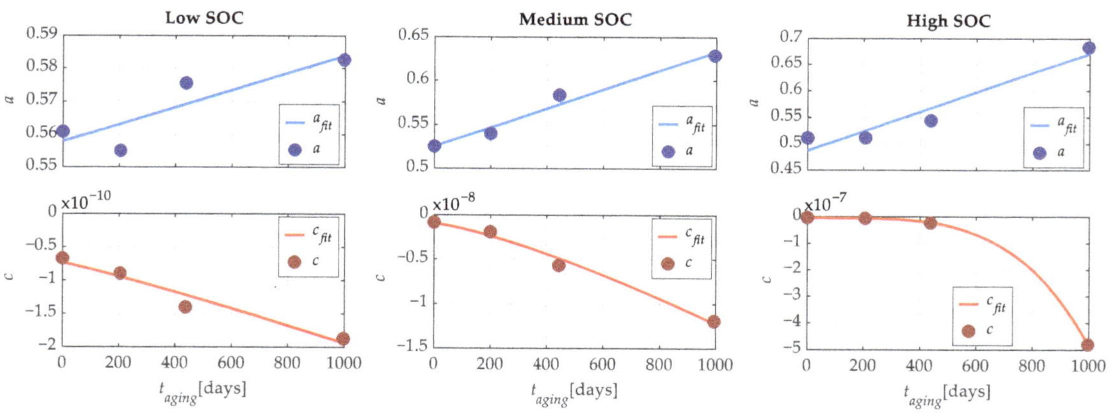

Figure 11. Parameters a and c as a function of the calendar time (with b and d fixed) and their fitting functions for the characterization subset.

Table 3. Coefficients of the parameters a and c.

SOC Level	α_a [V/Days]	β_a [V]	α_c [V/Days]	β_c [V]	δ_c
Low SOC	2.580×10^{-5}	0.5580	-6.017×10^{-14}	1.1	-7.362×10^{-11}
Medium SOC	1.072×10^{-4}	0.5258	-1.420×10^{-12}	1.3	-9.443×10^{-10}
High SOC	1.833×10^{-4}	0.4875	-4.808×10^{-19}	4.0	-4.551×10^{-9}

4.2. Model Validation

The validation of the model was assessed by evaluating the absolute error ($err_{OCV,abs}$) and relative percentage error ($err_{OCV,rel\%}$) between the estimated discharge OCV-q curves and the corresponding experimental curves, using all twelve experimental tests at different calendar times for each SOC level. They were evaluated as follows:

$$err_{OCV,abs} = \left| E^{\text{mod}}(q) - E^{\text{exp}}(q) \right| \tag{8}$$

$$err_{OCV,rel\%} = \frac{err_{OCV,abs}}{E^{\text{exp}}(q)} \cdot 100 \tag{9}$$

where $E^{\text{exp}}(q)$ and $E^{\text{mod}}(q)$ are, respectively, the experimental and modeled values of the OCV-q curves. Moreover, to have an overall indicator of the goodness of fit, the R^2 was also calculated.

Figure 12 shows the maximum and mean values of the relative percentage error of the OCV-q curves as a function of calendar aging. From this figure, it is possible to note that the highest maximum relative percentage errors are about 5%, 7%, and 6.5% for the low, medium, and high SOC tests, respectively. Instead, the mean relative percentage errors for the low, medium, and high SOC tests are lower than 0.45%, 0.65%, and 0.9%, respectively, across all calendar aging levels. On the other hand, the relative percentage error can have different weights for different states of discharge because the OCV curve changes as a function of q. For this reason, Figure 13 shows the maximum and mean values of the absolute error. The maximum value of the latter is about 165 mV for the low SOC tests and 220 mV for the medium and high SOC tests, while the mean values of the absolute error are lower than 15 mV, 22 mV, 30 mV for the low, medium, and high SOC tests, respectively, across all aging levels. Finally, Figure 14 reports the experimental OCV-q curves compared with the ones obtained through the proposed model for all tests and SOC levels, while Figure 15 shows the related R^2. The latter are larger than 0.9763, 0.9542, and 0.9490 for the low, medium, and high SOC tests, respectively.

The knowledge of the variation law of the OCV-q curve as a function of calendar aging can be employed to estimate the actual battery capacity. This, in turn, allows the correction of the SOC estimation and updating of the SOH in terms of capacity fade. In particular, the actual battery capacity, C_a, can be calculated by equating the (2) to the minimum cut-off voltage E_{min} of the battery. This is due to the fact that when the open circuit voltage E of the battery equals its minimum cut-off voltage, the absolute state of discharge q is equal to the actual battery capacity. The percentage relative error of the capacity was evaluated as follows:

$$err_{capacity,rel\%} = \frac{|q_{mod}(E_{\text{min}}) - q_{exp}(E_{\text{min}})|}{q_{exp}(E_{\text{min}})} \cdot 100 \tag{10}$$

where q_{mod} (E_{min}) and q_{exp} (E_{min}) are the estimated and experimental actual battery capacities, respectively.

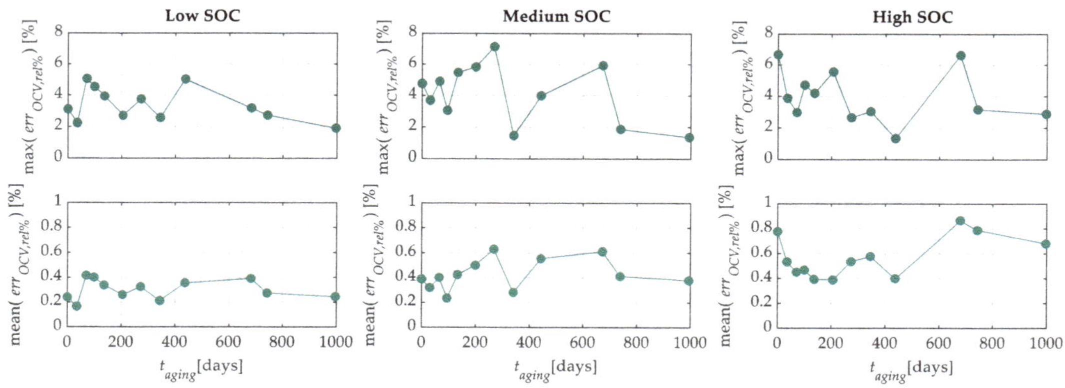

Figure 12. Mean and maximum percentage relative error as a function of the calendar time.

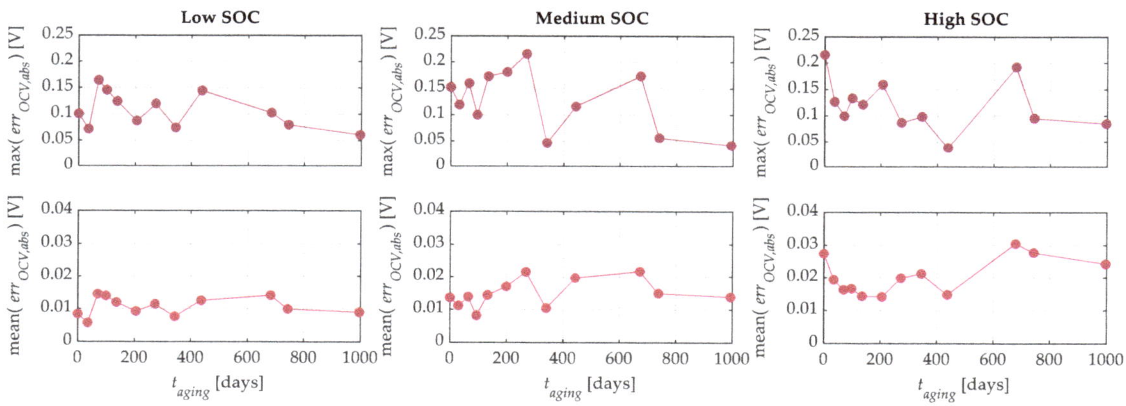

Figure 13. Mean and maximum absolute error as a function of the calendar time.

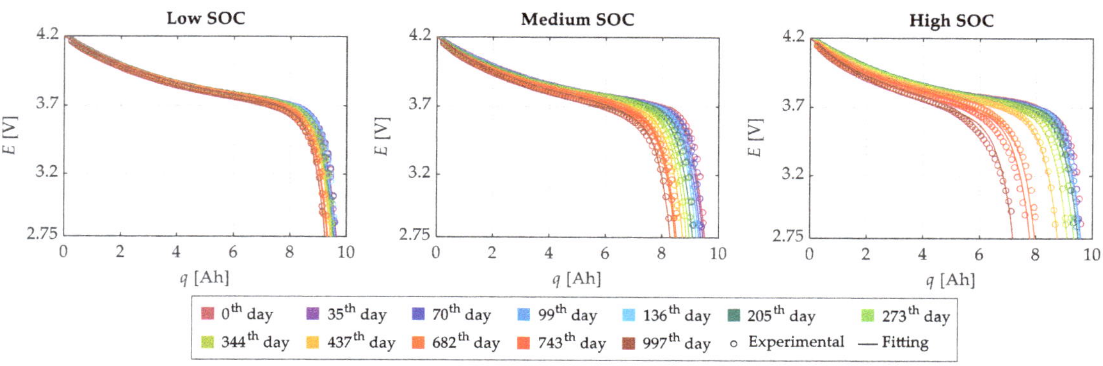

Figure 14. Experimental OCV-q curves and modeled ones for all the experimental tests.

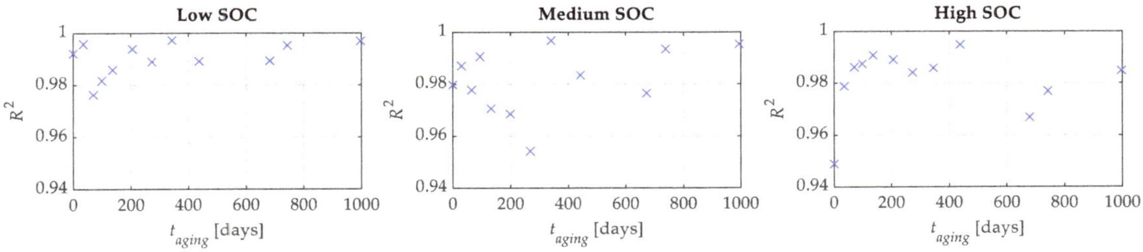

Figure 15. Coefficients of determination R^2 for all the experimental tests.

Theoretically, the minimum cut-off voltage for the battery used in this work would be 2.75 V. However, due to the elimination of the voltage drop over the internal resistance, the actual cut-off voltage was higher and depended on the amplitude of that voltage drop itself. Nonetheless, since the last experimental point of each discharge OCV-q curve is close to the minimum cut-off voltage, this experimental point was considered as E_{min}, and the related value of the actual battery capacity, C_a, was compared with the modeled one. Figure 16 shows the capacity percentage errors for the three SOC levels. From this figure, it is possible to recognize that the capacity error is lower than 1.5% for all the aging and SOC levels.

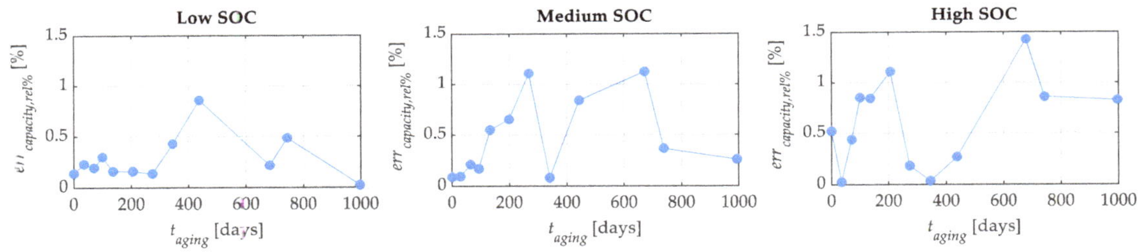

Figure 16. Capacity estimation error as a function of the calendar time.

Finally, once the actual capacity of the battery is obtained, it becomes possible to correct the SOC estimation according to (4) and update the value of SOH in terms of capacity fade as follows:

$$SOH = \left(1 - \frac{C_a}{C_i}\right) \cdot 100 \tag{11}$$

where C_i is the initial battery capacity. The results highlight the significant impact of storing the battery at a high SOC, as it degrades much faster. Specifically, the battery stored at a high SOC reached the end of its life in less than three years, whereas the one stored at a low SOC experienced minimal capacity fade.

5. Conclusions

In this work, a simple analytical function composed of two exponential terms and a constant one with five parameters was used to model the behavior of the discharge OCV-q curves of three $LiCoO_2$ batteries of the same batch stored at three different levels of SOC (low, medium, and high levels) for different calendar aging levels. Twelve discharge OCV-q experimental curves were performed over about 1000 days at different calendar times.

The proposed model was characterized using a subset of four discharge OCV-q experimental curves at the 0th, 205th, 437th, and 997th day. Firstly, the proposed analytical model, being nonlinear in its parameters, was fitted using the nonlinear regression least square method. Moreover, one of the parameters, namely f, was expressed as a function of the other two parameters, a and c, with the constraint that all the OCV-q curves started from the maximum cut-off voltage of 4.2 V. This reduced the number of parameters to be found to four. The fitting procedure yielded modeled data for the three SOC levels that showed good agreement with the experimental ones. This agreement was further confirmed by the high value of the R^2 for all the tests. The four parameters a, b, c, and d were reported as a function of the calendar time. On the other hand, the parameters a, b, and d did not exhibit a well-defined trend. Following the analysis performed in [33], we considered the value of the parameters b and d fixed at their mean values obtained previously for all the SOC levels. Therefore, the analytical model became linear in its parameters, and the fitting procedure was performed again using the linear least square method. In this way, the parameters a and c could be fitted using simple analytical expressions for the three SOC levels, and the OCV-q model considering the calendar aging was obtained.

Afterward, the model was validated through all twelve experimental OCV-q curves, and the absolute error, percentage relative error, and coefficients of determination R^2 were evaluated as indicators of the goodness of fit. Through the analysis of the results, it is possible to recognize that the mean relative percentage errors of the low, medium, and high SOC tests are, respectively, lower than 0.45%, 0.65%, and 0.9% for all the calendar aging levels. Additionally, the R^2 values are greater than 0.9763, 0.9542, and 0.9490 for the low, medium, and high SOC tests, respectively.

Finally, the proposed model can also be used to estimate the battery capacity for all storage SOC and aging levels. The estimated battery capacity values were compared with the corresponding experimental ones, and the capacity relative percentage error was calculated. This error is lower than 1.5% for all the tests. The battery capacity obtained can then be used to correct the SOC estimation and evaluate the capacity fade. Based on the results, it is possible to confirm that the worst storage condition for these kinds of LiBs occurs at high SOCs. Indeed, the battery stored at a high SOC reached its end of life before three years. On the other hand, the batteries stored at lower SOCs are still usable considering the same time span. This can be due to a notable imbalance that arises between the electrode and electrolyte interface caused by a high SOC. As a consequence, secondary chemical reactions such as corrosion, electrolyte decomposition, and SEI decomposition take place, resulting in the depletion of lithium inventory and active material.

In light of the above, it is possible to claim that the proposed calendar aging model is simple yet quite accurate in modeling the OCV-q curve. Moreover, it can be valuable for correcting SOC estimations and evaluating the actual battery capacity as an indicator of the SOH.

Author Contributions: Conceptualization, S.B. and S.C.; methodology, S.B. and S.C.; validation, S.B. and S.C.; writing—original draft preparation, S.B. and S.C.; writing—review and editing, S.B. and S.C.; supervision, L.P. and L.C. All authors have read and agreed to the published version of the manuscript.

Funding: This research received no external funding.

Data Availability Statement: Data sharing is not applicable.

Conflicts of Interest: The authors declare no conflict of interest.

References

1. Divakaran, M.A.; Minakshi, M.; Bahri, P.A.; Paul, S.; Kumari, P.; Divakaran, A.M.; Manjunatha, K.N. Rational Design on Materials for Developing next Generation Lithium-Ion Secondary Battery. *Prog. Solid State Chem.* **2021**, *62*, 100298. [CrossRef]
2. Nunotani, K.; Yoshida, F.; Kamiya, Y.; Daisho, Y.; Abe, K.; Kono, M.; Matsuo, H. Development and Performance Evaluation of Lithium Iron Phosphate Battery with Superior Rapid Charging Performance—Second Report: Evaluation of Battery Capacity Loss Characteristics. In Proceedings of the 2011 IEEE Vehicle Power and Propulsion Conference, Chicago, IL, USA, 6–9 September 2011; pp. 1–4.
3. Barcellona, S.; Brenna, M.; Foiadelli, F.; Longo, M.; Piegari, L. Analysis of Ageing Effect on Li-Polymer Batteries. *Sci. World J.* **2015**, *2015*, 979321. [CrossRef] [PubMed]

4. Barcellona, S.; Brenna, M.; Foiadelli, F.; Longo, M.; Piegari, L. Battery Lifetime for Different Driving Cycles of EVs. In Proceedings of the 2015 IEEE 1st International Forum on Research and Technologies for Society and Industry Leveraging a better Tomorrow (RTSI), Torino, Italy, 16–18 September 2015; pp. 446–450.
5. Mawonou, K.S.R.; Eddahech, A.; Dumur, D.; Beauvois, D.; Godoy, E. Improved State of Charge Estimation for Li-Ion Batteries Using Fractional Order Extended Kalman Filter. *J. Power Sources* **2019**, *435*, 226710. [CrossRef]
6. Li, Y.; Wang, C.; Gong, J. A Multi-Model Probability SOC Fusion Estimation Approach Using an Improved Adaptive Unscented Kalman Filter Technique. *Energy* **2017**, *141*, 1402–1415. [CrossRef]
7. Ning, B.; Xu, J.; Cao, B.; Wang, B.; Xu, G. A Sliding Mode Observer SOC Estimation Method Based on Parameter Adaptive Battery Model. *Energy Procedia* **2016**, *88*, 619–626. [CrossRef]
8. Ouyang, Q.; Chen, J.; Wang, F.; Su, H. Nonlinear Observer Design for the State of Charge of Lithium-Ion Batteries. *IFAC Proc. Vol.* **2014**, *47*, 2794–2799. [CrossRef]
9. Gan, L.; Yang, F.; Shi, Y.F.; He, H.L. Lithium-Ion Battery State of Function Estimation Based on Fuzzy Logic Algorithm with Associated Variables. *IOP Conf. Ser. Earth Environ. Sci.* **2017**, *94*, 012133. [CrossRef]
10. Jiao, M.; Wang, D.; Qiu, J. A GRU-RNN Based Momentum Optimized Algorithm for SOC Estimation. *J. Power Sources* **2020**, *459*, 228051. [CrossRef]
11. Zhou, W.; Zheng, Y.; Pan, Z.; Lu, Q. Review on the Battery Model and SOC Estimation Method. *Processes* **2021**, *9*, 1685. [CrossRef]
12. Ren, X.; Liu, S.; Yu, X.; Dong, X. A Method for State-of-Charge Estimation of Lithium-Ion Batteries Based on PSO-LSTM. *Energy* **2021**, *234*, 121236. [CrossRef]
13. Li, N.; Zhang, Y.; He, F.; Zhu, L.; Zhang, X.; Ma, Y.; Wang, S. Review of Lithium-Ion Battery State of Charge Estimation. *Glob. Energy Interconnect.* **2021**, *4*, 619–630. [CrossRef]
14. Xiong, R.; Yu, Q.; Wang, L.Y.; Lin, C. A Novel Method to Obtain the Open Circuit Voltage for the State of Charge of Lithium Ion Batteries in Electric Vehicles by Using H Infinity Filter. *Appl. Energy* **2017**, *207*, 346–353. [CrossRef]
15. Xiong, R.; He, H.; Guo, H.; Ding, Y. Modeling for Lithium-Ion Battery Used in Electric Vehicles. *Procedia Eng.* **2011**, *15*, 2869–2874. [CrossRef]
16. Sundaresan, S.; Devabattini, B.; Kumar, P.; Pattipati, K.; Balasingam, B. Tabular Open Circuit Voltage Modelling of Li-Ion Batteries for Robust SOC Estimation. *Energies* **2022**, *15*, 9142. [CrossRef]
17. Pillai, P.; Sundaresan, S.; Kumar, P.; Pattipati, K.R.; Balasingam, B. Open-Circuit Voltage Models for Battery Management Systems: A Review. *Energies* **2022**, *15*, 6803. [CrossRef]
18. He, H.; Xiong, R.; Zhang, X.; Sun, F.; Fan, J. State-of-Charge Estimation of the Lithium-Ion Battery Using an Adaptive Extended Kalman Filter Based on an Improved Thevenin Model. *IEEE Trans. Veh. Technol.* **2011**, *60*, 1461–1469. [CrossRef]
19. Sun, D.; Yu, X.; Wang, C.; Zhang, C.; Huang, R.; Zhou, Q.; Amietszajew, T.; Bhagat, R. State of Charge Estimation for Lithium-Ion Battery Based on an Intelligent Adaptive Extended Kalman Filter with Improved Noise Estimator. *Energy* **2021**, *214*, 119025. [CrossRef]
20. Zhao, S.; Duncan, S.R.; Howey, D.A. Observability Analysis and State Estimation of Lithium-Ion Batteries in the Presence of Sensor Biases. *IEEE Trans. Control. Syst. Technol.* **2017**, *25*, 326–333. [CrossRef]
21. Xiong, R.; He, H.; Sun, F.; Zhao, K. Online Estimation of Peak Power Capability of Li-Ion Batteries in Electric Vehicles by a Hardware-in-Loop Approach. *Energies* **2012**, *5*, 1455–1469. [CrossRef]
22. Ta, K.P.; Newman, J. Proton Intercalation Hysteresis in Charging and Discharging Nickel Hydroxide Electrodes. *J. Electrochem. Soc.* **1999**, *146*, 2769–2779. [CrossRef]
23. Baccouche, I.; Jemmali, S.; Manai, B.; Omar, N.; Amara, N. Improved OCV Model of a Li-Ion NMC Battery for Online SOC Estimation Using the Extended Kalman Filter. *Energies* **2017**, *10*, 764. [CrossRef]
24. Zhang, C.; Jiang, J.; Zhang, L.; Liu, S.; Wang, L.; Loh, P. A Generalized SOC-OCV Model for Lithium-Ion Batteries and the SOC Estimation for LNMCO Battery. *Energies* **2016**, *9*, 900. [CrossRef]
25. Weng, C.; Sun, J.; Peng, H. A Unified Open-Circuit-Voltage Model of Lithium-Ion Batteries for State-of-Charge Estimation and State-of-Health Monitoring. *J. Power Sources* **2014**, *258*, 228–237. [CrossRef]
26. Zhang, R.; Xia, B.; Li, B.; Cao, L.; Lai, Y.; Zheng, W.; Wang, H.; Wang, W.; Wang, M. A Study on the Open Circuit Voltage and State of Charge Characterization of High Capacity Lithium-Ion Battery Under Different Temperature. *Energies* **2018**, *11*, 2408. [CrossRef]
27. Dong, G.; Wei, J.; Zhang, C.; Chen, Z. Online State of Charge Estimation and Open Circuit Voltage Hysteresis Modeling of LiFePO4 Battery Using Invariant Imbedding Method. *Appl. Energy* **2016**, *162*, 163–171. [CrossRef]
28. Hu, Y.; Yurkovich, S.; Guezennec, Y.; Yurkovich, B.J. Electro-Thermal Battery Model Identification for Automotive Applications. *J. Power Sources* **2011**, *196*, 449–457. [CrossRef]
29. Wang, L.; Lu, D.; Liu, Q.; Liu, L.; Zhao, X. State of Charge Estimation for LiFePO4 Battery via Dual Extended Kalman Filter and Charging Voltage Curve. *Electrochim. Acta* **2019**, *296*, 1009–1017. [CrossRef]
30. Tong, S.; Klein, M.P.; Park, J.W. On-Line Optimization of Battery Open Circuit Voltage for Improved State-of-Charge and State-of-Health Estimation. *J. Power Sources* **2015**, *293*, 416–428. [CrossRef]
31. Farmann, A.; Sauer, D.U. A Study on the Dependency of the Open-Circuit Voltage on Temperature and Actual Aging State of Lithium-Ion Batteries. *J. Power Sources* **2017**, *347*, 1–13. [CrossRef]

32. Tian, J.; Xiong, R.; Shen, W.; Sun, F. Electrode Ageing Estimation and Open Circuit Voltage Reconstruction for Lithium Ion Batteries. *Energy Storage Mater.* **2021**, *37*, 283–295. [CrossRef]
33. Barcellona, S.; Codecasa, L.; Colnago, S.; Piegari, L. Cycle Aging Effect on the Open Circuit Voltage of Lithium-Ion Battery. In Proceedings of the International Conference on Electrical Systems for Aircraft, Railway, Ship Propulsion and Road Vehicles & International Transportation Electrification Conference (ESARS-ITEC), Venice, Italy, 28–31 March 2023.
34. Kim, Y.H.; Ha, H.D. Design of Interface Circuits with Electrical Battery Models. *IEEE Trans. Ind. Electron.* **1997**, *44*, 81–86. [CrossRef]
35. Dürr, M.; Cruden, A.; Gair, S.; McDonald, J.R. Dynamic Model of a Lead Acid Battery for Use in a Domestic Fuel Cell System. *J. Power Sources* **2006**, *161*, 1400–1411. [CrossRef]
36. Chan, H.L. A New Battery Model for Use with Battery Energy Storage Systems and Electric Vehicles Power Systems. In Proceedings of the 2000 IEEE Power Engineering Society Winter Meeting. Conference Proceedings (Cat. No.00CH37077), Singapore, 23–27 January 2000; Volume 1, pp. 470–475.
37. Randles, J.E.B. Kinetics of Rapid Electrode Reactions. *Discuss. Faraday Soc.* **1947**, *1*, 11. [CrossRef]
38. Zhang, S.S.; Xu, K.; Jow, T.R. Electrochemical Impedance Study on the Low Temperature of Li-Ion Batteries. *Electrochim. Acta* **2004**, *49*, 1057–1061. [CrossRef]
39. Buller, S.; Thele, M.; DeDoncker, R.W.A.A.; Karden, E. Impedance-Based Simulation Models of Supercapacitors and Li-Ion Batteries for Power Electronic Applications. *IEEE Trans. Ind. Appl.* **2005**, *41*, 742–747. [CrossRef]
40. Barré, A.; Deguilhem, B.; Grolleau, S.; Gérard, M.; Suard, F.; Riu, D. A Review on Lithium-Ion Battery Ageing Mechanisms and Estimations for Automotive Applications. *J. Power Sources* **2013**, *241*, 680–689. [CrossRef]
41. Birkl, C.R.; Roberts, M.R.; McTurk, E.; Bruce, P.G.; Howey, D.A. Degradation Diagnostics for Lithium Ion Cells. *J. Power Sources* **2017**, *341*, 373–386. [CrossRef]

energies

MDPI

Article

Evaluation of the Power Generation Impact for the Mobility of Battery Electric Vehicles

Javier Rey and Lázaro V. Cremades *

Department of Project and Construction Engineering, Universitat Politècnica de Catalunya, 08028 Barcelona, Spain
* Correspondence: lazaro.cremades@upc.edu

Abstract: European institutions have decided to ban the sale of Internal Combustion Vehicles (ICEVs) in the EU from 2035. This opens a possible scenario in which, in the not-too-distant future, all vehicles circulating in Europe are likely to be Battery Electric Vehicles (BEVs). The Spanish vehicle fleet is one of the oldest and has the lowest percentage of BEVs in Europe. The aim of this study is to evaluate the hypothetical scenario in which the current mobility of ICEVs is transformed into BEVs, in the geographical area of the province of Barcelona and in Spain in general. The daily electricity consumption, the required installation capacity of wind and solar photovoltaic energies, and the potential reduction of NO_x and particulate matter (PM) emissions are estimated. The daily emission reduction would be about 314 tons of NO_x and 17 tons of PM in Spain. However, the estimated investment required in Spain to generate the additional electricity from renewable sources would be enormous (over EUR 25.4 billion), representing, for example, 5.5% of the total national budget in 2022.

Keywords: ICEVs; BEVs; mobility; electricity generation; wind energy; solar photovoltaic energy; renewable energy sources; vehicle fleet; pollutant emissions

Citation: Rey, J.; Cremades, L.V. Evaluation of the Power Generation Impact for the Mobility of Battery Electric Vehicles. *Energies* **2023**, *16*, 5006. https://doi.org/10.3390/en16135006

Academic Editor: Simone Barcellona

Received: 23 May 2023
Revised: 24 June 2023
Accepted: 26 June 2023
Published: 28 June 2023

1. Introduction

The 2015 Paris summit, COP21, ended with an agreement between all industrialized nations [1]. This agreement calls on the European Union (EU) to try to avoid a global temperature increase of more than 2 °C compared to pre-industrial levels. To achieve these goals, the European institutions have proposed to achieve climate neutrality, i.e., zero carbon dioxide (CO_2) emissions, so that no more greenhouse gases (GHG) are produced by human activity.

In order to contribute to these objectives, it is necessary to regulate vehicle transport. Member States have committed themselves to taking appropriate measures to achieve this objective, which is why the successive environmental permits for vehicle transport in the EU are becoming increasingly restrictive for internal combustion engine vehicles (ICEVs). However, the ultimate goal is to ban ICEVs in the medium and long term. Currently, in the absence of a final agreement with all Member States, European institutions have decided to ban the sale of internal combustion vehicles in the EU from 2035 [2]. Although this only affects the sale of new vehicles, major European cities are also restricting the circulation of ICEVs in low emission zones [3].

It is therefore foreseeable that the current vehicle fleet will be transformed at high speed in the coming years, and it is therefore important to study and analyze the following questions: Is it possible to transform the current vehicle fleet into a battery electric vehicle fleet? And if it is possible, how can the transition be made and what are the socio-economic costs?

Virtually since the popularization of the automobile at the beginning of the 20th century, all ground transportation has been powered by internal combustion engines. However, this form of mobility generates negative externalities that were not considered until the last few decades. ICEVs emit gaseous and particulate pollutants from the combustion process [4].

Successive environmental regulations have restricted the maximum value of each of the emissions or introduced a new restriction that was not foreseen in the previous regulations.

The EU has adopted and applied a series of environmental regulations, known as EURO regulations, which affect the homologation of vehicles, in an attempt to reduce the pollutant emissions of ICEVs. The EURO 6d regulation is currently in effect [5–7]. Each revision of these regulations has further restricted the maximum level of pollutant emissions and introduced new, more demanding type-approval tests based on real-world driving, such as the WLTP procedure or the Real Driving Emissions tests [5].

In order to comply with these regulations, vehicle manufacturers have developed technological improvements to meet the EURO regulations, such as particulate filters or Selective Catalytic Reduction (SCR) catalysts using AdBlue [8]. Now, with the foreseeable introduction of the next update of the EURO regulation, EURO 7 [5], European manufacturers are declaring that it will be impossible to comply with the new restrictions and make the necessary investments to adapt the existing technology while keeping it profitable, even more so when the sale of internal combustion vehicles is finally banned in the EU from 2035. It is therefore foreseeable that the introduction of electric vehicles will accelerate in the coming years.

There are several types of electric vehicle (Figure 1):

- Battery electric vehicle (BEV): They store chemical energy in a battery, which provides the electrical energy for consumption by the electric motor, which converts the electrical energy into mechanical energy. It is currently the main alternative to the conventional ICEV and is the main pillar of the new mobility. This is due to the fact that the use of the battery electric vehicles is considered to produce no polluting gases or particles, although this depends on the origin of the electrical energy and the entire life cycle of the battery electric vehicle.
- Hybrid electric vehicle (HEV): They combine an internal combustion engine with an electric motor. The internal combustion engine can operate in two ways. The combustion engine provides mechanical power directly to the transmission or to an electric generator. The generator feeds a battery, which feeds an electric motor. The electric motor provides mechanical power to the transmission. HEVs have seen significant development and uptake in recent years due to their lower fuel consumption, combined with a competitive price compared to the ICEV.
- Plug-in hybrid electric vehicle (PHEV): They are essentially HEVs that allow electricity to be supplied to the battery directly from the grid. The battery is smaller than that of a BEV. This makes it possible to provide driving modes using only the electric motor supplied by the battery, and therefore driving modes that are a priori free of polluting emissions. The main problem with PHEVs is that there is no guarantee that the user will recharge the battery and not constantly use the vehicle powered by the combustion engine, which is effectively equivalent to an ICEV.
- Fuel cell electric vehicle (FCEV): They use the electrical energy generated in a fuel cell, which uses the chemical energy stored in a pressurized tank, mainly from hydrogen. This electrical energy powers a smaller battery than in the BEV, to ultimately provide mechanical power to the drivetrain through the electric motor. FCEVs can become another alternative for zero-emission mobility and coexist with the BEV in the medium and long term. Currently, the purchase price of existing FCEVs, such as the Toyota Mirai [9] or the Hyundai NEXO [10], is high and the technology to produce green hydrogen from clean electricity is not yet developed enough to compete with ICEVs and BEVs. Finally, the refueling infrastructure is very poorly developed and free mobility with an FCEV is practically impossible [11].

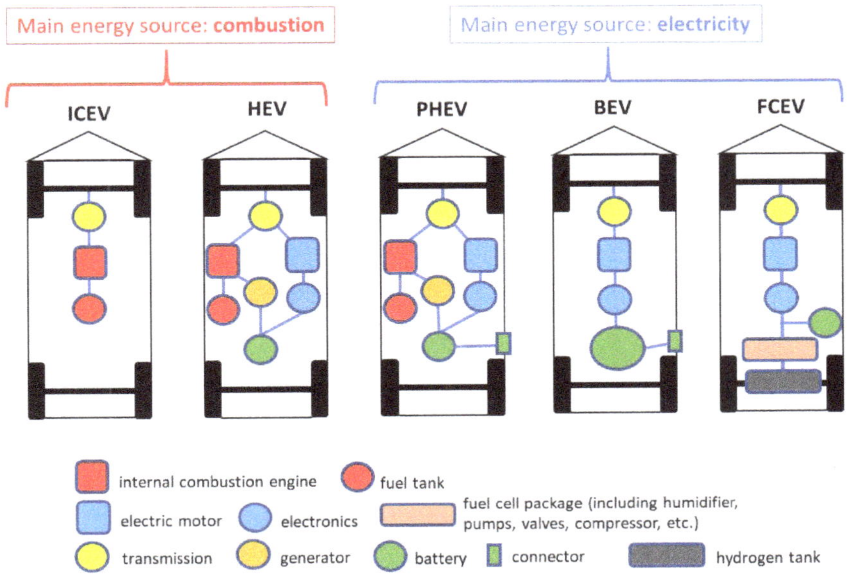

Figure 1. Types of vehicles (internal combustion and electric) and their main components. Modified from [12].

For the end user, the main difference between an ICEV and a BEV is the provision and storage of the energy needed to operate the vehicle. This difference is currently the main advantage of the ICEV, and therefore the main disadvantage of the BEV. In order to drive the BEV, the vehicle's battery needs to be charged. There are currently four possible modes of charging from the grid [13,14]:

- Mode 1. The vehicle is directly connected to the conventional grid without the need for any additional special equipment or systems. This mode is very practical for small vehicles such as bicycles or mopeds, but is not recommended for commercial vehicles.
- Mode 2. This mode provides a slow charge. This type of charging is single-phase with a voltage of 230 V and a maximum power of 3.7 kW. The BEV is connected to the mains via the appropriate plug/adapter to ensure the safety of the charging process.
- Mode 3. This mode provides semi-fast charging. The electric vehicle is connected to the alternating current grid via a dedicated BEV charging outlet. The most commonly used plug for this type of charging is Type 2 [15]. This mode allows single-phase or three-phase charging. Single-phase connections charge at 7.4 kW and three-phase connections charge at 22 kW.
- Mode 4. Its charging power is equal to or greater than 50 kW, allowing "super fast" and "ultra fast" charging. The latter is not recommended for daily charging, as it can damage the battery if used regularly. It is specifically designed for outdoor public use stations and could be similar to a gas station, where the vehicle can be recharged during long trips or in specific situations where passengers are short of time.

Mode 4 charging uses direct current (DC), as opposed to the previous modes that use alternating current (AC). The most commonly used plug for this type of charging is currently the CCS Combo, which combines a Type 2 plug with two extra terminals to allow DC power to pass through [15].

The main objective of this paper is to study the feasibility of transforming the current mobility of the vehicle fleet in Spain based on the use of fossil fuels (gasoline and diesel) in internal combustion engines (ICEVs) to a mobility based on the use of electricity from renewable sources for battery electric vehicles (BEVs). This feasibility is analyzed from a techno-economic and environmental point of view.

There are several papers in the literature that have examined the impact on the reduction of GHG and/or non-GHG emissions by substituting ICEVs with BEVs in specific regions, such as [16,17]. There are also those that performed life cycle analyses comparing ICEVs with BEVs, such as [18–20]. But none of them explored the feasibility of transforming the actual mobility of ICEVs into BEVs and its impact on electricity generation in Spain, as presented in this paper.

2. Method

In order to study the feasibility of this change, an in-depth analysis of daily mobility at different geographical levels and of the type of vehicle used for each trip was carried out, making it possible to estimate the amount of additional electrical energy that would need to be generated to make a fleet of vehicles made up entirely of BEVs feasible. However, to achieve a true environmental transition in vehicle transportation, the electricity needed to power a fleet of BEVs must come from renewable sources. This would require the installation of new solar and wind power plants, which would need to be located and installed somewhere in the territory, along the associated transportation network.

The following steps were taken to conduct the study (Figure 2):

(1) Collect available information on charging stations, vehicle fleet, actual mobility, certified electricity consumption of BEVs, electricity generation, and pollutant emissions of ICEVs according to vehicle age.
(2) Estimate the daily electricity consumption that would be required for the mobility of the vehicles if they were all BEVs.
(3) Determine the hourly availability for charging these vehicles based on the daily mobility data and energy source. Since photovoltaic (PV) solar energy can only be generated during daylight hours, the percentage of daily use of this source can be defined. The rest of the energy is assumed to come from wind.
(4) Estimate the required installation capacity of wind turbines and PV panels to provide the energy needed to charge the BEVs. Also estimate the investment required.
(5) Estimate the reduction in pollutant emissions from replacing ICEVs with BEVs.

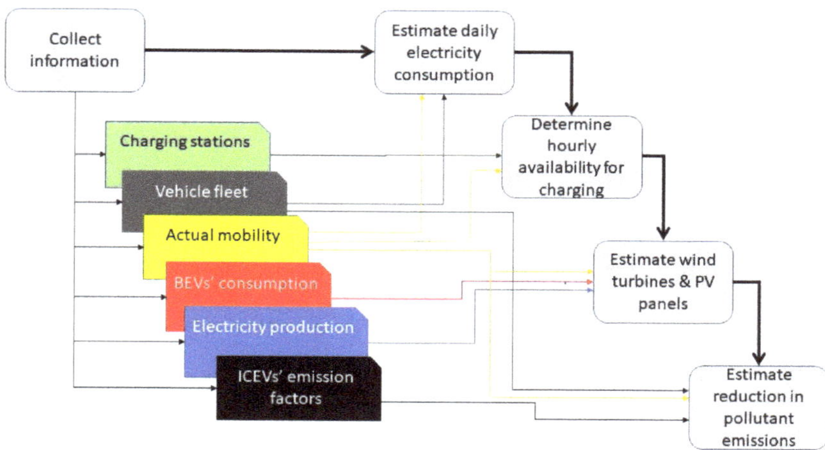

Figure 2. Schematic of the methodological steps followed in this study.

The fleet information was used to estimate the emission reduction of the ICEVs that would be replaced by BEVs, based on their age. However, the number of ICEVs to be replaced by BEVs was determined by the actual mobility. Vehicle mobility trips are assumed to correspond to the proportion of such vehicles in the fleet, both in terms of category and age.

The study, analysis and proposal to be carried out in this work mainly focused on the province of Barcelona (PB) (Spain), as there is a survey of actual mobility that is regularly carried out for this region [21]. However, the composition and age of the vehicle fleet have been obtained from data for Spain as a whole.

The results obtained for PB can be extrapolated to the whole of Spain, assuming that the mobility pattern is similar.

The PB covers mobility that takes place in the region bordering the city of Barcelona. At the end of 2022, the PB had a population of 5,727,615 inhabitants [22] in an area of 7726 km^2, which corresponds to a population density of 741.2 inhabitants/km^2, of which 1,636,193 inhabitants correspond to the city of Barcelona, with a population density of 15,992.2 inhabitants/km^2, which acts as a dynamic center of socio-economic activity.

In this paper, clean electrical energy is defined as energy produced by renewable energy sources. Renewable energy is defined as energy that is consumed more slowly than it is produced. Therefore, in this study, only energy from solar and wind sources is considered to be clean electricity.

Available data from actual wind and PV solar farms in Spain were extrapolated to estimate the area and investment costs required to meet the estimated energy demand.

The optimal use of a BEV implies a usage model with recharging at a dedicated charging point and at a recharging rate that does not imply a sudden degradation of the battery system. For this reason, in the hypothetical case study of a ground transport BEV, it was assumed that the vehicle would not be charged in the usual way at a fast charging station.

3. Basic Data

3.1. Charging Points

In order to increase the use of BEVs, it is necessary to develop a network of charging stations with public access, in addition to the private charging points that each user can install at home.

Currently, the two most powerful public charging stations in Spain are located in the Basque Country and have been installed by Repsol [23]. The fact that a fuel company is installing charging stations shows a clear commitment to electrification. These four-terminal installations have a capacity of 400 kW, which corresponds to a charging time of between 5 and 10 min, similar to the refueling time of an ICEV.

However, according to the Spanish Association of Automobile and Truck Manufacturers (ANFAC) [24], the public access charging network in Spain is growing very slowly and is poorly distributed throughout the country. At the beginning of 2021, there were 11,517 public access charging points in Spain, 83% of which had power of less than 22 kW, which does not allow fast charging. While in the EU as a whole the average number of charging points per million inhabitants is 573, in Spain it is only 245. This is in line with the current low presence of BEVs in Spain.

3.2. Vehicle Fleet Composition

From the studies and statistics published by the *Dirección General de Tráfico* (DGT in Spanish) [25], it is possible to know the composition of the current vehicle fleet in each of the Spanish municipalities and, therefore, in the country as a whole.

On 31 December 2022, the total number of main categories of motor vehicles in Spain was 34,304,426, divided as follows:

- Passenger cars: 25,222,554 (73.5%);
- Motorcycles: 4,006,804 (11.7%);
- Vans: 2,617,145 (7.6%);
- Trucks: 2,457,923 (7.2%).

The DGT statistics also allowed us to classify each vehicle type according to its age, from 2022 backwards (Table 1), which allowed us to know its pollution potential according to the existing environmental approval in place at the time of its registration.

Table 1. Age of the Spanish vehicle fleet by vehicle type in 2022 [26].

Vehicle	Age	Gasoline	% [1]	Diesel	% [1]	Others	% [1]	Total	% [2]
Passenger cars	<5 years	3,242,568	62.7	1,739,699	33.6	190,474	3.7	5,172,741	*20.5*
	5–9 years	1,813,269	38.6	2,859,712	60.8	26,896	0.6	4,699,877	*18.6*
	10–14 years	1,150,761	29.1	2,802,644	70.8	3542	0.1	3,956,947	*15.7*
	15–19 years	1,854,694	32.0	3,944,656	68.0	2511	0.0	5,801,861	*23.0*
	>19 years	3,689,507	66.0	1,899,161	34.0	2460	0.0	5,591,128	*22.2*
Motorcycles	<5 years	817,978	95.9	1398	0.2	33,988	4.0	853,364	*21.3*
	5–9 years	612,144	99.0	1853	0.3	4625	0.7	618,622	*15.4*
	10–14 years	644,651	99.5	922	0.1	2478	0.4	648,051	*16.2*
	15–19 years	810,876	99.8	1186	0.1	118	0.0	812,180	*20.3*
	>19 years	1,073,538	99.9	645	0.1	404	0.0	1,074,587	*26.8*
Vans	<5 years	49,376	9.2	470,241	87.1	19,986	3.7	539,603	*20.6*
	5–9 years	13,425	3.4	373,018	95.4	4400	1.1	390,843	*14.9*
	10–14 years	13,616	5.2	249,505	94.5	795	0.3	263,916	*10.1*
	15–19 years	36,964	7.9	430,543	91.9	778	0.2	468,285	*17.9*
	>19 years	302,203	31.7	651,928	68.3	367	0.0	954,498	*36.5*
Trucks	<5 years	11,108	3.7	275,904	92.8	10,405	3.5	297,417	*12.1*
	5–9 years	5511	2.1	252,210	97.1	2106	0.8	259,827	*10.6*
	10–14 years	3995	1.3	293,101	98.2	1331	0.4	298,427	*12.1*
	15–19 years	18,085	2.3	777,168	97.7	402	0.1	795,655	*32.4*
	>19 years	38,597	4.8	767,888	95.2	112	0.0	806,597	*32.8*
Total		16,202,866	47.2	17,793,382	51.9	308,178	0.9	34,304,426	

[1] Percentage of each fuel type in its age group; [2] percentage contribution of each age for the type of vehicle

Table 1 shows how the share of diesel passenger cars has fallen significantly in recent years, from 60 to 70% of the fleet to just 34% for cars less than 5 years old. This is largely due to the reputational crisis suffered by the diesel engine as a result of the so-called "dieselgate" in September 2015. It should also be noted that the sales of alternative vehicles ("Others" column in Table 1) compared to conventional vehicles have started to increase in recent years, but still represent a very small number. Furthermore, the age of the fleet is remarkable, as the number of vehicles 10 years old or more accounts for more than 60% in all categories.

Within PB, the total fleet of main vehicles in 2022 was 3,560,977, as follows [25]:

- Passenger cars: 2,417,620 (67.9%);
- Motorcycles: 696,678 (19.6%);
- Vans: 241,125 (6.8%);
- Trucks: 205,554 (5.8%).

Age and energy type distribution can be assumed to be similar to that of the Spanish fleet.

3.3. Mobility in the Barcelona Province

According to the latest published survey on mobility in the Barcelona metropolitan area in 2021, there were 16,909,491 trips per working day in PB, which corresponds to an average of 3.5 trips per inhabitant per day [21,27]. These trips were made on foot (46.5%) or by bicycle/scooter (2.5%), public transport (14%), or private transport (37%), the latter including cars (86%), motorcycles (10.6%), and vans/trucks (3.4%). The total number of trips in 2021 was lower than it was in 2019 (19,259,471), in the pre-pandemic COVID-19 situation, but the distribution of trips by mode was similar [21].

On the other hand, the average distance of each trip by private vehicle in PB was as follows [27]: 8.9 km by car; 5.8 km by motorcycle, and 15.2 km by van/truck.

Multiplying each trip by the average distance per trip gives the total distance traveled per working day by each type of vehicle (Table 2).

Table 2. Total daily distance traveled by type of vehicle in Barcelona province.

Private Transport	Trips	km/Trip	Distance (km)
Car	5,370,628	8.9	47,798,589
Motorcycle	664,715	5.8	3,855,347
Van/Truck	214,745	15.2	3,264,124
Total	6,250,088		54,918,060

The total distance traveled by private vehicles was 54,918,060 km, which corresponds to a total of 9.6 km traveled per inhabitant per working day in a private vehicle.

Figure 3 shows the hourly distribution of private transport mobility during a weekday in PB. It follows a pattern of occupational mobility (work and/or study), as the largest number of trips was concentrated between 7 and 9 a.m. and 5 and 7 p.m., coinciding with the departure and return home.

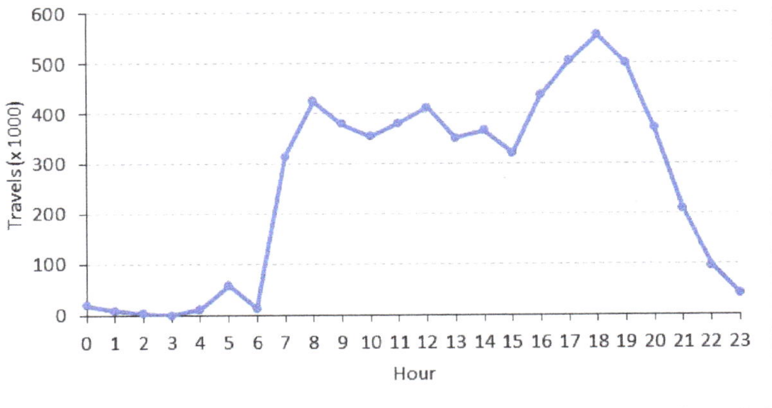

Figure 3. Private transportation trips in PB during the course of a workday. Based on [27].

3.4. Electric Vehicle Power Consumption

In order to know the energy consumption of a fleet made up entirely of electric vehicles, the homologated consumption of models representative of the type of transport can be used as a basis for calculation. From the lowest to the highest electric energy consumption per kilometre traveled, the following models were selected:

- Motorcycles: Silence S01 [28]. This electric motorcycle is the number 1 seller on the Spanish market, with sales of 1433 units in 2022. It has a range of 133 km with a battery of 5.6 kWh of stored energy, giving an energy consumption per 100 km of 4.2 kWh/100 km.
- Passenger cars: Tesla Model 3 [29]. This car is the number 1 in the Spanish electric vehicle market, with sales of 2677 units in 2022. It has an approved energy consumption of 16.2 kWh/100 km.
- Vans: Peugeot Expert [30]. It is a mid-size van and therefore covers the entire range of this type of vehicle. In 2022, 722 units were sold on the Spanish market. It has an energy consumption of 20.5 kWh/100 km, slightly higher than that of a car.
- Trucks: DAF LF Electric [31]. The disadvantage of the battery electric vehicle market for trucks is the significant energy consumption required to move such a large

and heavy vehicle. Therefore, to achieve a homologated range of 280 km, a battery with 254 kWh of stored energy is required, resulting in an energy consumption of 110.2 kWh/100 km, more than five times higher than that of the reference van.

To disaggregate the van and truck mobility data in Table 2, we have assumed that 55% are vans and 45% are trucks. This proportion is about the same proportion as their share in the PB fleet (see Section 3.2).

3.5. Electric Power Generation

In 2022, a total of 276,316 GWh of electricity was produced in Spain, corresponding to an average daily production of 757 GWh [32]. Up to 44% of this energy came from renewable sources, i.e., around 330 GWh per day, mainly from wind, photovoltaics, and hydropower (Table 3).

Table 3. Electricity production by technology in Spain by 2022 [32].

Generation Type	Source Category	Energy Generated (GWh)	%
Combined cycle	Pollutant	60,652	24.66
Wind	Renewable	59,805	22.14
Nuclear	No emissions *	55,984	20.26
Photovoltaic	Renewable	27,283	10.08
Hydropower	Renewable	17,860	6.46
Co-generation	Pollutant	17,732	6.43
Coal	Pollutant	7687	2.81
Other renewable sources	Renewable	4646	1.69
Solar thermal	Renewable	4123	1.49
Turbine pumping	Renewable	3776	1.37
Diesel engines	Pollutant	2548	0.92
Non-renewable wastes	Pollutant	1761	0.69
Steam turbine	Pollutant	1207	0.44
Renewable wastes	Renewable	739	0.32
Gas turbine	Pollutant	657	0.24
Hydro-wind	Renewable	23	0.01
TOTAL		276,316	100.00

* but radioactive wastes

3.6. Pollutant Emission Factors for ICEVs

To estimate the emissions of ICEVs, we used the European Environment Agency emission factors for road transport according to the Tier 2 methodology, which take into account the technology or legislation (age) of the vehicle [33]. We only considered the emissions of nitrogen oxides (NO_x) and particulate matter (PM). Table 4 shows the emission factors expressed in grams of pollutant emitted per kilometer driven, used in this study for both gasoline and diesel vehicles. The passenger car factors corresponded to mid-size passenger cars. The truck factors corresponded to trucks between 7.5 and 16 tons. And the motorcycle factors corresponded to four-stroke motorcycles between 250 and 750 cm^3.

In the absence of factors, the same values are assumed for gasoline trucks as for vans. Similarly, the PM factors for diesel motorcycles are assumed to be the same as for cars.

Table 4. NO_x and PM emission factors (in g/km) for ICEVs (European gasoline and diesel vehicles) taken from [33]. Values in italics are estimates used in this study.

Technology/Regulation	Passenger Cars		Vans		Trucks		Motorcycles	
	NO_x [a]	PM [b]	NO_x [a]	PM [b]	NO_x [a]	PM [b]	NO_x [a]	PM [b,c]
Gasoline								
ECE (until 1977)	2.53	0.0022	*3.09*	0.0023	*3.09*	*0.0023*	*0.233*	*0.2*
ECE (1978–1980)	2.40	0.0022	*3.09*	0.0023	*3.09*	*0.0023*	*0.233*	*0.2*
ECE (1981–1985)	2.51	0.0022	*3.09*	0.0023	*3.09*	*0.0023*	*0.233*	*0.2*
ECE (1985–1992)	2.66	0.0022	3.09	0.0023	*3.09*	*0.0023*	*0.233*	*0.2*
Euro 1 (1992–1996)	0.485	0.0022	0.563	0.0023	*0.563*	*0.0023*	0.233	0.2
Euro 2 (1996–2000)	0.255	0.0022	0.23	0.0023	*0.23*	*0.0023*	0.477	0.08
Euro 3 (2000–2005)	0.097	0.0011	0.129	0.0011	*0.129*	*0.0011*	0.317	0.04
Euro 4 (2005–2010)	0.061	0.0011	0.064	0.0011	*0.064*	*0.0011*	0.194	0.04
Euro 5 (2010–2016)	0.061	0.0014	0.064	0.0014	*0.064*	*0.0014*	0.194	0.01
Euro 6 until 2016	0.061	0.0014	0.064	0.0012	*0.064*	*0.0012*	*0.194*	*0.01*
Euro 6 2017–2019	0.061	0.0016	0.064	0.0012	*0.064*	*0.0012*	*0.194*	*0.01*
Euro 6 2020+	0.061	0.0016	0.064	0.0012	*0.064*	*0.0012*	*0.194*	*0.01*
Diesel								
Conventional (until 1992)	0.546	0.2209	0.87	0.356	8.92	0.3344	*0.546*	*0.2209*
Euro 1 (1992–1996)	0.690	0.0842	0.69	0.117	5.31	0.201	*0.690*	*0.0842*
Euro 2 (1996–2000)	0.716	0.0548	0.716	0.117	5.5	0.104	*0.716*	*0.0548*
Euro 3 (2000–2005)	0.773	0.0391	0.77	0.0783	4.3	0.0881	*0.773*	*0.0391*
Euro 4 (2005–2010)	0.58	0.0314	0.58	0.0409	2.65	0.0161	*0.58*	*0.0314*
Euro 5 (2010–2016)	0.55	0.0021	0.55	0.001	1.51	0.0161	*0.55*	*0.0021*
Euro 6 until 2016	0.45	0.0015	0.45	0.0009	0.291	0.0008	*0.45*	*0.0015*
Euro 6 2017–2019	0.35	0.0015	0.35	0.0009	0.291	0.0008	*0.35*	*0.0015*
Euro 6 2020+	0.17	0.0015	0.17	0.0009	0.291	0.0008	*0.17*	*0.0015*

ECE = Economic Commission for Europe; [a] NO_x = nitrogen oxides, expressed as NO_2; [b] PM = particulate matter, expressed as $PM_{2.5}$; [c] values for quadbikes.

4. Results

The total energy needed to cover the mobility in PB with a fleet composed exclusively of BEVs can be estimated by multiplying the homologated power consumption of each vehicle type selected in Section 3.4 by the total distance traveled by each vehicle in PB. The results are shown in Table 5.

Table 5. Daily energy consumption estimated for a fleet of BEVs in Barcelona province.

Private Transport	Distance (km)	Power Consumption (kWh/100 km)	Total Daily Consumption (kWh)
Cars	47,798,589	16.2	7,743,371
Motorcycles	3,855,347	4.2	162,310
Vans	1,795,268	20.5	368,030
Trucks	1,468,856	110.2	1,619,267
TOTAL	54,918,060		9,892,978

In summary, the daily electrical energy required to switch from fossil-fuel-based ground transportation to clean electric transportation would be about 10 GWh. This amount would require a 3% increase in clean electricity generation. This increase seems more than affordable for Spain as a whole, with the exception of hydropower.

Traditionally, hydropower has been very important in electricity generation. The main problem with this energy source is that due to the drought situation in 2022, the total production of hydroelectricity in Spain has decreased by 39.7% compared to 2021. The foreseeable increase in droughts and the abandonment of the construction of hydroelectric dams in Spain means that an increase in hydroelectric capacity has not been considered.

Therefore, it seems reasonable to consider an increase in electricity generation capacity with only wind and PV solar energies.

Unlike nuclear power and combined cycle power plants, wind and solar energies cannot be controlled by existing human technology, but depend on the weather. This fact means that the installation of new electricity generation mechanisms must take into account: (1) the hourly schedules for charging BEVs, and (2) the hourly production of electricity for each technology.

The main characteristic of the BEV is the slowness of recharging the vehicle compared to refueling an ICEV. Since the BEV needs to be parked for a relatively long time to recharge, we propose relating the period of electricity generation needed to recharge the BEV fleet inversely proportionally to the mobility schedule in PB (Figure 3). That is, we assume that the fewer trips are made in PB, the more likely it is that the BEV will be recharging batteries in the standard recharge mode. With this assumption, the distribution of electric power throughout the day would be as shown in Figure 4. Therefore, most of the additional electricity generation to charge the batteries of the BEVs that would replace the current fleet would have to be generated at night.

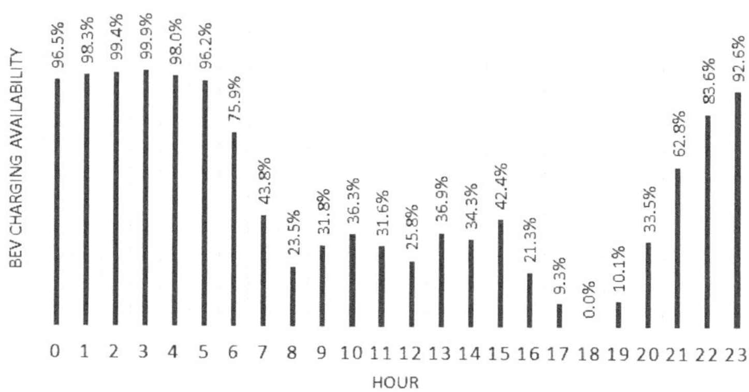

Figure 4. Estimated availability of electric power to recharge a BEV in PB.

A proposal to increase wind and PV solar power generation capacity is presented below.

4.1. PV Solar Energy

Solar energy reaches its daily production peak in Barcelona between 10 a.m. and 5 p.m. [34]. In winter, approximately 90% of the total energy is produced in this hourly window. However, this is also the season with the lowest total production. In summer, about 70% is produced in this time window [34]. Therefore, it is estimated that 75% of the total energy production on an annual average is in this time window.

From Figure 4, it can be deduced that there would be an average of 29.7% charge availability in this period. To achieve this charging scenario, it is necessary to find a compromise between the recommendation of charging the BEV with non-fast charging and the availability of using PV solar energy. Therefore, in order to maximize the possibilities of producing 100% of the electrical energy from clean sources, it seems reasonable to consider that 30% of this energy can come from PV solar power.

The installed PV capacity in Spain in 2022 was 19,348 MW [35]. Therefore, its load factor L can be calculated as follows:

$$L = \frac{E}{P \cdot t} \tag{1}$$

where E is the total annual energy generated by photovoltaic source, i.e., 27,283 GWh (see Table 3); P is the installed PV capacity, and t is the number of hours per year available for

PV generation, i.e., 2557 h. Applying Equation (1) with these values gives a load factor of 55.1%.

Then, the additional PV power, P_A, to be installed for only BEV charging in PB during the period between 10 a.m. and 5 p.m., i.e., 7 h, can be estimated as follows:

$$P_A = \frac{f \cdot E_C}{L \cdot t_c} \quad (2)$$

where f is the fraction of energy from PV solar power, i.e., 0.3; E_C is the total daily consumption of energy, i.e., 9893 MWh (see Table 5); and t_c is the number of charging hours, i.e., 7 h. Applying Equation (2) with these values gives P_A = 770 MW.

As a reference, there are currently 196 PV park projects in Catalonia, with the construction of PV modules planned to occupy 4977 hectares (ha), with a total PV solar power of 2888 MW [36]. This corresponds to an average power of 0.58 MW/ha. Applying the same ratio to the additional power of 770 MW, at least 1327 ha of land would need to be covered with solar panels.

As a reference for estimating the investment required to install this amount of electricity, we have extrapolated the budget of the Ancar II PV solar farm [37]. This farm, located in the province of Teruel, has a nominal capacity of 41.58 MW, generated by 116,032 PV generation modules covering an area of 106.53 ha. The budget for the Ancar II PV solar park is EUR 21.2 M, of which the cost of the solar panels represents 62% [37]. Assuming the same cost per hectare, an investment of around EUR 265 M would be required to install 1327 ha.

If we extrapolate these calculations to the whole of Spain, assuming that the mobility of private transport in Spain as a whole follows a behavior similar to that of PB, as shown in Figure 3, it would be necessary to occupy an area of approximately 14,000 ha with an investment of around EUR 2800 M.

4.2. Wind Energy

The remaining 70% of the electrical energy to be produced would come from wind.

The installed capacity of wind energy in Spain in 2022 was 29,417 MW [35]. When we applied Equation (1) to the data for wind power (i.e., E = 59,805 GWh, P = 29,417 MW, t = 8766 h), the load factor of the entire wind power generation grid was only 23.2%.

As shown in Figure 4, there is virtually 100% hourly availability for BEV charging between 0 and 6 a.m. Therefore, the additional wind power to be installed for only BEV charging in PB can be estimated by applying Equation (2) with f = 0.7, E_C = 9893 MWh, L = 0.232, and t_c = 6 h. The result is P_A = 4976 MW, or about 5 GW.

There are currently 45 wind farm projects in Catalonia (region in the northeast of Spain where PB is located), with 207 wind turbines in the pipeline, representing a total capacity of 1184 MW [36]. This corresponds to an average capacity of 5.72 MW per wind turbine. Applying the same ratio to the additional 5 GW of wind power, at least 870 turbines would need to be installed.

To estimate the budget of this proposal, we can extrapolate the actual budgets of existing wind farms. For example, the budget of the Cabigordo wind farm in the province of Teruel (Spain) is EUR 32.6 M, of which the main budget item is the purchase and installation of the wind turbines, which amounts to EUR 22 M [38]. This wind farm consists of nine wind turbines with a total capacity of 50 MW. Considering a similar cost for the wind turbines, the total cost of just installing the 870 wind turbines, without considering all the other costs of the project, would be approximately EUR 2127 M.

At the country level, assuming the same mobility pattern as in PB, it is estimated that approximately 9240 turbines would need to be installed at a cost of around EUR 22,600 M.

4.3. Reduction of Pollutant Emissions

Table 6 shows the estimated NO_x and PM emissions in kg/day that would be avoided if the ICEVs currently circulating in PB were replaced by BEVs. For this purpose, it was assumed that the mobility envisaged in Figure 3 is composed of ICEVs of the four categories analyzed (cars, motorcycles, vans, and trucks) and grouped in five age periods (up to 1993, 1994–2008, 2009–2013, 2014–2018, and 2019–2022) in the proportions that exist in the current Spanish fleet (see Table 1). The emission factors in Table 4 have been applied to each category by averaging the factors for the specified periods. These factors are multiplied by the daily distance traveled by each category of ICEV, as shown in Table 5.

Table 6. Estimated pollutant emissions from ICEVs in PB (kg/day).

Vehicle	Age	Period	Gasoline		Diesel	
			NO_x	PM	NO_x	PM
Passenger cars	<5 years	2019–2022	375	10	1045	9
	5–9 years	2014–2018	210	5	1546	6
	10–14 years	2009–2013	133	3	1232	37
	15–19 years	1994–2008	789	6	2424	184
	>19 years	up to 1993	15,283	15	4321	1067
Motorcycles	<5 years	2019–2022	153	2	134	1
	5–9 years	2014–2018	114	1	265	1
	10–14 years	2009–2013	120	4	350	10
	15–19 years	1994–2008	231	33	538	41
	>19 years	up to 1993	283	186	638	158
Vans	<5 years	2019–2022	2	0	6	0
	5–9 years	2014–2018	1	0	4	0
	10–14 years	2009–2013	1	0	5	0
	15–19 years	1994–2008	6	0	17	2
	>19 years	up to 1993	553	0	162	49
Trucks	<5 years	2019–2022	0	0	2	0
	5–9 years	2014–2018	0	0	2	0
	10–14 years	2009–2013	7	0	5	0
	15–19 years	1994–2008	3	0	48	1
	>19 years	up to 1993	62	0	164	6
Total			18,325	265	12,910	1572

According to these calculations, more than 31 tons of NO_x and almost two tons of PM would be avoided daily.

Assuming that the mobility pattern throughout Spain is similar to that of PB, it can be estimated that the daily emission reduction would be about 314 tons of NO_x and 17 tons of PM in Spain.

5. Discussion

The methodology used could be applied to other regions or countries where the vehicle fleet is mainly composed of ICEVs and for which disaggregated mobility information is available.

To understand the scale of the challenge of installing the required wind power capacity, we can compare it with the current situation of existing wind farms. The total number of wind turbines currently operating in Catalonia is 846 [36]. These turbines are mainly located in four areas where the wind tends to blow with greater intensity and frequency. As a result, project proposals in Catalonia are concentrated in these areas [39]. However, this may foreshadow a land use problem, as the best sites are limited. For this reason, there is a first wind farm project, called Tramuntana Park, directly in the sea, on the coast of the Gulf of Roses, which foresees an initial installation of 35 wind turbines [40]. This would make it possible to take advantage of the high wind speeds in this area, although it is expected to have a negative impact on the ecosystem.

In terms of the estimated demand for solar energy, the 1327 ha of land required is equivalent to 1858 soccer fields, or 13.1% of the surface area of the city of Barcelona. The requirements to meet this demand appear to be lower than those for wind energy because, on the one hand, the surface area to be installed is relatively small, since only 30% of the electrical energy would be covered by PV solar energy, and, on the other hand, there are not as many restrictions on the location of the solar panels.

In the case of wind energy, the occupied area would be much smaller. For example, the total area occupied by nine wind turbines and foundations at the Cabigordo wind farm site is 65,413 m^2 [38]. We can then estimate that 870 turbines would occupy an area of about 63 ha. The impact on the land would therefore be small compared to that of a PV solar installation with a similarly rated output. However, the environmental impact generated by wind turbines is not harmless, as it mainly affects birds [41].

The cost per unit of installed power is lower for PV solar farms (EUR 0.34 M/MW) than for wind farms (EUR 0.43 M/MW). The main issue in achieving an effective use of clean energy for a BEV fleet is the need to ensure the production of clean energy at night. For this reason, it is essential to prioritize electrical energy produced by wind power, as it is not possible to produce electrical energy from PV solar sources during the night.

In addition to this investment in electricity generation, there is the cost of the infrastructure required to distribute and install the charging points. This cost is very variable because it depends on many factors, such as the price of the chosen charging point, the charging power, the distance to the meter or electrical panel, and the location of the charging point, as well as the auxiliary work that needs to be carried out to supply the charging point. However, as a guideline, the cost of installing one charging point can range from one to several thousand euros [42].

In terms of environmental benefits, the significant reduction in pollutant emissions that would be achieved during vehicle operation is noteworthy. The estimated values are relatively higher than those published in other studies, such as the one carried out for the urban areas of Berlin and Stuttgart [17]. This is probably due to the high age of the Spanish vehicle fleet, which is one of the oldest in Europe, with an average age of more than 13.5 years, compared to the European average of 11.5 years [43]. Likewise, the Spanish fleet has one of the lowest percentages of BEVs (3.8% [44]) in Europe (average 14.2% [45]).

In conclusion, if the assumptions made in the calculations are accepted, the resulting estimates allow us to state that the impact of the mobility transformation would be positive in terms of reducing pollutant emissions during the service life of BEVs. However, the environmental impact of the rest of the life cycle of the vehicle (extraction of raw materials, production, and recycling) would have to be taken into account.

The estimated investment required in Spain to generate the additional electricity from wind and PV solar power is enormous (more than 25.4 billion euros), representing, for example, 5.5% of the total national budget in 2022. On the other hand, the installation of the estimated quantities of wind turbines and solar panels, although important, would not be able to meet the electricity needs alone, unless they are accompanied by other auxiliary sources to meet demand when the weather is unfavorable.

Estimated energy requirements could be reduced if batteries with higher energy densities than those currently available are developed in the future. Another possibility for the future, perhaps in the longer term, is to advance the development of hydrogen production/use technology from renewable sources to make the use of FCEVs competitive.

Author Contributions: Conceptualization, J.R. and L.V.C.; methodology, J.R. and L.V.C.; validation, L.V.C.; formal analysis, J.R. and L.V.C.; investigation, J.R. and L.V.C.; data curation, J.R.; writing—original draft preparation, L.V.C.; writing—review and editing, J.R. and L.V.C.; visualization, J.R. and

L.V.C.; supervision, L.V.C.; project administration, L.V.C. All authors have read and agreed to the published version of the manuscript.

Funding: This research received no external funding.

Data Availability Statement: Not applicable.

Conflicts of Interest: The authors declare no conflict of interest.

References

1. United Nations Climate Change. The Paris Agreement. 2023. Available online: https://unfccc.int/process-and-meetings/the-paris-agreement (accessed on 19 April 2023).
2. News European Parliament. EU Ban on the Sale of New Petrol and Diesel Cars from 2035 Explained. 2023. Available online: https://www.europarl.europa.eu/news/en/headlines/economy/20221019STO44572/eu-ban-on-sale-of-new-petrol-and-diesel-cars-from-2035-explained (accessed on 23 April 2023).
3. Eurocities. Low Emission Zones: Challenges and Solutions. 2021. Available online: https://eurocities.eu/latest/low-emission-zones-challenges-and-solutions/ (accessed on 29 April 2023).
4. Krajinska, A. Electric Vehicles Are Far Better than Combustion Engine Cars When it Comes to Air Pollution. Here's Why. 2021. Available online: https://www.transportenvironment.org/discover/electric-vehicles-are-far-better-than-combustion-engine-cars-when-it-comes-to-air-pollution-heres-why/ (accessed on 29 April 2023).
5. EUR-Lex. Commission Regulation (EU) 2017/1151 Supplementing Regulation (EC) No 715/2007 of the European Parliament and of the Council on Type-Approval of Motor Vehicles with Respect to Emissions from Light Passenger and Commercial Vehicles (Euro 5 and Euro 6). Available online: http://publications.europa.eu/resource/cellar/7d1c640d-62d8-11e7-b2f2-01aa75ed71a1.0006.02/DOC_1 (accessed on 29 April 2023).
6. Motorway. Euro 6 Compliance and Emission Standards—The 2023 Guide. 2023. Available online: https://motorway.co.uk/sell-my-car/guides/euro-6-emission-standards-compliance (accessed on 29 April 2023).
7. BUYACAR. What Are Euro 6 Emissions Standards? *Why You Need to Know*. 2022. Available online: https://www.buyacar.co.uk/cars/euro-6-emissions-standard-cars/522/what-are-euro-6-emissions-standards-why-you-need-to-know (accessed on 29 April 2023).
8. Brzezanski, M.; Sala, R. The influence of AdBlue dosage on the process of selective catalytic reduction of nitrogen oxides. *ResearchGate* **2013**, *154*, 1032–1037.
9. Toyota. 2023 MIRAI. 2023. Available online: https://www.toyota.com/mirai/ (accessed on 29 April 2023).
10. Hyundai. NEXO. 2023. Available online: https://www.hyundai.com/worldwide/en/eco/nexo/because-of-you (accessed on 29 April 2023).
11. Albatayneh, A.; Juaidi, A.; Jaradat, M.; Manzano-Agugliaro, F. Future of Electric and Hydrogen Cars and Trucks: An Overview. *Energies* **2023**, *16*, 3230. [CrossRef]
12. McKinsey & Company. EVolution. *Electric Vehicles in Europe: Gearing Up for A New Phase?* 2014. Available online: https://www.mckinsey.com/featured-insights/europe/electric-vehicles-in-europe-gearing-up-for-a-new-phase (accessed on 29 April 2023).
13. Schneider Electric. Electric Vehicle and EV Charging Fundamentals. 2021. Available online: https://www.electrical-installation.org/enwiki/Electric_Vehicle_and_EV_charging_fundamentals (accessed on 29 April 2023).
14. go-e. The 4 EV Charging Modes You Need to Know about. 2022. Available online: https://go-e.com/en/magazine/charging-modes-electric-car (accessed on 29 April 2023).
15. Wallbox. EV Charging Connector Types: What You Need to Know. 2023. Available online: https://wallbox.com/en_catalog/faqs-plug-types (accessed on 29 April 2023).
16. Lewison, T. Full fuel cycle emissions reductions through the replacement of ICEVs with BEVs. In *Climate Change 2-Canadian Technology Development Conference*; Canadian Nuclear Society: Toronto, ON, Canada, 2001.
17. Fournier, G.; Boos, A.; Wörner, R.; Jaroudi, I.; Morozova, I.; Nemoto, E. Substituting individual mobility by mobility on demand using autonomous vehicles—A sustainable assessment simulation of Berlin and Stuttgart. *Int. J. Automot. Technol. Manag.* **2020**, *20*, 369–407. [CrossRef]
18. Li, M.; Zhang, X.; Li, G. A comparative assessment of battery and fuel cell electric vehicles using a well-to-wheel analysis. *Energy* **2016**, *94*, 693–704. [CrossRef]
19. Choma, E.; Ugaya, C. Environmental impact assessment of increasing electric vehicles in the Brazilian fleet. *J. Clean. Prod.* **2017**, *152*, 497–507. [CrossRef]
20. Held, M.; Schücking, M. Utilization effects on battery electric vehicle life-cycle assessment: A case-driven analysis of two commercial mobility applications. *Transp. Res. Part D Transp. Environ.* **2019**, *75*, 87–105. [CrossRef]

21. Autoritat del Transport Metropolità. Enquesta de Mobilitat en Dia Feiner (EMEF). 2023. Available online: https://www.atm.cat/comunicacio/publicacions/emef (accessed on 11 May 2023). (In Catalan).
22. Padron.com.es. Población de Barcelona (Provincia). 2023. Available online: https://padron.com.es/barcelona/ (accessed on 14 May 2023). (In Spanish).
23. Repsol. Repsol Opens the Most Powerful Electric Vehicle Charging Station in Europe. 2019. Available online: https://www.repsol.com/en/press-room/press-releases/2019/repsol-opens-the-most-powerful-electric-vehicle-charging-station-in-europe.cshtml (accessed on 29 April 2023).
24. ANFAC. Mapa de Infraestructuras de Recarga de Acceso Público en España Para 2021–2030. 2021. Available online: https://anfac.com/publicaciones/mapa-de-infraestructuras-de-recarga-de-acceso-publico-en-espana-a-2021-2030/ (accessed on 29 April 2023).
25. Dirección General de Tráfico. Parque de Vehículos—Tablas Estadísticas 2022. 2023. Available online: https://www.dgt.es/menusecundario/dgt-en-cifras/dgt-en-cifras-resultados/dgt-en-cifras-detalle/?id=00866 (accessed on 29 April 2023). (In Spanish).
26. Dirección General de Tráfico. Parque de Vehículos—Tabla Estadística Auxiliar 2022. 2023. Available online: https://www.dgt.es/menusecundario/dgt-en-cifras/dgt-en-cifras-resultados/dgt-en-cifras-detalle/?id=00867 (accessed on 2 May 2023). (In Spanish).
27. Institut d'Estudis Regionals i Metropolitans de Barcelona. Enquesta de Mobilitat en Dia Feiner 2021 (EMEF 2021). 2022. Available online: https://omc.cat/documents/662112/1025112/EMEF+2021_Informe+SIMMB_juny2022+%281%29.pdf/17ff87aa-7005-193e-ad0e-e08330798c53?t=1666599585086 (accessed on 12 May 2023). (In Catalan).
28. Silence. Silence S01. 2023. Available online: https://www.silence.eco/lang/en/s01/ (accessed on 11 May 2023).
29. Tesla. Model 3. 2023. Available online: https://www.tesla.com/en_eu/model3 (accessed on 11 May 2023).
30. Peugeot. Peugeot e-Expert. 2023. Available online: https://www.peugeot.es/gama/expert-furgon.html (accessed on 11 May 2023).
31. DAF. DAF LF Electric. 2023. Available online: https://www.daf.es/es-es/descubra-los-camiones-daf/alternativas-de-combustibles-y-cadenas-cinematicas/vehiculos-electricos-de-bateria/daf-lf-electric (accessed on 11 May 2023). (In Spanish).
32. Red Eléctrica de España. Generación Total. 2023. Available online: https://www.sistemaelectrico-ree.es/informe-del-sistema-electrico/generacion/generacion-de-energia-electrica/generacion-total-de-energia-electrica (accessed on 15 May 2023). (In Spanish).
33. European Environment Agency. EMEP/EEA air pollutant emission inventory guidebook 2019. In *European Monitoring and Evaluation Programme (EMEP)*; European Environment Agency (EEA): Copenhagen, Denmark, 2019.
34. Red Eléctrica de España. Península—Seguimiento de la Demanda de Energía Eléctrica. 2023. Available online: https://demanda.ree.es/visiona/peninsula/demandaqh/total/2023-03-07 (accessed on 16 May 2023). (In Spanish).
35. Red Eléctrica de España. Potencia Instalada. 2023. Available online: https://www.sistemaelectrico-ree.es/informe-del-sistema-electrico/generacion/potencia-instalada (accessed on 16 May 2023). (In Spanish).
36. Generalitat de Catalunya. Visor Sobre el Estado de Tramitación Ambiental de los Proyectos de Energías Renovables en Catalunya. 2023. Available online https://mediambient.gencat.cat/es/detalls/Articles/visor (accessed on 16 May 2023). (In Spanish).
37. Gobierno de Aragón. Planta Fotovoltaica Ancar II en los Término Municipales de Alfambra y Orrios (Teruel). 2023. Available online: https://www.aragon.es/-/planta-fotovoltaica-ancar-ii (accessed on 16 May 2023).
38. Gobierno de Aragón. Parque Eólico Cabigordo en los Términos Municipales de Corbalán, Cedrillas y El Pobo (Teruel). 2023. Available online: https://www.aragon.es/-/parque-eolico-cabigordo?p_l_back_url=%2Fbuscador%3Fq%3Deolico%2Bcabigordo%26category-td%3D466682%26category%3D465154%26type%3Dcom.liferay.journal.model.JournalArticle (accessed on 16 May 2023). (In Spanish).
39. Generalitat de Catalunya. Hipermapa. *Sol·Licituds de Parcs Eòlics*. 2023. Available online: https://sig.gencat.cat/visors/hipermapa_antic.html#param=param&color=vermell&background=topo_ICC_grisos&BBOX=134915.697674,4485000,659084.302326,4752000&layers=ENERGIA_PARCSEOLICS_MULTIPUNT,ENERGIA_PARCSEOLICSSOLARS_LIN,ENERGIA_PARCSSOLARS_MULTIPOLI (accessed on 16 May 2023).
40. Parc Tramuntana. Parc Tramuntana. *Una Oportunidad Azul Para la Sostenibilidad Del Empordà i Girona*. 2023. Available online: https://parctramuntana.com/es/inicio/ (accessed on 16 May 2023). (In Spanish).
41. Miao, R.; Ghosh, P.N.; Khanna, M.; Wang, W.; Rong, J. Effect of wind turbines on bird abundance: A national scale analysis based on fixed effects models. *Energy Policy* **2019**, *132*, 357–366. [CrossRef]
42. Chargeguru. Cuánto Cuesta la Instalación de un Punto de Recarga Para Vehículo Eléctrico? 2023. Available online: https://chargeguru.com/es/2021/09/17/cual-es-el-precio-de-la-instalacion-de-un-punto-de-recarga-para-vehiculo-electrico/ (accessed on 15 May 2023). (In Spanish).
43. ANFAC. La Antigüedad del Parque de Automóviles en España Sigue al Alza y Alcanza Los 13,5 Años de Edad Media. 2022. Available online: https://anfac.com/actualidad/la-antiguedad-del-parque-de-automoviles-en-espana-sigue-al-alza-y-alcanza-los-135-anos-de-edad-media/ (accessed on 15 May 2023). (In Spanish).

44. ANFAC. Vehículo Electrificado. *Informe Anual 2022*. 2023. Available online: https://anfac.com/wp-content/uploads/2023/04/2022_Informe_electrificados-Anual.pdf (accessed on 15 May 2023). (In Spanish).
45. Rodríguez, M. Electric Vehicle Sales in Europe in the First Quarter of 2023. 2023. Available online: https://www.electromaps.com/en/blog/electric-vehicle-sales-europe-first-quarter-2023 (accessed on 15 May 2023).

energies

MDPI

Article

Degradation Prediction and Cost Optimization of Second-Life Battery Used for Energy Arbitrage and Peak-Shaving in an Electric Grid

Rongheng Li, Ali Hassan, Nishad Gupte, Wencong Su and Xuan Zhou *

Department of Electrical and Computer Engineering, University of Michigan-Dearborn, Dearborn, MI 48128, USA; rongheng@umich.edu (R.L.); alihssn@umich.edu (A.H.); nishadg@umich.edu (N.G.); wencong@umich.edu (W.S.)
* Correspondence: xuanzhou@umich.edu

Abstract: With the development of the electric vehicle industry, the number of batteries that are retired from vehicles is increasing rapidly, which raises critical environmental and waste issues. Second-life batteries recycled from automobiles have eighty percent of the capacity, which is a potential solution for the electricity grid application. To utilize the second-life batteries efficiently, an accurate estimation of their performance becomes a crucial portion of the optimization of cost-effectiveness. Nonetheless, few works focus on the modeling of the applications of second-life batteries. In this work, a general methodology is presented for the performance modeling and degradation prediction of second-life batteries applied in electric grid systems. The proposed method couples an electrochemical model of the battery performance, a state of health estimation method, and a revenue maximization algorithm for the application in the electric grid. The degradation of the battery is predicted under distinct charging and discharging rates. The results show that the degradation of the batteries can be slowed down, which is achieved by connecting numbers of batteries together in parallel to provide the same amount of required power. Many works aim for optimization of the operation of fresh Battery Energy Storage Systems (BESS). However, few works focus on the second-life battery applications. In this work, we present a trade-off between the revenue of the second-life battery and the service life while utilizing the battery for distinct operational strategies, i.e., arbitrage and peak shaving against Michigan's DTE electricity utility's Dynamic Peak Pricing (DPP) and Time of Use (TOU) tariffs. Results from case studies show that arbitrage against the TOU tariff in summer is the best choice due to its longer battery service life under the same power requirement. With the number of retired batteries set to increase over the next 10 years, this will give insight to the retired battery owners/procurers on how to increase the profitability, while making a circular economy of EV batteries more sustainable.

Keywords: second-life battery; electricity grid application; electrochemical modeling; degradation prediction; battery operational strategy

Citation: Li, R.; Hassan, A.; Gupte, N.; Su, W.; Zhou, X. Degradation Prediction and Cost Optimization of Second-Life Battery Used for Energy Arbitrage and Peak-Shaving in an Electric Grid. *Energies* **2023**, *16*, 6200. https://doi.org/10.3390/en16176200

Academic Editor: Simone Barcellona

Received: 31 July 2023
Revised: 19 August 2023
Accepted: 21 August 2023
Published: 26 August 2023

1. Introduction

With the widespread use of electric vehicles, huge numbers of batteries are employed. As the battery is charged and discharged continuously, the capacity degrades gradually. When it reaches a certain level, i.e., 80% of the original state of health, the battery should be retired from the electric vehicles [1,2]. With the continuous development of the electric vehicle industry, the number of batteries retired from vehicles is increasing rapidly. The expense of disassembling and recycling these batteries is relatively high, which raises a substantial issue on how to handle them appropriately. A recent work that utilizes an indicator, i.e., global warming potential, demonstrates that reusing the existing electric vehicle battery in a secondary application provides a significant environmental benefit compared to manufacturing a new battery for the same purpose [3]. By appropriate assessment and treatment, the retired batteries can be utilized for other applications. Such reused batteries are referred to as second-life batteries [4–6]. The implementation of second-life batteries is also beneficial for protecting the environment and saving money [7].

Among the various applications of the second-life battery, it is found to be suitable for employment as an energy storage system in the electricity grid system [8]. Energy storage systems have proven to be a game changer for the integration of renewable energy and the stability of modern power systems [9]. Their use in the grid can be for various purposes. In [10], G. Fitzgerald et al. mentioned thirteen different uses for BESSs in general. Main grid utilization of BESSs is for energy arbitrage [11–13], frequency regulation [14–16], peak shaving [17–19], or power smoothing [20–22]. The second-life batteries have been deployed by industrial consortiums as discussed in [23–25]. In [26], Zhang et al. proposes a remaining useful life prediction methodology using a deep learning integrated approach. In [27], Xiofan et al. discusses a power processing methodology for power converters for power optimization of second-life batteries. Many research studies [28–30] focus on the aging of a fresh energy storage system, however, there is a clear research gap in predicting the degradation of second-life batteries. Therefore, accurate battery performance estimation becomes essential in the choice of operational grid strategies. In the short-term range, correct estimation of the battery charging and discharging responses can enhance the stabilization of the grid system [31]. For the long-term consideration, accurate state of health prediction of the second-life battery is beneficial for reducing the cost and improving the grid system efficiency [32].

For distinct working conditions, i.e., operational strategies and user loads, the cost-effectiveness of the batteries can be optimized with appropriate configurations, i.e., series and parallel connections [33]. Compared to costly and time-consuming experimental testing, numerical modeling is a better approach to speed up the simulation of the operating process [34]. The three main categories of modeling methods are the physics-based electrochemical models [35], the electrical equivalent circuit models [36,37], and the data-driven models [38,39]. The advantages and limitations of these approaches are summarized in Table 1. The physics-based model shows the mechanisms of electrochemical reactions, which are particularly important in predicting degradation. Among various models, the Doyle-Fuller-Newman (DFN) model attracts great attention due to its good accuracy and suitability for diverse working conditions [40]. For engineering practice, the P2D model is simplified to a more concise form [41,42]. Xu et al. presented an electrochemical-thermal-capacity model that minimizes capacity fade and reduces the temperature rise to prevent a thermal runaway [43]. Song et al. developed an electro-chemo-mechanical model, which couples the mechanical and electrochemical factors [44]. The framework of the P2D model is implemented in commercial software, such as COMSOL [45], and open-source codes, i.e., DUALFOIL [46], LIONSIMBA [47], and PyBaMM [48].

Table 1. A summary of the battery performance modeling approaches, including their benefits and limitations.

Category	Benefits	Limitations
Physics-based electrochemical models [35,40–43]	Reveal the physical phenomenon behind the battery; suitable for wide scenario ranges	Complex to implement; time-consuming
Electrical equivalent circuit models [36,37]	Robust; easy to implement	Less accurate
Data-driven models [38,39]	Moderate accuracy; knowledge of the underlying process is not required	Accuracy depends on the quality of training data; usually requires a large set of charging and discharging data

A number of major degradation mechanisms are proposed to accurately model the battery aging process. They are solid electrolyte interphase (SEI) layer growth, lithium plating, particle cracking, and active material losses. Safri et al. developed a solvent-decomposition reaction model to simulate the growth of solid electrolyte interphase at the anode [49]. Single et al. revealed the mechanism of SEI formation [50]. Their work demonstrated that the diffusion of neutral radicals is the cause of long-term SEI growth. Luo et al. revealed the mechanism of SEI formation [51]. They simulated the battery degradation under various depths of discharge, state-of-charge swing ranges, and temperatures. O'Kane et al. coupled four degradation mechanisms in the electrochemical model [52]. They reported that five distinct pathways can result in end-of-life, which depends on how the cell is charged and discharged.

This article outlines a technique for predicting the performance and degradation of second-life batteries utilized in electric grid systems. The approach consists of an electrochemical model of the battery's performance, a health monitoring method, and an algorithm to reduce costs for grid applications. The governing equations of the electrochemical process and the degradation mechanisms are demonstrated. Different charging and discharging rates are utilized to predict battery degradation. The study shows that the connection of batteries in parallel can slow down the degradation, but this option requires a balance between the cost of the battery and its lifetime. Different charging and discharging strategies are considered in the research, including DPP and TOU. The strategy with both a longer battery life and mediate cost is depicted.

The main contributions of this work are summarized as follows:

- A methodology that couples the DFN electrochemical model and a revenue maximization algorithm is introduced to model the performance of a second-life battery in the application of the electric grid.
- Simulations of distinct configurations demonstrate that a trade-off between the revenue of the battery and the service life should be optimized.
- Through the implementation of distinct battery operational strategies, the study illustrates that the arbitrage against the TOU tariff in summer is the optimal solution among various combinations due to its longer battery service life while providing the same amount of power.

2. Methodology

The structure of a battery under the P2D model configuration is displayed in Figure 1. The cell consists of three regions: the positive electrode, the separator, and the negative electrode. The active materials are modeled as spherical particles that fill in the positive and negative electrodes. Two dimensions are considered in the P2D model. The x-axis is defined in the direction perpendicular to these layers to account for the diffusion and migration of Li ions in the liquid phase. The origin of the r-axis is at the center of the solid sphere, which describes the diffusion of Li ions in the active materials.

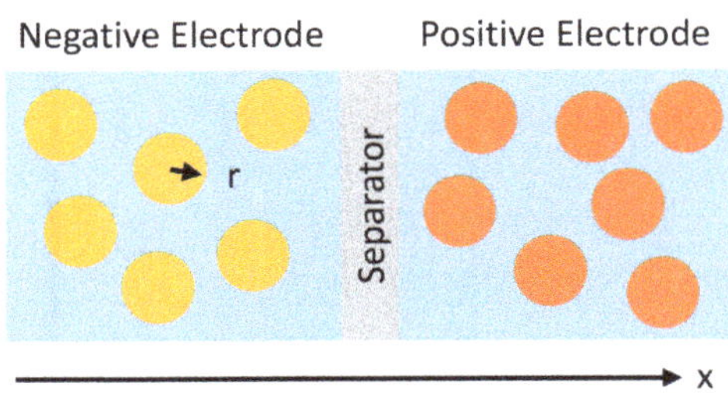

Figure 1. Structure configuration of the P2D model of the Li-ion battery.

2.1. P2D Models

The electrochemical process described in P2D models is governed by the following five partial differential equations. They are the equations for: charge conservation in the homogeneous solid, mass conservation in the homogeneous solid, mass conservation in the homogeneous electrolyte, charge conservation in the homogeneous electrolyte, and lithium movement between the solid and electrolyte phases [40,53,54].

2.1.1. Mass Conservation in Solid

The lithium diffusion inside the solid particles is driven by the gradient of its concentration, which is governed by Fick's second law as:

$$\frac{\partial c_s(x,r,t)}{\partial t} = \frac{1}{r^2}\frac{\partial}{\partial r}\left(D_s r^2 \frac{\partial c_s\,(x,r,t)}{\partial r}\right) \tag{1}$$

where $c_s(x,t,r)$ and D_s are the concentration and the diffusion coefficient of lithium in the solid phase, respectively. The initial condition is:

$$c_s(x,r,t)|t\ = 0\ = C_{s,0} \tag{2}$$

The boundary conditions are:

$$D_s \left.\frac{\partial c_s\,(x,r,t)}{\partial r}\right|_{r=0} = 0 \tag{3}$$

$$D_s \left.\frac{\partial c_s\,(x,r,t)}{\partial r}\right|_{r=R_s} = \frac{j(x,t)}{F} \tag{4}$$

where $c_{s,0}$, R_s, F, and $j(x,t)$ are the initial concentration of lithium, the radius of solid-state particles, the Faraday constant, and the current density of the surface particle's electrochemical reaction rate, respectively.

2.1.2. Mass Conservation in Electrolyte

The movement of lithium ions inside the electrolyte is governed by the diffusion and migration process, which is described by:

$$\varepsilon_e \frac{\partial c_e(x,t)}{\partial_t} = \frac{\partial}{\partial x}\left(D_e^{eff}\frac{\partial c_e(x,t)}{\partial_x}\right) + a_s\left(1 - t_+^0\right)\frac{j(x,t)}{F} \tag{5}$$

with $c_e(x,t)$, ε_e, and t_+^0 being the lithium ion concentration, the electrolyte volume fraction, and the lithium ion transfer number, respectively. D_e^{eff} is the effective diffusion coefficient in the electrolyte and is expressed as:

$$D_e^{eff} = \varepsilon_e^{brugg} D_e \tag{6}$$

where ε_e^{brugg} is the Bruggeman correction coefficient. D_e is the electrolyte diffusion coefficient. The specific surface area of the solid particles is calculated as:

$$a_s = \frac{3\varepsilon_s}{R_s} \tag{7}$$

where ε_s is the solid phase volume fraction.

The corresponding boundary conditions are:

$$\left.\frac{\partial c_e(x,t)}{\partial_x}\right|_{x=0} = \left.\frac{\partial c_e(x,t)}{\partial_x}\right|_{x=L} = 0 \tag{8}$$

$$D_{e,neg}^{eff}\left.\frac{\partial c_e(x,t)}{\partial_x}\right|_{x=x_{neg}} = D_{e,sep}^{eff}\left.\frac{\partial c_e(x,t)}{\partial_x}\right|_{x=x_{neg}^+} \tag{9}$$

$$c_e(x,t)|_{x=x_{neg}^-} = c_e(x,t)|_{x=x_{neg}^+} \tag{10}$$

$$D_{e,sep}^{eff} \frac{\partial c_e(x,t)}{\partial_x}\bigg|_{x=x_{sep}} = D_{e,pos}^{eff} \frac{\partial c_e(x,t)}{\partial_x}\bigg|_{x=x_{sep}^+} \tag{11}$$

$$c_e(x,t)|_{x=x_{sep}^-} = c_e(x,t)|_{x=x_{sep}^+} \tag{12}$$

where $x = 0$ represents the boundary between the current collector and the negative electrode; $x = x_{neg}$ is the boundary between the negative electrode and the separator; $x = x_{sep}$ defines the boundary between the separator and the positive electrode. In addition, the superscripts "−" and "+" denote the negative and positive portions of the battery.

2.1.3. Charge Conservation in Solid

Charge conservation shows that there is rarely a loss of net charge. Ohm's law can be seen in the electrode solid-phase potential distribution, which can be stated as follows:

$$-\frac{\partial i_s(x,t)}{\partial x} = \frac{\partial}{\partial x}\left(\sigma_S^{eff} \frac{\partial}{\partial x} \varnothing_s(x,t)\right) = a_s j(x,t) \tag{13}$$

where $i_s(x,t)$ is the electrical current density in the solid phase, $\phi_s(x,t)$ is the potential present in the solid phase, and σ_S^{eff} is the material's effective electrical conductivity in the solid state defined as:

$$\sigma_S^{eff} = \varepsilon_s^{bu} \sigma_s \tag{14}$$

with σ_s denoting the material's conductivity in the solid phase.

The boundary conditions are as follows:

$$-\sigma_S^{eff} \frac{\partial}{\partial x} \varnothing_s(x,t)\bigg|_{x=0} = -\sigma_S^{eff} \frac{\partial}{\partial x} \varnothing_s(x,t)\bigg|_{x=L} = \frac{I_{app}(t)}{A_{cell}} \tag{15}$$

where cell A is the electrode area and $I_{app}(t)$ is the charge/discharge current value of the external circuit when the battery is functioning.

2.1.4. Charge Conservation in Electrolyte

The change is conserved in the liquid phase. The lithium ions are intercalated and deintercalated inside the solid particles, which is governed by:

$$-\frac{\partial i_e(x,t)}{\partial x} = \frac{\partial}{\partial x}\left(k_e^{eff} \frac{\partial}{\partial x} \varnothing_e(x,t)\right) + \frac{\partial}{\partial x}\left(k_D^{eff} \frac{\partial}{\partial x} lnc_e(x,t)\right) = -a_s j(x,t) \tag{16}$$

where $i_e\ (x,t)$, $c_e(x,t)$, $\phi_e(x,t)$, and k_e^{eff} are the ionic current density, the lithium concentration, the potential, and the effective ionic conductivity in the electrolyte. Then:

$$k_e^{eff} = \varepsilon_e^{brugg} K_e \tag{17}$$

with K_e being the ionic conductivity of the electrolyte. The effective diffusion conductivity of the electrolyte is expressed as:

$$k_D^{eff} = \frac{2RTk_e^{eff}}{F}\left(1 + \frac{d\ lnf_\pm}{d\ lnc_e}\right)\left(t_+^0 - 1\right) \tag{18}$$

with R, T, and $f_\pm$ being the gas constant, the temperature, and the molar activity coefficients of the electrolyte. The boundary conditions are:

$$\frac{\partial \varnothing_e(x,t)}{\partial_x}\bigg|_{x=0} = \frac{\partial \varnothing_e(x,t)}{\partial_x}\bigg|_{x=L} = 0 \tag{19}$$

$$i_e(x,t)|_{x=x_{neg}^-} = i_e(x,t)|_{x=x_{neg}^+} = \frac{I_{app}(t)}{A_{cell}} \tag{20}$$

$$\varnothing_e(x,t)|_{x=x_{neg}^-} = \varnothing_e(x,t)|_{x=x_{neg}^+} \tag{21}$$

$$i_e(x,t)|_{x=x_{sep}^-} = i_e(x,t)|_{x=x_{sep}^+} = \frac{I_{app}(t)}{A_{cell}} \tag{22}$$

$$\varnothing_e(x,t)|_{x=x_{sep}^-} = \varnothing_e(x,t)|_{x=x_{sep}^+} \tag{23}$$

2.1.5. Electrochemical Kinetics Equation

To link the surface over potential $\eta(x,t)$ with the electrochemical reaction rate $j(x,t)$, the Bulter–Volmer kinetic equation is employed.

$$j(x,t) = i_0\left[exp\left(\frac{\alpha_a F}{RT}\eta(x,t)\right) - exp\left(\frac{-\alpha_c F}{RT}\eta(x,t)\right)\right] \tag{24}$$

with i_0, α_a, and α_c being the exchange current density, the transfer coefficient of the anode, and the transfer coefficient of the cathode, respectively. The overpotential is expressed as:

$$\eta(x,t) = \varnothing_s(x,t) - \varnothing_e(x,t) - E_j^{ref}(\theta_j(x,t)) - j(x,t)R_{SEI} \tag{25}$$

with R_{SEI} being the resistance of the SEI film, and E_j^{ref} being the equilibrium potential of the electrodes. In addition:

$$\theta_j(x,t) = \frac{c_{s,surf(x,t)}}{c_{s,max,j}}, j = neg, pos \tag{26}$$

where $c_{s,surf}(x,t)$ stands for the lithium concentration of the active material surface; $c_{s,\ max}$ is the maximal lithium concentration of electrodes. The exchange current can be calculated as:

$$i_0 = Fk_0\left(c_{s,max} - c_{s,surf}\right)^{\alpha_a} c_{s,surf}{}^{\alpha_c} c_e{}^{\alpha_a} \tag{27}$$

with k_0 as the electrochemical reaction rate constant.

2.2. State of Health

For the long-term lifespan, four mechanisms can be utilized for the prediction of the state of health. In this paper, the interstitial-diffusion-limited SEI growth model is implemented [51]. The four mechanisms are the solid electrolyte interphase layer growth, lithium plating, particle cracking, and active material losses.

$$j_{inter}^* = -\frac{F^* D_{Li,i}^*}{L_\sigma^*} c_{Li,i,0}^* \exp\left(-\frac{F^*}{R_g^* T^*}(\varnothing_{s,n}^* - \varnothing_{s,e}^*)\right) \tag{28}$$

Here, j_{inter}^* represents the interfacial current density, $D_{Li,i}^*$ is the diffusivity of lithium ions in the inner SEI, $\varnothing_{s,n}^*$ is the negative electrode potential, and $c_{Li,i,0}^*$ is the inner SEI's concentration of lithium ion interstitials when $\varnothing_{s,n}^* = \varnothing_{e,n}^*$. Here, $\varnothing_{e,n}^*$ is the true electrolyte potential, F^* shows Faraday's constant, R_g^* is the universal gas constant, and T^* is the reference temperature. The lithium plating is considered irreversible in this work. For the lithium plating portion, the irreversible model is selected. The SEI on cracks and loss of active material losses are not considered in this work. The description and derivation of the governing equations for the degradation mechanisms are well demonstrated in [48–52,55]. The default settings of PyBaMM are utilized for the rest simulations.

2.3. Grid Optimization

This section will present an optimization problem for the LG M50 cylindrical cell used in a pack acting as an energy storage system to provide power to the grid for two management strategies: energy arbitrage and peak shaving. They are under extensive investigation on the demand side in the area of power systems. Arbitrage involves exploiting temporal price variations in the electricity market by purchasing electricity during periods of lower prices and selling it during periods of higher prices, thereby maximizing economic gains. On the other hand, peak shaving aims to strategically decrease electricity consumption during peak demand periods to smooth out load curves and reduce grid stress. The advancement of sustainable energy systems is aided by these strategies, which play a crucial role in enhancing grid stability, cost-effectiveness, and the integration of renewable energy sources.

The mathematical statement of the optimization problem is described in the following equations.

$$max\sum_{t=1}^{96} E_{dch}(t)C_{pk}(t) - E_{ch}(t)C_{ofpk}(t) \tag{29}$$

s.t.

$$C_{battery}(t) = C_{battery}(t-1) + E_{ch}(t)\eta_{ch} - E_{dch}(t)\eta_{dch} \tag{30}$$

$$SOC(t=1) = SOC(t=96) \tag{31}$$

$$E_{ch}(t) \leq E_{ch\_max}(t) * 0.95 * u(t)\ \forall u \in 0,1 \tag{32}$$

$$E_{dch}(t) \leq E_{dch\_max}(t) * 0.95 * u(t)\ \forall u \in 0,1 \tag{33}$$

$$SOC_{min} \leq SOC \leq SOC_{max} \tag{34}$$

$$E_{dch}(t) * u(t) + E_{ch}(t) * (1-u(t)) = 0 \\ where \begin{cases} u(t) = 1,\ t = 44\ to\ 80 \\ u(t) = 0,\ t = 1\ to\ 43\ or\ 81\ to\ 96 \end{cases} \tag{35}$$

In Equation (30), E_{dch} is the discharge energy in kWh from the battery, E_{ch} is the charging energy in kWh into the battery, C_{pk} is the cost of energy in USD/kWh for the peak hours, and C_{ofpk} is the cost of energy in USD/kWh for the off-peak hours, respectively. η_{ch} and η_{dch} are the charge and discharge efficiencies taken as 0.95 (95%). The variable $u(t)$ is a binary variable that makes sure that the battery either charges or discharges at one 15-min time interval. The t represents 15-min intervals in a 24-h period, i.e., ($25 \times 15 = 96$). The cost function maximizes the profit earned by selling the power (revenue) subtracted from the money incurred by charging the battery (cost). Constraint (31) calculates the capacity $C_{battery}$ of the battery at each time step t, i.e., $C_{battery}$ decreases if the battery is discharged or increases if it is charged. Equation (31) sets the state of charge SOC at $t = 1$ at the start of the day equal to the SOC at $t = 96$ at the end of the day. Equations (32)–(34) keep the charge energy, discharge energy, and state of charge within the maximum physical limit of the battery, respectively. Equation (35) is an additional constraint only for the peak-shaving strategy that forces the battery to discharge during the peak demand hours. The optimization problem is an MILP formulated using the abstract model in Pyomo 6.6.0 and solved using the CPLEX optimization solver.

The load profile of Southeast Michigan on a typical weekday is given in Figure 2, which shows the load profile for one day of residential customers in Michigan. The load peaks at around 12 p.m. and dips around 8 p.m. This is known as peak load for which utilities must fire up the peak power plants, which is both expensive and inefficient. However, with recent advances, energy storage systems can be used to provide the stored power to the grid to meet the peak load, which is known as peak shaving.

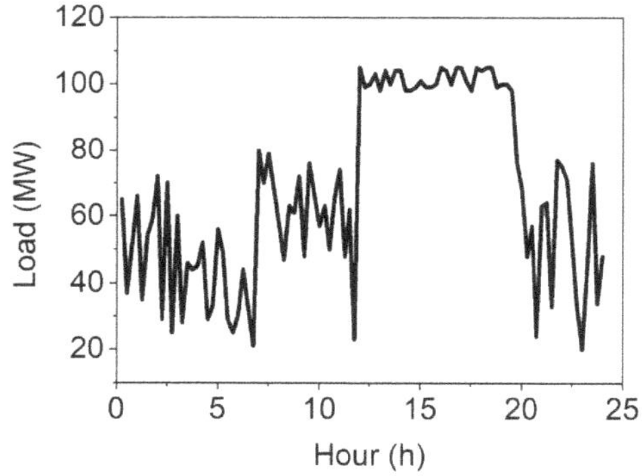

Figure 2. The load profile of Southeast Michigan on a typical weekday.

The optimization problem in this work is solved for the two different residential tariffs of a DTE utility in Michigan: Time of Use (TOU) and Dynamic Peak Pricing (DPP) rate as shown in Table 2. The concept of TOU rate structures is to divide the day into different periods, each with varying electricity prices, which encourages consumers to switch their electricity consumption to off-peak hours when prices are lower. On the other hand, the DPP rate dynamically adjusts electricity prices in real time depending on the supply and demand conditions of the network, favoring a more flexible and reactive consumption behavior.

Table 2. DTE TOU and DPP tariff.

Tariff Type	On-Peak Hours	Mid-Peak Hours	Off-Peak Hours	Summer (Jun through Sep) (¢/kWh (USD))		Summer (Jun through Sep) (¢/kWh (USD))	
				On-Peak	Off-Peak	On-Peak	Off-Peak
Time of Use (TOU)	Monday–Friday 3 p.m. to 7 p.m.	---	Monday–Friday 12 a.m.–3 p.m., 7 p.m.–12 a.m. and all-day Saturday, Sunday	7.941	4.828	5.560	4.828
Dynamic Peak Pricing (DPP)	Monday–Friday 3 p.m. to 7 p.m.	Monday–Friday 7 a.m. to 3 p.m. and 7 p.m. to 11 p.m.	11 p.m. to 7 a.m. and all-day Saturday, Sunday	Same rates all around the year			
					On-Peak	Mid-Peak	Off-Peak
					12.658	5.486	1.184

3. Results

3.1. Model Validation

To validate the proposed calculation method, we have conducted simulations using the LG M50 cylindrical cell. The detailed values of battery parameters were taken from the reported works [52,56]. To represent different operating conditions, three distinct discharging rates were selected, which are 0.05C, 1C, and 2C. The voltage responses obtained from our model, which was implemented using PyBaMM v23.5, were compared with experimental data available online and depicted in Figure 3. Due to the significant difference in discharging timespans between 0.05C and the other rates, their responses were plotted separately in Figure 3a. The model exhibited good agreement with the experimental data for the discharging curve. Minor deviations between simulation results and experimental data were observed for higher discharging rates, i.e., 1C and 2C in Figure 3b, yet remained within acceptable tolerances.

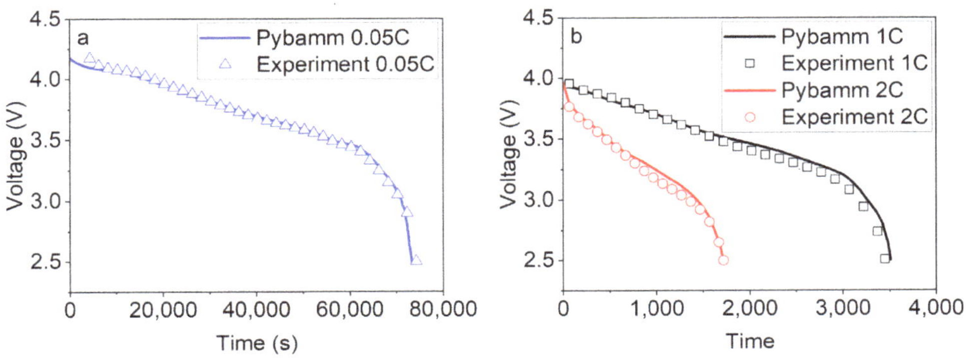

Figure 3. The voltage responses of the battery at (**a**) 0.05C and (**b**) 1C and 2C.

With the short time performance of the battery validated, the long-term degradation estimation is also validated. The following cycling protocol is utilized, which is the same as a testing report. Firstly, the cell is charged at a constant current of 0.33C to 4.2 V. Then, the cell is discharged at the same constant rate of 2.85 V. The above protocol is repeated for 1000 cycles with the discharged capacity displayed in Figure 4. As depicted in the figure, the discharge capacity drops to 4.10 A.h, which is around 80% of the initial capacity. It is well agreed with the reported testing results.

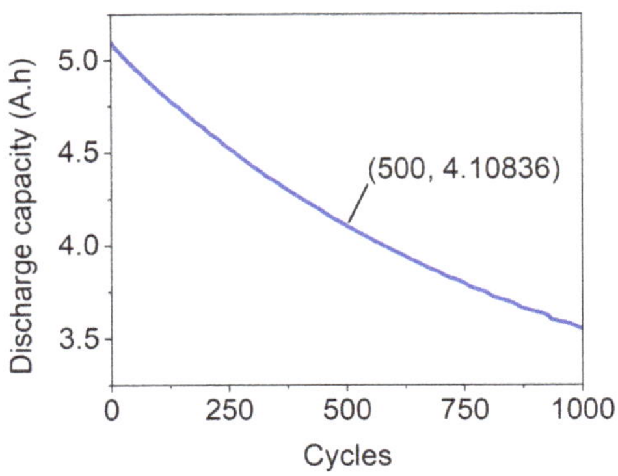

Figure 4. The discharge capacity of the battery for 1000 charging and discharging cycles.

3.2. Optimization for Distinct Strategies

Using the second-life battery cell in a grid would need a scheduled charge and discharge, which is calculated by an optimization problem explained in Section 2. The charge and discharge energy (kWh) of the battery cell is optimized using an optimization problem formulation explained in Section 2.3. while following the constraints. Figure 5a,b shows the charging and discharging energy (kWh) of a battery cell for the DPP tariff for arbitrage and peak-shaving. The battery charges and discharges aggressively for DPP-Arbitrage during hours 8 to 16 while for DPP—Peak-Shaving the charge and discharge trend is concentrated from hours 19 to 22. Likewise, Figure 6a,b show charge and discharge energy (kWh) calculated for the TOU tariff for summer arbitrage and peak-shaving. Figure 7a,b provides similar results for the TOU tariff for winter. The overall charge and discharge profile is rigorous for TOU summer peak shaving as compared to the rest. In terms of battery degradation, such a profile will lessen cell life. For longer life, the charging and discharging of the battery needs to be less rigorous for TOU peak shaving during winters. The peaks are widely distributed with a uniform discharge during the peak hours.

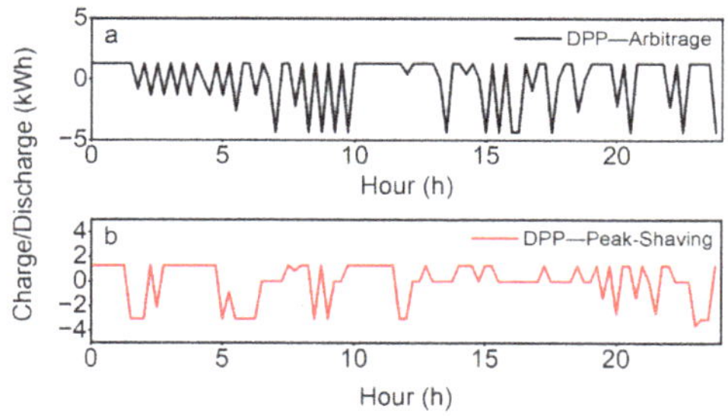

Figure 5. (**a**) Charge and discharge energy (kWh) for arbitrage against DPP tariff; (**b**) charge and discharge energy (kWh) for peak shaving against DPP tariff.

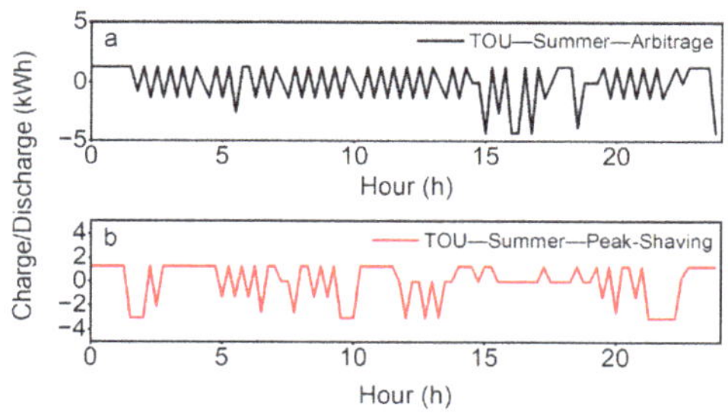

Figure 6. (**a**) Charge and discharge energy (kWh) for arbitrage against TOU tariff (summer); (**b**) charge and discharge energy (kWh) for peak shaving against TOU tariff (summer).

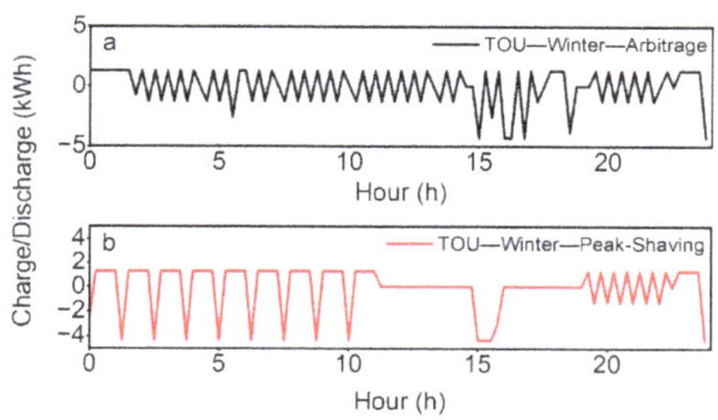

Figure 7. (**a**) Charge and discharge energy (kWh) for arbitrage against TOU tariff (winter); (**b**) charge and discharge energy (kWh) for peak shaving against TOU tariff (winter).

3.3. Battery Cell Capacity

To further expand on the results above, Figures 8–10 show the battery cell's capacity in kWh for DPP and TOU tariffs. Figure 8 shows that the battery capacity profile for DPP peak shaving is more rigorous as compared to DPP arbitrage. For Figure 9, the battery capacity profile for TOU peak shaving in summer is more rigorous than TOU arbitrage in summer. The battery capacity profile for TOU winter peak shaving is more rigorous than

TOU winter arbitrage. Overall, the battery capacity profile for TOU winter arbitrage and peak shaving looks the smoothest as compared to all other cases, which is in accordance with the charge and discharge profile results above.

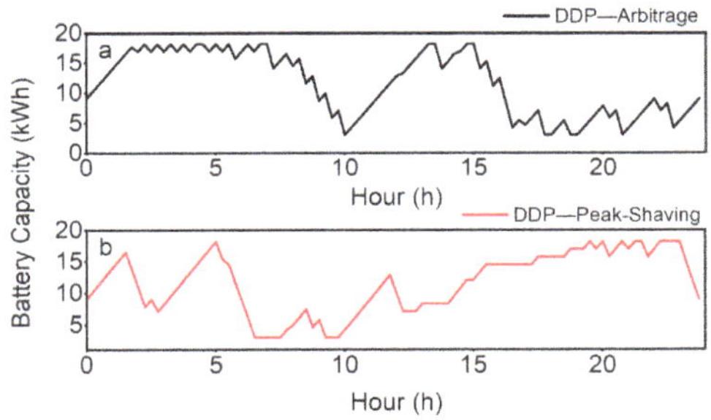

Figure 8. (**a**) Battery capacity (kWh) for 1 day (15 min interval) for DPP arbitrage; (**b**) battery capacity (kWh) for 1 day (15 min interval) for DPP peak shaving.

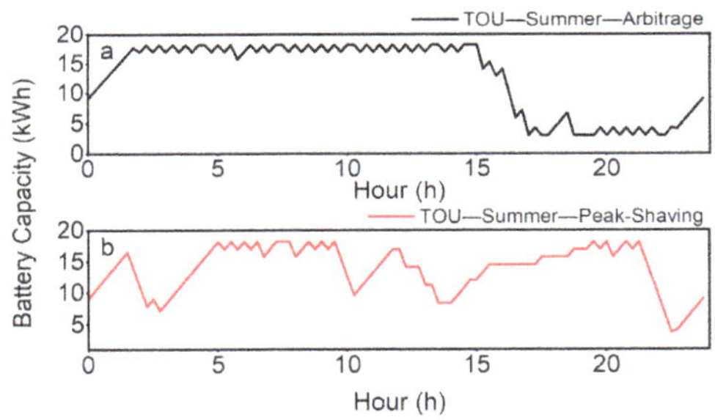

Figure 9. (**a**) Battery capacity (kWh) for 1 day (15 min interval) for TOU (summer) EA; (**b**) battery capacity (kWh) for 1 day (15 min interval) for TOU (summer) peak shaving.

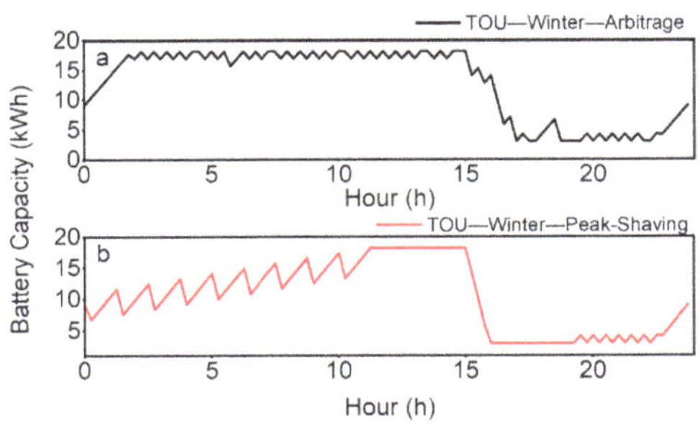

Figure 10. (**a**) Battery capacity (kWh) for 1 day (15 min interval) for TOU (winter) arbitrage; (**b**) battery capacity (kWh) for 1 day (15 min interval) for TOU (winter) peak shaving.

3.4. Usage Days for Different Operational Strategies

In this part, the usage days of two different operational strategies are investigated. Case 1 simulates the configuration in that the required power is supplied using one LG M50 cell, while case 2 is for the same power supplied using two identical cells. For each case, the battery cell(s) starts from 80% of its original state of health, justifying it as a second-life battery use. It will be terminated at 64%, which is 80% of the starting point of the second-life use. Figure 11 shows the cell life (usage days) calculated for two different configurations.

The first configuration considers only one cell providing (discharge) and receiving (charge) energy/power to and from the grid, respectively, against DPP and TOU tariffs for EA and peak shaving. Conversely, the second configuration considers the same for two battery cells. The bar graph shows that cell life is the shortest in the first configuration where the peak shaving is provided by one cell for summer TOU. This is because the cell charges and discharges aggressively for this scenario as compared to the other cases, resulting in heavy cell degradation. The cell life is the longest for the arbitrage case against the DPP tariff. For configuration 2, again the cell life is lowest for peak shaving for TOU (summer), while the cell life is longest for TOU (winter) peak shaving.

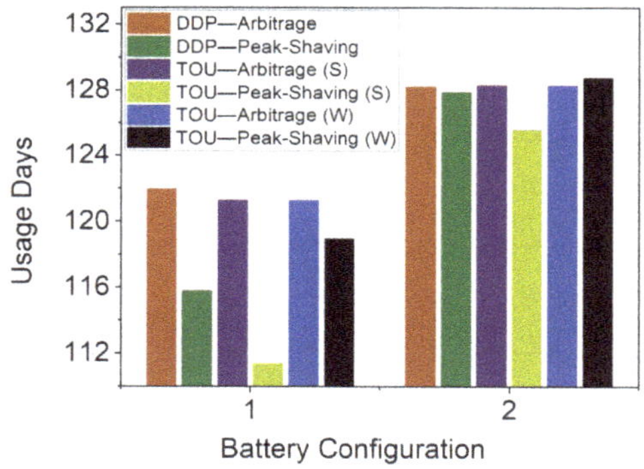

Figure 11. Usage days when required power is provided by one vs. two battery cells.

Table 3 shows the revenue (in USD) calculated by optimizing the cost function. The profit is at a maximum (USD 21.25) for the arbitrage for TOU in summer. However, the degradation for the same battery cell as shown above was the maximum. This again shows that excessive charge and discharge may earn revenue but will also degrade the battery fast. Therefore, the battery needs to generate revenue while at the same time degrading slowly for long life. Considering the results from Figure 11 and Table 3, it is recommended to use the second-life battery pack made of used LG M50 cells for arbitrage against the TOU tariff in summer.

Table 3. DTE TOU and DPP tariff.

Scenario	Profit per 24 h (USD)
DPP—Arbitrage	0.7067
DPP—Peak Shaving	1.77
TOU—Arbitrage (Summer)	20.08
TOU—Peak Shaving (Summer)	21.25
TOU—Arbitrage (Winter)	0.32
TOU—Peak Shaving (Winter)	0.32

4. Conclusions

A method for predicting the performance and deterioration of second-life batteries deployed in electric grid systems is presented in this paper. The approach incorporates an electrochemical model to evaluate battery performance, a health monitoring technique, and a cost-reduction algorithm designed for grid applications. The method predicts battery degradation based on various charging and discharging rates, which reveals that parallel battery connections can mitigate degradation, despite necessitating a balance between battery cost and longevity. Various charging and discharging strategies, including energy arbitrage and peak-shaving against DPP and TOU tariffs of Michigan's DTE utility, are explored. Note that the proposed methodology utilizes the SEI layer growth and lithium plating mechanisms for battery degradation prediction to demonstrate the idea of its implementation in the performance estimation of second-life batteries. More complex mechanisms, i.e., the particle cracking model, loss of active material model, and the coupling of these mechanisms, could be employed to further improve the estimation accuracy. This methodology offers an appropriate framework for the analysis of the second-life battery in grid applications. It could be extended further for electric grid systems with photovoltaic panels and wind turbines. Moreover, combined with detailed cost estimation approaches for the battery, i.e., replacement and depreciation models, a more comprehensive computational framework could be developed for the revenue optimization of the electric grid system.

Author Contributions: Writing—original draft, R.L.; Grid Optimization, A.H.; Governing Equations of Battery, N.G.; Supervision, W.S. and X.Z. All authors have read and agreed to the published version of the manuscript.

Funding: This research was funded by the Research Initiation & Development Grant (FY23 Winter Cycle) of the University of Michigan—Dearborn Office of Research.

Data Availability Statement: The data are available from the corresponding author upon reasonable request.

Acknowledgments: We are grateful to acknowledge the comments and suggestions from the editor and reviewers for their help in improving the quality of this paper.

Conflicts of Interest: The authors declare no conflict of interest.

References

1. Skeete, J.P.; Wells, P.; Dong, X.; Heidrich, O.; Harper, G. Beyond the EVent Horizon: Battery Waste, Recycling, and Sustainability in the United Kingdom Electric Vehicle Transition. *Energy Res. Soc. Sci.* **2020**, *69*, 101581. [CrossRef]
2. Wang, L.; Wang, X.; Yang, W. Optimal Design of Electric Vehicle Battery Recycling Network—From the Perspective of Electric Vehicle Manufacturers. *Appl. Energy* **2020**, *275*, 115328. [CrossRef]
3. Ioakimidis, C.S.; Murillo-Marrodán, A.; Bagheri, A.; Thomas, D.; Genikomsakis, K.N. Life Cycle Assessment of a Lithium Iron Phosphate (LFP) Electric Vehicle Battery in Second Life Application Scenarios. *Sustainability* **2019**, *11*, 2527. [CrossRef]
4. Hua, Y.; Liu, X.; Zhou, S.; Huang, Y.; Ling, H.; Yang, S. Toward Sustainable Reuse of Retired Lithium-Ion Batteries from Electric Vehicles. *Resour. Conserv. Recycl.* **2021**, *168*, 105249. [CrossRef]
5. Hua, Y.; Zhou, S.; Huang, Y.; Liu, X.; Ling, H.; Zhou, X.; Zhang, C.; Yang, S. Sustainable Value Chain of Retired Lithium-Ion Batteries for Electric Vehicles. *J. Power Sources* **2020**, *478*, 228753. [CrossRef]
6. Chen, W.; Liang, J.; Yang, Z.; Li, G. A Review of Lithium-Ion Battery for Electric Vehicle Applications and Beyond. *Energy Procedia* **2019**, *158*, 4363–4368. [CrossRef]
7. Haram, M.H.S.M.; Lee, J.W.; Ramasamy, G.; Ngu, E.E.; Thiagarajah, S.P.; Lee, Y.H. Feasibility of Utilising Second Life EV Batteries: Applications, Lifespan, Economics, Environmental Impact, Assessment, and Challenges. *Alex. Eng. J.* **2021**, *60*, 4517–4536. [CrossRef]
8. White, C.; Thompson, B.; Swan, L.G. Repurposed Electric Vehicle Battery Performance in Second-Life Electricity Grid Frequency Regulation Service. *J. Energy Storage* **2020**, *28*, 101278. [CrossRef]
9. Sahoo, S.; Timmann, P. Energy Storage Technologies for Modern Power Systems: A Detailed Analysis of Functionalities, Potentials, and Impacts. *IEEE Access* **2023**, *11*, 49689–49729. [CrossRef]
10. Fitzgerald, G.; Mandel, J.; Morris, J.; Touati, H. *The Economics of Battery Energy Storage How Multi-Use, Customer-Sited Batteries Deliver the Most Services and Value to Customers and the Grid the Economics of Battery Energy Storage*; Rocky Mountain Institute: Snowmass, CO, USA, 2015.
11. Aaslid, P.; Korpas, M.; Belsnes, M.M.; Fosso, O.B. Stochastic Optimization of Microgrid Operation With Renewable Generation and Energy Storages. *IEEE Trans. Sustain. Energy* **2022**, *13*, 1481–1491. [CrossRef]
12. Gu, B.; Mao, C.; Liu, B.; Wang, D.; Fan, H.; Zhu, J.; Sang, Z. Optimal Charge/Discharge Scheduling for Batteries in Energy Router-Based Microgrids of Prosumers via Peer-to-Peer Trading. *IEEE Trans. Sustain. Energy* **2022**, *13*, 1315–1328. [CrossRef]
13. Wu, X.; Zhao, J.; Conejo, A.J. Optimal Battery Sizing for Frequency Regulation and Energy Arbitrage. *IEEE Trans. Power Deliv.* **2022**, *37*, 2016–2023. [CrossRef]
14. Hassan, A.; Altin, M.; Bingöl, F. Inertia and Droop Controller for a Modern Variable Speed Wind Turbine to Provide Frequency Control in a Microgrid. *Politek. Derg.* **2020**, *23*, 771–777. [CrossRef]

15. Guan, M. Scheduled Power Control and Autonomous Energy Control of Grid-Connected Energy Storage System (ESS) with Virtual Synchronous Generator and Primary Frequency Regulation Capabilities. *IEEE Trans. Power Syst.* **2022**, *37*, 942–954. [CrossRef]
16. Hassan, A. Frequency Control in an Isolated Power System with High Penetration of Wind Power. Ph.D. Thesis, Izmir Institute of Technology, İzmir, Turkey, 2019.
17. Li, X.; Wang, L.; Yan, N.; Ma, R. Cooperative Dispatch of Distributed Energy Storage in Distribution Network with PV Generation Systems. *IEEE Trans. Appl. Supercond.* **2021**, *31*, 1–4. [CrossRef]
18. Argiolas, L.; Stecca, M.; Elizondo, L.R.; Soeiro, T.B.; Bauer, P. Optimal Battery Energy Storage Dispatch in Energy and Frequency Regulation Markets While Peak Shaving an EV Fast Charging Station. *IEEE Open Access J. Power Energy* **2022**, *9*, 374–385. [CrossRef]
19. Li, X.; Cao, X.; Li, C.; Yang, B.; Cong, M.; Chen, D. A Coordinated Peak Shaving Strategy Using Neural Network for Discretely Adjustable Energy-Intensive Load and Battery Energy Storage. *IEEE Access* **2020**, *8*, 5331–5338. [CrossRef]
20. Rasool, S.; Muttaqi, K.M.; Sutanto, D. A Multi-Filter Based Dynamic Power Sharing Control for a Hybrid Energy Storage System Integrated to a Wave Energy Converter for Output Power Smoothing. *IEEE Trans. Sustain. Energy* **2022**, *13*, 1693–1706. [CrossRef]
21. Fan, F.; Kockar, I.; Xu, H.; Li, J. Scheduling Framework Using Dynamic Optimal Power Flow for Battery Energy Storage Systems. *CSEE J. Power Energy Syst.* **2022**, *8*, 271–280. [CrossRef]
22. Abdalla, A.A.; El Moursi, M.S.; El-Fouly, T.H.M.; Al Hosani, K.H. A Novel Adaptive Power Smoothing Approach for PV Power Plant with Hybrid Energy Storage System. *IEEE Trans. Sustain. Energy* **2023**, *14*, 1457–1473. [CrossRef]
23. Europe's Largest Energy Storage System Now Live at the Johan Cruijff Arena. Available online: https://global.nissannews.com/en/releases/europes-largest-energy-storage-system-now-live-at-the-johan-cruijff-arena (accessed on 15 April 2021).
24. Schimpe, M.; Piesch, C.; Hesse, H.C.; Paß, J.; Ritter, S.; Jossen, A. Power Flow Distribution Strategy for Improved Power Electronics Energy Efficiency in Battery Storage Systems: Development and Implementation in a Utility-Scale System. *Energies* **2018**, *11*, 533. [CrossRef]
25. Second Life Energy Storage: VHH and MAN Testing Use of Second Life of Batteries for EBus Charging Station. Available online: https://press.mantruckandbus.com/second-life-energy-storage-vhh-and-man-testing-use-of-secondlife-of-batteries-for-ebus-charging-station (accessed on 30 June 2020).
26. Chen, Z.; Chen, L.; Shen, W.; Xu, K. Remaining Useful Life Prediction of Lithium-Ion Battery via a Sequence Decomposition and Deep Learning Integrated Approach. *IEEE Trans. Veh. Technol.* **2022**, *71*, 1466–1479. [CrossRef]
27. Cui, X.; Ramyar, A.; Contreras, V.; Judge, G.; Siegel, J.; Stefanopoulou, A.; Avestruz, A.T. Lite-Sparse Hierarchical Partial Power Processing for Parallel Batteries in Heterogeneous Energy Storage Systems. In Proceedings of the 2021 IEEE 22nd Workshop on Control and Modelling of Power Electronics, COMPEL 2021, Cartegena, Colombia, 2–5 November 2021; Institute of Electrical and Electronics Engineers Inc.: New York, NY, USA, 2021.
28. Li, S.; Zhao, P.; Gu, C.; Li, J.; Huo, D.; Cheng, S. Aging Mitigation for Battery Energy Storage System in Electric Vehicles. *IEEE Trans. Smart Grid* **2023**, *14*, 2152–2163. [CrossRef]
29. Fan, F.; Xu, Y.; Zhang, R.; Wan, T. Whole-Lifetime Coordinated Service Strategy for Battery Energy Storage System Considering Multi-Stage Battery Aging Characteristics. *J. Mod. Power Syst. Clean Energy* **2022**, *10*, 689–699. [CrossRef]
30. Zhang, L.; Yu, Y.; Li, B.; Qian, X.; Zhang, S.; Wang, X.; Zhang, X.; Chen, M. Improved Cycle Aging Cost Model for Battery Energy Storage Systems Considering More Accurate Battery Life Degradation. *IEEE Access* **2022**, *10*, 297–307. [CrossRef]
31. Janota, L.; Králík, T.; Knápek, J. Second Life Batteries Used in Energy Storage for Frequency Containment Reserve Service. *Energies* **2020**, *13*, 6396. [CrossRef]
32. Steckel, T.; Kendall, A.; Ambrose, H. Applying Levelized Cost of Storage Methodology to Utility-Scale Second-Life Lithium-Ion Battery Energy Storage Systems. *Appl. Energy* **2021**, *300*, 117309. [CrossRef]
33. Wang, B.; Ji, C.; Wang, S.; Sun, J.; Pan, S.; Wang, D.; Liang, C. Study of Non-Uniform Temperature and Discharging Distribution for Lithium-Ion Battery Modules in Series and Parallel Connection. *Appl. Therm. Eng.* **2020**, *168*, 114831. [CrossRef]
34. Wang, Y.; Tian, J.; Sun, Z.; Wang, L.; Xu, R.; Li, M.; Chen, Z. A Comprehensive Review of Battery Modeling and State Estimation Approaches for Advanced Battery Management Systems. *Renew. Sustain. Energy Rev.* **2020**, *131*, 110015. [CrossRef]
35. Zhang, Q.; Wang, D.; Yang, B.; Cui, X.; Li, X. Electrochemical Model of Lithium-Ion Battery for Wide Frequency Range Applications. *Electrochim. Acta* **2020**, *343*, 136094. [CrossRef]
36. Nejad, S.; Gladwin, D.T.; Stone, D.A. A Systematic Review of Lumped-Parameter Equivalent Circuit Models for Real-Time Estimation of Lithium-Ion Battery States. *J. Power Sources* **2016**, *316*, 183–196. [CrossRef]
37. Seaman, A.; Dao, T.S.; McPhee, J. A Survey of Mathematics-Based Equivalent-Circuit and Electrochemical Battery Models for Hybrid and Electric Vehicle Simulation. *J. Power Sources* **2014**, *256*, 410–423. [CrossRef]
38. Wang, Y.; Yang, D.; Zhang, X.; Chen, Z. Probability Based Remaining Capacity Estimation Using Data-Driven and Neural Network Model. *J. Power Sources* **2016**, *315*, 199–208. [CrossRef]
39. Wei, J.; Dong, G.; Chen, Z. Remaining Useful Life Prediction and State of Health Diagnosis for Lithium-Ion Batteries Using Particle Filter and Support Vector Regression. *IEEE Trans. Ind. Electron.* **2018**, *65*, 5634–5643. [CrossRef]
40. Doyle, M.; Fuller, T.; Newman, J. Modeling of Galvanostatic Charge and Discharge of the Lithium/Polymer/Insertion Cell. *J. Electrochem. Soc.* **1993**, *140*, 1526–1533. [CrossRef]

41. Dao, T.S.; Vyasarayani, C.P.; McPhee, J. Simplification and Order Reduction of Lithium-Ion Battery Model Based on Porous-Electrode Theory. *J. Power Sources* **2012**, *198*, 329–337. [CrossRef]
42. Han, X.; Ouyang, M.; Lu, L.; Li, J. Simplification of Physics-Based Electrochemical Model for Lithium Ion Battery on Electric Vehicle. Part II: Pseudo-Two-Dimensional Model Simplification and State of Charge Estimation. *J. Power Sources* **2015**, *278*, 814–825. [CrossRef]
43. Xu, M.; Wang, R.; Zhao, P.; Wang, X. Fast Charging Optimization for Lithium-Ion Batteries Based on Dynamic Programming Algorithm and Electrochemical-Thermal-Capacity Fade Coupled Model. *J. Power Sources* **2019**, *438*, 227015. [CrossRef]
44. Song, X.; Lu, Y.; Wang, F.; Zhao, X.; Chen, H. A Coupled Electro-Chemo-Mechanical Model for All-Solid-State Thin Film Li-Ion Batteries: The Effects of Bending on Battery Performances. *J Power Sources* **2020**, *452*, 227803. [CrossRef]
45. Cai, L.; White, R.E. Mathematical Modeling of a Lithium Ion Battery with Thermal Effects in COMSOL Inc. Multiphysics (MP) Software. *J. Power Sources* **2011**, *196*, 5985–5989. [CrossRef]
46. FORTRAN Programs for the Simulation of Electrochemical Systems. Available online: http://www.cchem.berkeley.edu/jsngrp/fortran.html (accessed on 15 July 2023).
47. Torchio, M.; Magni, L.; Gopaluni, R.B.; Braatz, R.D.; Raimondo, D.M. LIONSIMBA: A Matlab Framework Based on a Finite Volume Model Suitable for Li-Ion Battery Design, Simulation, and Control. *J. Electrochem. Soc.* **2016**, *163*, A1192–A1205. [CrossRef]
48. Sulzer, V.; Marquis, S.G.; Timms, R.; Robinson, M.; Chapman, S.J. Python Battery Mathematical Modelling (PyBaMM). *J. Open Res. Softw.* **2021**, *9*, 14. [CrossRef]
49. Safari, M.; Morcrette, M.; Teyssot, A.; Delacourt, C. Multimodal Physics-Based Aging Model for Life Prediction of Li-Ion Batteries. *J. Electrochem. Soc.* **2009**, *156*, A145. [CrossRef]
50. Single, F.; Latz, A.; Horstmann, B. Identifying the Mechanism of Continued Growth of the Solid–Electrolyte Interphase. *ChemSusChem* **2018**, *11*, 1950–1955. [CrossRef] [PubMed]
51. Luo, G.; Zhang, Y.; Tang, A. Capacity Degradation and Aging Mechanisms Evolution of Lithium-Ion Batteries under Different Operation Conditions. *Energies* **2023**, *16*, 4232. [CrossRef]
52. O'Kane, S.E.J.; Ai, W.; Madabattula, G.; Alonso-Alvarez, D.; Timms, R.; Sulzer, V.; Edge, J.S.; Wu, B.; Offer, G.J.; Marinescu, M. Lithium-Ion Battery Degradation: How to Model It. *Phys. Chem. Chem. Phys.* **2022**, *24*, 7909–7922. [CrossRef]
53. Doyle, M.; Newman, J. Comparison of Modeling Predictions with Experimental Data from Plastic Lithium Ion Cells. *J. Electrochem. Soc.* **1996**, *143*, 1890–1903. [CrossRef]
54. Plett, G.L. *Battery Management Systems, Volume I: Battery Modeling*; Artech House: London, UK, 2015; Volume 1, p. 343.
55. Marquis, S.G. Long-Term Degradation of Lithium-Ion Batteries. Ph.D. Thesis, University of Oxford, Oxford, UK, 2020.
56. Chen, C.-H.; Brosa Planella, F.; O'Regan, K.; Gastol, D.; Widanage, W.D.; Kendrick, E. Development of Experimental Techniques for Parameterization of Multi-Scale Lithium-Ion Battery Models. *J. Electrochem. Soc.* **2020**, *167*, 080534. [CrossRef]

energies

MDPI

Article

HPPC Test Methodology Using LFP Battery Cell Identification Tests as an Example

Tadeusz Białoń [1,2], Roman Niestrój [1,2], Wojciech Skarka [3,4,*] and Wojciech Korski [2]

1 Department of Electrical Engineering and Computer Science, Faculty of Electrical Engineering, Silesian University of Technology, 44-100 Gliwice, Poland; tadeusz.bialon@polsl.pl (T.B.); roman.niestroj@polsl.pl (R.N.)
2 Łukasiewicz Research Network—Institute of Innovative Technologies EMAG, 40-189 Katowice, Poland; wojciech.korski@emag.lukasiewicz.gov.pl
3 Department of Fundamentals of Machinery Design, Faculty of Mechanical Engineering, Silesian University of Technology, 44-100 Gliwice, Poland
4 Bumech S.A., 40-389 Katowice, Poland
* Correspondence: wojciech.skarka@polsl.pl

Abstract: The aim of this research was to create an accurate simulation model of a lithium-ion battery cell, which will be used in the design process of the traction battery of a fully electric load-hull-dump vehicle. Discharge characteristics tests were used to estimate the actual cell capacity, and hybrid pulse power characterization (HPPC) tests were used to identify the Thevenin equivalent circuit parameters. A detailed description is provided of the methods used to develop the HPPC test results. Particular emphasis was placed on the applied filtration and optimization techniques as well as the assessment of the quality and the applicability of the acquired measurement data. As a result, a simulation model of the battery cell was created. The article gives the full set of parameter values needed to build a fully functional simulation model. Finally, a charge-depleting cycle test was performed to verify the created simulation model.

Keywords: lithium-ion iron phosphate (LFP) battery; hybrid pulse power characterization (HPPC); Thevenin equivalent circuit

Citation: Białoń, T.; Niestrój, R.; Skarka, W.; Korski, W. HPPC Test Methodology Using LFP Battery Cell Identification Tests as an Example. *Energies* **2023**, *16*, 6239. https://doi.org/10.3390/en16176239

Academic Editor: Simone Barcellona

Received: 30 June 2023
Revised: 20 August 2023
Accepted: 24 August 2023
Published: 28 August 2023

1. Introduction

The research described here aims to create an accurate simulation model of a battery cell, which will be used in the design process of the traction battery of a fully electric load-hull-dump (LHD) vehicle. The model-based design (MBD) method [1,2] was used to create numerical models of vehicle subassemblies in order to test them by means of simulation. The MBD method is frequently used to design vehicles and mobile robots [3], as well as to design manned and unmanned aerial vehicles [4,5]. The practical importance of model-based techniques in energy storage analysis and design is also underlined in [6–10].

The most basic element of a traction battery is a single cell. Its equivalent circuit, describing the static and dynamic properties, is the starting point for creating a simulation model. In practice, many different forms of battery cell equivalent circuits are used [11–21], taking into account various physical and chemical phenomena. The appropriate choice of model depends on its intended use and the method by which its parameters will be identified.

Battery cell models are often used in the form of an equivalent circuit [11–13,16,22–25], and many authors emphasize the advantages of this method. The first advantage is the simplicity of the model, most often composed of resistors and capacitors, and a voltage source to represent the battery's OCV [15–17,26]. The battery's electrical properties are described by their characteristics dependent on the state of charge (SOC) of the cell. Another advantage is the flexibility. The complexity of the equivalent circuit (i.e., number of RC pairs in the Thevenin model) can be adjusted according to the desired model accuracy and applied parameter identification method. To represent battery cell properties over a wide frequency range, an additional inductance may be introduced [27,28]. In this work it is omitted, because the inductance value cannot be identified with the HPPC tests. The third important advantage is the composability. Introducing additional elements to the model, phenomena such as self-discharge can be taken into consideration [18]. The equivalent circuit element characteristics can also be extended with the thermal and aging model, creating a multi-physical model [29].

The equivalent circuit model approach also shows high fidelity in simulating battery performance characteristics [30–33]. Model fidelity, which measures how closely a model or simulation mimics the state and behavior of a real-world item, is crucial in MBD.

For the purposes of battery design using the MBD method, models describing the dependence of the open-circuit voltage (OCV) characteristics on SOC [11,18,34,35] and the dynamic properties of the cell with one or two time constants are most often used. The Thevenin circuit [17,20,25,36–38] is such model and was used in the research described here.

In general, it is difficult to estimate battery parameters quickly and accurately from input–output cycling data [23,24,39]; therefore, special identification tests must be used. The pulse charge or discharge test [14,40–42] or hybrid pulse power characterization (HPPC) tests [43–47], combined with charge and discharge characteristics [44,48,49], are the most commonly used. These tests (performed once) can reflect the properties of the cell for the current state of health (SOH), so they cannot identify the effects of cell aging [21] or changes in its parameters during long-term operation. They are also unable to identify self-discharge effects [50]. These effects, however, have little impact on the basic operational properties of the battery and are usually neglected in the design process with the MBD method.

Identification of resistor–capacitor (RC) parameters of the Thevenin equivalent circuit depends on the HPPC impulse voltage approximation quality with an exponential function [40,48,49,51] or multi-exponential function [35,47,48]. However, usually two exponential terms are used. Approximation may be performed by optimization. Deterministic optimization methods may be used [21,44,52]. However, in this case, the optimization result depends on the starting point of the algorithm, which is not always easy to choose. However, this problem does not occur in population-based metaheuristic algorithms [53]. Such algorithms, i.e., genetic algorithms [17,54,55], particle swarm optimization (PSO) [21,52,56], and others [57], are also used for HPPC results processing. PSO was also applied in the research described here.

This article details the step-by-step process of preparing HPPC tests and processing their results. Typical technical problems, including those resulting from the physical properties of lithium-ion iron phosphate (LFP) cells, are discussed and methods of solving them are proposed. LFP battery cells have a lower energy density than the most popular electromobility applications of nickel manganese cobalt (NMC) cells [58,59], but they ensure greater safety of use due to much lower susceptibility to thermal runaway [59,60]. LFP battery cells also perform more favorably in terms of product sustainability [61,62].

The article describes the issues that are a continuation of the research described in the article [63].

The novelties are as follows:

- An optimization-based battery cell time constant identification algorithm is implemented in software written by the authors.
- An HPPC-based method for OCV vs. SOC characteristic determination is established.
- Other contributions of the article are as follows:
- This paper gives the values of all parameters necessary to build a fully parameterized mathematical model of the cell.
- The paper explains the HPPC test development methodology step by step. In the literature, usually only the results of HPPC are given, but the process of obtaining them is not described. This paper fills that gap.

- The paper discusses potential flaws in the HPPC test results. Not every HPPC pulse recorded during measurements is suitable for further analysis and must be omitted. In the literature, this problem is hardly commented on. This paper fills that gap.
- The paper applies edge detection techniques in the analysis of the HPPC test results.
- The paper remarks on battery cell true capacity experimental estimation.

2. Materials and Methods

The general research methodology is schematically presented in Figure 1.

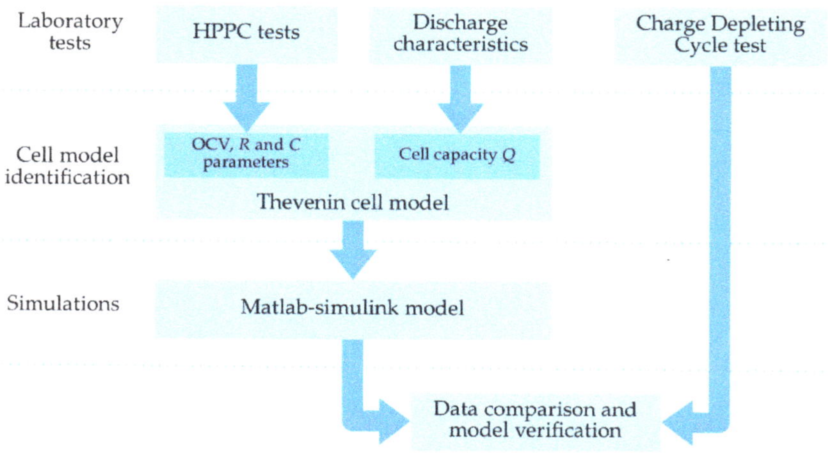

Figure 1. Research methodology overview.

The research was divided into three main stages: laboratory tests, identification of cell parameters based on test results, and creation of a simulation model and simulations (Figure 1). In the laboratory phase, tests were performed that were the basis for model identification and its subsequent verification. Identification of the parameters of the cell model was carried out using software written by the authors, using innovative optimization algorithms based on particle swarm optimization (PSO) and the Levenberg–Marquardt method. Known signal processing techniques, such as edge detection and filtering of measurement data, were also used in an original way. As a result, a simulation model was created in the MATLAB/Simulink environment, using the Simscape Electrical library.

The tests were carried out for the LFP ($LiFePO_4$) battery cell with the rated parameters given in Table 1. The following laboratory tests were carried out: discharge characteristics to estimate the actual cell capacity, HPPC tests to identify the equivalent circuit parameters, and a charge-depleting cycle (CDC) test [64] to verify the identified mathematical model.

Table 1. ThunderSky Winston LFP040AHA cell nominal parameters.

Parameter	Value
Capacity Q_n	40 Ah
Energy density	82.5 Wh/kg
Voltage (min./nominal/max.)	2.5/3.3/4.0
Current (typical/max. discharge)	20 A (0.5C [1])/400 A (10C [1])

[1] Battery cell C-rating, based on nominal capacity: 1C = 40 A.

The tests were carried out in the laboratory setup shown in Figure 2. The voltage at the cell terminals and at the shunt was recorded using a National Instruments NI 6251 M A series data acquisition device was equipped with a 16-bit analog-to-digital converter.

The sampling rate was 135 Hz during HPPC tests and 100 Hz during the CDC test. The main element of the setup was a programmable power supply with a load function ITECH IT6522C, additionally equipped with a dedicated power dissipater module IT-E502. This set enables both power supply and active load up to 3000 W and 120 A DC. The power supply may operate in constant current (CC) and constant voltage (CV) modes. The battery cell was operated in CC mode, in accordance with the given reference current test profile, only if the cell voltage value was within the tolerable limits given in Table 1. When the cell voltage reached the minimum or maximum value, the power supply was switched into CV mode, in which the current was limited to keep the voltage within specified limits. Measurements were carried out at an ambient temperature of about 22 °C, with deviations up to 1 °C.

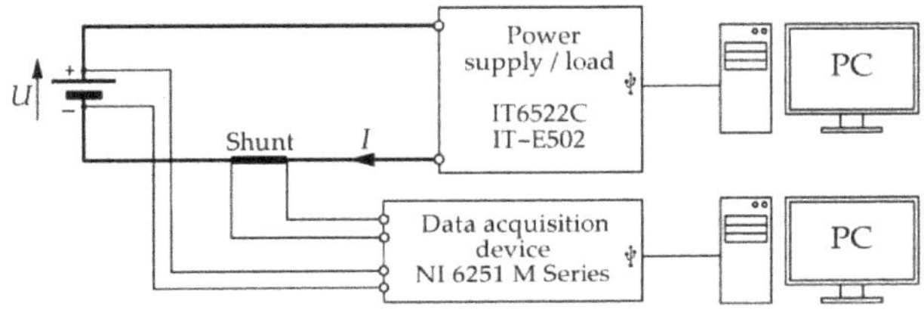

Figure 2. Laboratory setup overview.

The simulation model created during the research directly reflects the structure of the laboratory setup. Not only a battery cell model based on Thevenin's equivalent circuit (Section 3.1) was created, but also an active power supply model was created, including the implementation of CC/CV mechanisms. This makes it possible to compare the simulation results with the laboratory CDC test results. In both cases, i.e., in the real power supply and its simulation model, the same profile of the reference current was implemented.

3. Results

The aim of the research was to create an accurate simulation model of an LFP battery cell. The basis of the model was the equivalent circuit described in Section 3.1, the parameters of which were determined on the basis of HPPC test results as described in Section 3.3. The cell capacity set in the simulation model was determined by the methods described in Section 3.2. The method of verifying the identified model is described in Section 3.4.

3.1. Battery Cell Equivalent Circuit

A mathematical model of the battery cell in the form of a Thevenin equivalent circuit [21,37,38,46,65–67] was used (Figure 3). The circuit contains two RC pairs, thus simulating two time constants of the dynamic model [65,66,68,69]:

$$\tau_1 = R_1 C_1,\ \tau_2 = R_2 C_2, \tag{1}$$

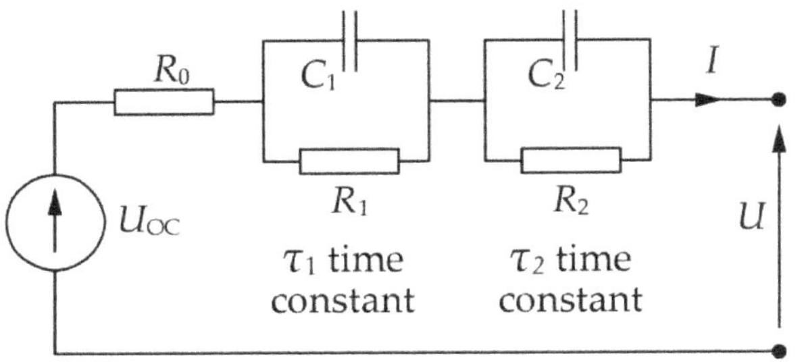

Figure 3. Thevenin equivalent circuit of the battery cell.

Thevenin models are usually used with one to five RC pairs and their corresponding time constants. In the case of LFP cells, the first (smallest) time constant has values of a few seconds, whereas the second has values of tens of seconds. The third time constant is measured in tens of minutes, etc. HPPC tests can only identify the first two RC pairs (see Section 3.3), so the others are omitted [22].

All the resistances, capacities as well as the OCV (U_{OC} in Figure 3) depend on the *SOC* of the cell [31], which is estimated on the basis of the cell current [22,37,51,57,67,68,70]:

$$SOC = SOC_0 - \frac{1}{Q}\int_0^t I d\tau, \tag{2}$$

where SOC_0 is the initial *SOC* of the cell, and Q is the cell capacity. Note that the actual cell capacity depends on many factors, such as temperature and SOH of the cell, and is usually different from the rated one, Q_n. Here, it was estimated based on the measurement results as described in Section 3.2.

Determination of the OCV vs. *SOC* characteristic is described in Section 3.3.2. The determination of the dependence of RC parameters on *SOC* is described in Sections 3.3.4 and 3.3.5.

3.2. Capacity and State of Charge Estimation

The actual capacity of the cell is usually different from the nominal one and is crucial from the point of view of correct parameterization of the created mathematical model. The correctness of determining the *SOC* depends on this, and more precisely, it is necessary to determine the value of Q in Formula (2), corresponding to $SOC = 1$.

By definition, the charge drawn from the battery is equal to the integral of the current over time [71]:

$$Q = \int_0^t I d\tau. \tag{3}$$

In geometric interpretation, Q is the area under the current waveform. However, recording of the battery discharge current can be made at different values of I, and different assumptions as to the operating conditions. Two different methods were used here. The first was a measurement based on the discharge characteristics [44,48,49]. The second, proposed by the authors, was the use of current waveforms recorded during HPPC tests [63].

Discharge tests were performed at four different current values (0.5 C, 1 C, 2 C, and 3 C) by recording the current and voltage. The tests were started with the cell charged to the maximum voltage (Table 1) and after several hours of relaxation. The current was recorded while the laboratory setup was operating in the CC mode, and the charge taken under these conditions was called Q_{CC} (Figure 4).

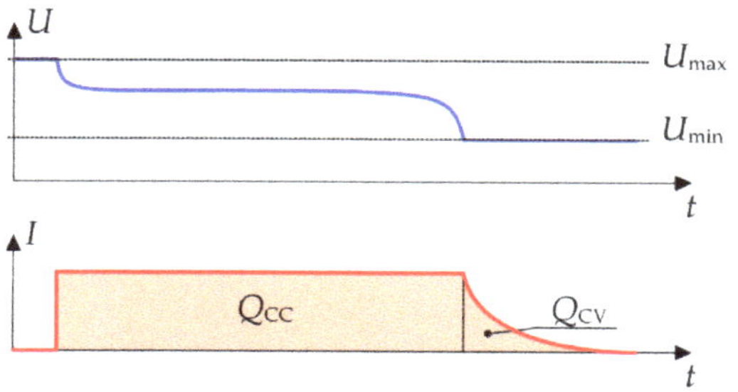

Figure 4. Cell capacity estimation based on discharge characteristic.

When the cell voltage reached the minimum value, and the system went into CV mode, the recording continued until the current completely dropped to 0. The charge determined under these conditions was called Q_{CV}. The results obtained during the tests are summarized in Table 2, and the recorded transients are shown in Figure 5.

Table 2. Cell capacity measurement results based on discharge characteristics.

Relative Discharge Current	Total Discharge Q [Ah]	Discharge in CC Mode Q_{CC} [Ah]	Discharge in CV Mode Q_{CV} [Ah]
0.5C	47.71	46.30	1.407
1C	47.71	45.78	1.934
2C	47.70	45.21	2.495
3C	47.62	45.41	2.212

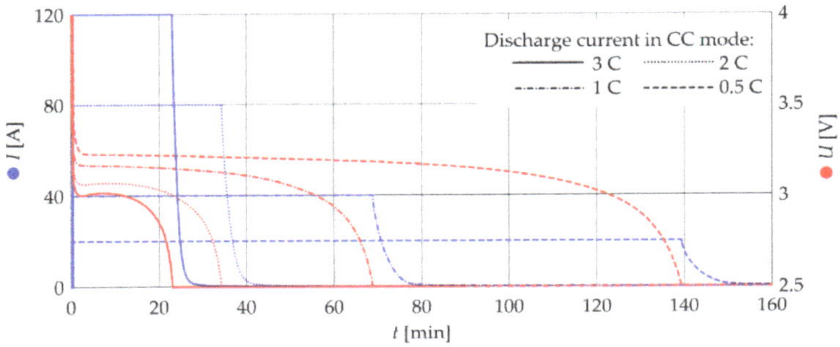

Figure 5. Discharge characteristics for cell capacity estimation.

Table 2 shows that the total charge taken from the cell during the discharge tests was slightly dependent on the current. However, the proportion between the Q_{CC} and Q_{CV} values changes. The higher the discharge current, the lower the Q_{CC} and the higher the Q_{CV}.

Then, the total charge taken from the cell during the HPPC tests (described in Section 3.3) was determined by integrating the currents recorded in each of the 18 tests and summing the results. The value of Q = 50.71 Ah was obtained (see Table 3 in Section 3.3).

Table 3. HPPC tests and impulses summary.

HPPC Test No.	Impulse No., Type and Relative Current Value								ΔQ [Ah]	$\Delta Q/Q_n$ [%]	Q [Ah]
	1 0.5 C	2 0.5 C	3 1 C	4 1 C	5 2 C	6 2 C	7 3 C	8 3 C			
1	(−)	(+)	(−)	(+)	(−)	(+)	(−)	(+)	2.41	6.03	2.41
2	(−)	(+)	(−)	(+)	(−)	(+)	(−)	(+)	2.21	5.53	4.63
3	(−)	(+)	(−)	(+)	(−)	(+)	(−)	(+)	4.21	10.53	8.84
4	(−)	(+)	(−)	(+)	(−)	(+)	(−)	(+)	4.21	10.53	13.05
5	(−)	(+)	(−)	(+)	(−)	(+)	(−)	(+)	4.22	10.54	17.27
6	(−)	(+)	(−)	(+)	(−)	(+)	(−)	(+)	4.23	10.57	21.50
7	(−)	(+)	(−)	(+)	(−)	(+)	(−)	(+)	4.22	10.56	25.72
8	(+)	(−)	(+)	(−)	(+)	(−)	(+)	(−)	4.21	10.54	29.93
9	(+)	(−)	(+)	(−)	(+)	(−)	(+)	(−)	4.22	10.55	34.16
10	(+)	(−)	(+)	(−)	(+)	(−)	(+)	(−)	2.16	5.40	36.31
11	(+)	(−)	(+)	(−)	(+)	(−)	(+)	(−)	2.18	5.45	38.49
12	(+)	(−)	(+)	(−)	(+)	(−)	(+)	(−)	2.20	5.51	40.70
13	(+)	(−)	(+)	(−)	(+)	(−)	(+)	(−)	2.21	5.52	42.90
14	(+)	(−)	(+)	(−)	(+)	(−)	(+)	(−)	2.22	5.54	45.12
15	(+)	(−)	(+)	(−)	(+)	(−)	(+)	(−)	2.20	5.49	47.32
16	(+)	(−)	(+)	(−)	(+)	(−)	(+)	(−)	2.22	5.55	49.54
17	(+)	(−)	(+)	(−)	(+)	(−)	(+)	(−)	0.92	2.31	50.46
18	(+)	(−)	(+)	(−)	(+)	(−)	(+)	(−)	0.25	0.62	50.71

Impulses: ●—healthy, ●—trimmed, ●—distorted, (+)—charging, (−)—discharging.

It should be noted that all measured charge values (in discharge and HPPC tests) were greater than the nominal cell capacitance Q_n, but the differences between them were significant. Therefore, the question as to which of them should be treated as the final one (the total capacity of the cell) should be asked, which will be used in the created mathematical model. In order to find the answer, a number of simulations were carried out for all the values obtained and the results of the selected values are presented in Section 3.4. The best result was obtained for the value of Q = 45.7 Ah, calculated as the average of the Q_{CC} values for all four current values (average of the values from the third column of Table 2).

3.3. HPPC Tests

The basic idea of an HPPC test is to analyze the cell voltage response to a rectangular current pulse. This response is a multi-exponential waveform, the time constants of which should be determined in the identification process [17,44,46,48,69]:

$$U = U_{OC} - IR_0 - IR_1\left(1 - e^{-\frac{t}{\tau_1}}\right) - IR_2\left(1 - e^{-\frac{t}{\tau_2}}\right). \quad (4)$$

In order for the determination of the time constant to be possible and precise, the recorded voltage response should last several times longer than the expected length of the time constant [43]. So, the longer the recorded transient, the better. On the other hand, a single HPPC pulse should be as short as possible so as not to change the SOC of the battery, which results directly from (2). The HPPC pulse length used in practice is therefore a compromise between these two requirements. In the case of nickel manganese cobalt (NMC) cells, the first two time constants are relatively short [63] and do not exceed a dozen or so seconds, so pulses from 9 s to 18 s are sufficient. Usually 10 s pulses are used [16,43–45]. In the case of the considered LFP cell, the time constants are longer, so the duration time of the HPPC pulses was extended to 60 s.

In practice, HPPC profiles containing different numbers of pulses are used, but it is always an even number [16,44,55]. This is because the pulses always occur in pairs (a charging pulse with a discharging pulse), so that the series of pulses does not change the SOC of the cell. The number of pairs of pulses may be different, sometimes only one is used [69]. When there is more than one pair, then individual pairs differ in current values. Here, four pairs of pulses were used, successively with current values of 0.5 C, 1 C, 2 C, and 3 C (Figure 6). The greater the current value, the greater the voltage change in response to the impulse, so the easier it is to record (see Section 3.3.1). On the other hand, the greater the current value, the greater the SOC change during the pulse duration, which may cause the problems described in Section 3.3.3. The order of the pulses in the pair also matters. For high SOC values, the discharge pulse was used first. Starting with a charging pulse would risk increasing the cell voltage during the pulse duration, which for a high SOC value (close to 1) could cause the measurement system to switch from CC to CV mode and cut the pulse (see Section 3.3.3). For small SOC values (close to 0), for the same reason, the order was reversed with the charging pulse used first. The sequence change is seen in Table 3 after test number 7.

Another important consideration is the relaxation time between pulses. In principle, it should be much longer than the expected values of the time constants of the cell, so that before the next pulse occurs, the cell voltage has time to stabilize after the preceding pulse. However, due to the very large time constants of the examined LFP cell, it was difficult to meet this assumption. In the conducted tests, a relaxation time between pulses of 20 min was used (Figure 6).

The last element of the HPPC profile shown in Figure 6 is the discharge of the cell before the next HPPC test. The values of the cell equivalent circuit parameters change most rapidly for very small and very large SOC values, but for intermediate values ($SOC \approx 0.5$), they are almost constant. To capture the shape of the characteristics, a discharge of 0.05 Q_n was used for large SOC values, and then the interval was increased to 0.1 Q_n to return to 0.05 Q_n for small SOC values.

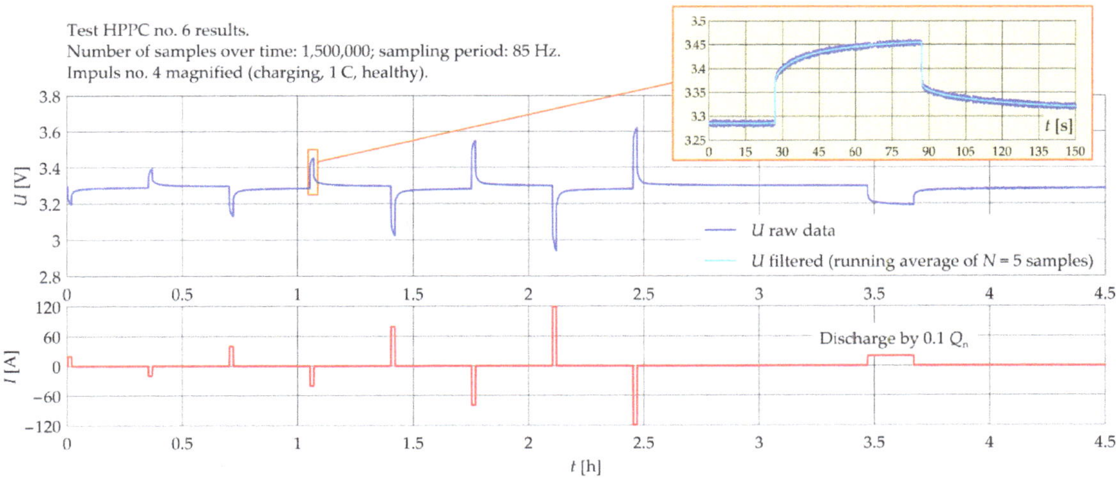

Figure 6. Exemplary HPPC test result.

Successive HPPC tests were carried out until the total charge taken from the battery during the test dropped significantly below the set discharge value (0.05 Q_n), which resulted from the fact that the discharge pulses in the profile were cut off by the CC/CV mechanism due to reaching the minimum voltage. All the performed tests are summarized in Table 3, where ΔQ is the charge taken from the cell during the whole HPPC test (including final discharge by 0.05/0.1 Q_n). For the last two tests (17 and 18), this value drops significantly, which means that the cell is already discharged. In Table 3, Q is the total value of the charge taken from the cell at the end of the given test, taking into account the charge taken in the preceding tests.

3.3.1. Filtering and Slope Detection

The recorded HPPC test transients shown in Figure 6 contain 1,500,000 samples over time. For this reason, precise localization of the beginnings and ends of HPPC pulses is a challenge. In addition, the recorded waveforms, in particular the voltages, contain a lot of noise that hinders further analysis. The high noise content results from the unfavorable proportion of the analyzed voltage changes to the measuring range of the data acquisition device. The voltage changes caused by an HPPC pulse range from ten to several tens of mV, whereas the measured voltage values reach up to 4 V. Therefore, a measuring transducer with a range of 10 V was used. Consequently, the analyzed changes constitute only a few percent of the measurement range, which, even with good quality converters and 16-bit sampling, results in a relatively large amount of noise.

Both problems, noise removal in the voltage transient and detection of the beginnings and ends of pulses based on the current transient, can be solved by using appropriate data-filtering techniques.

To detect the edges marking the beginnings and ends of the pulses, a method based on the analysis of the transient of the difference of two exponential averages was used:

$$r_i(\alpha) = \begin{cases} I_i & i = 1 \\ \alpha I_i + (1-\alpha) r_{i-1} & i > 1, \end{cases} \tag{5}$$

$$\Delta_i = r_i(\alpha_{\text{fast}}) - r_i(\alpha_{\text{slow}}). \tag{6}$$

In (5) and (6), I_i is the i-th sample of the current waveform, r_i is the i-th sample of the exponential moving average, α is the weight coefficient, and Δ_i is the i-th sample of the waveform difference. The principle of operation of the method is shown in Figure 7. The difference Δ between two waveforms averaged with different weight values α ($\alpha_{\text{slow}} = 0.02$,

$\alpha_{fast} = 0.1$) contains peaks at moments when there is a rapid change in the trend of the source waveform.

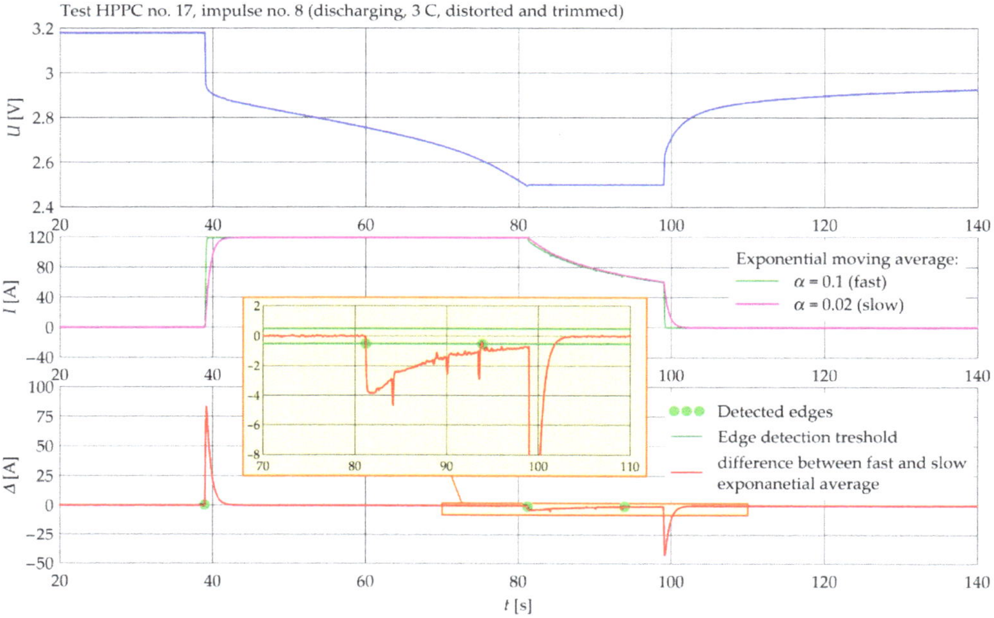

Figure 7. Edge detection method on the example of trimmed HPPC impulse.

An edge is detected when the value of Δ exceeds the set threshold, which was 0.5 here. It should be noted that the values of the weight coefficient α are selected according to the sampling frequency of the source waveform, and the threshold value should be selected according to the noise content and disturbances in the source waveform.

A running average of order $N = 5$ was used to filter out the noise from the voltage waveform:

$$U_{\text{filtered } i} = \frac{1}{2N+1} \sum_{k=i-N}^{i+N} U_k. \tag{7}$$

In (7), i is the sample number of the measured voltage U that corresponds to the $U_{\text{filtered } i}$ filtered voltage sample. The filtration consists of calculating the average for N samples preceding and following the sample with the number i.

This simple method gave good results due to the high sampling frequency of the recorded voltage waveform and the random character of the filtered noise. Order $N = 5$ was sufficient, and its low value introduced negligible distortion of the voltage waveform, having no significant impact on the subsequent identification of time constants. The filtration results for an exemplary HPPC pulse recorded at the smallest of the applied currents of 0.5 C (i.e., in conditions where the relative noise content is the highest) are shown in Figure 8.

3.3.2. OCV vs. *SOC* Characteristic

The OCV characteristic, represented by U_{OC} (i.e., the voltage source in the Thevenin equivalent circuit), is identified by measurement. The averaged charging and discharging characteristics may be used here [44,48,49]. However, this method has some disadvantages. The measured cell voltage contains not only the OCV but also the voltage drop at the impedance, which also depends on the *SOC*. Moreover, the measured charge and discharge capacities differ due to power losses. This makes it difficult to correlate them before the averaging.

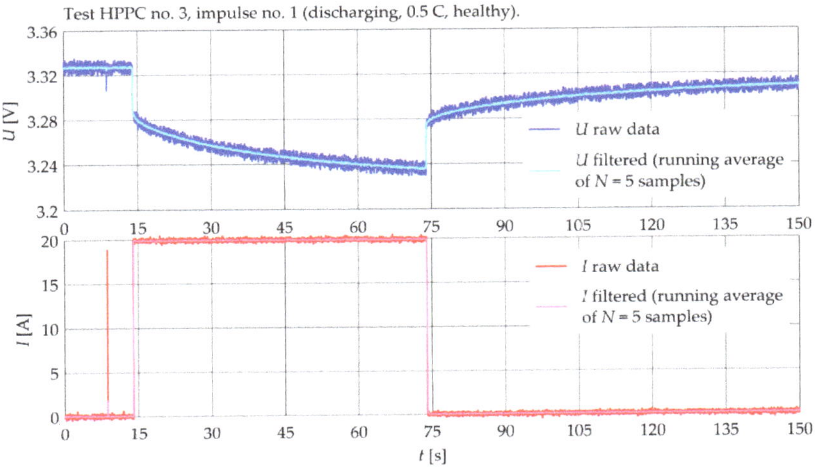

Figure 8. Exemplary HPPC impulse before and after data filtering. Near t = 10 s, a false peak generated by the control system of the active power supply is visible.

To avoid these problems, the authors proposed a method for determining the OCV characteristics based on the results of HPPC tests, consisting of averaging (over a 10 s time period) the voltage recorded in the no-current state before each pulse. The *SOC* value corresponding to the voltage obtained this way is calculated in reference to the total charge of all HPPC tests, that is, the Q value from the last row of Table 3.

A measurement-based OCV characteristic is too irregular to be directly applied in the cell mathematical model and must be approximated [11,70,72–76]. Choice of the appropriate approximating function is a further problem. Several types of functions were tested, but a log-linear exponential (LLE) function [11,75] gave the best result [63]. The LEE function has the following form:

$$U_{OC}(SOC) = a + b\ \ln(SOC + c) + d\ SOC + e^{e(SOC - f)}. \tag{8}$$

Its coefficients a to f were obtained by optimization with the particle swarm method (PSO) described in [63]. The resulting function plot and its coefficient values are given in Figure 9.

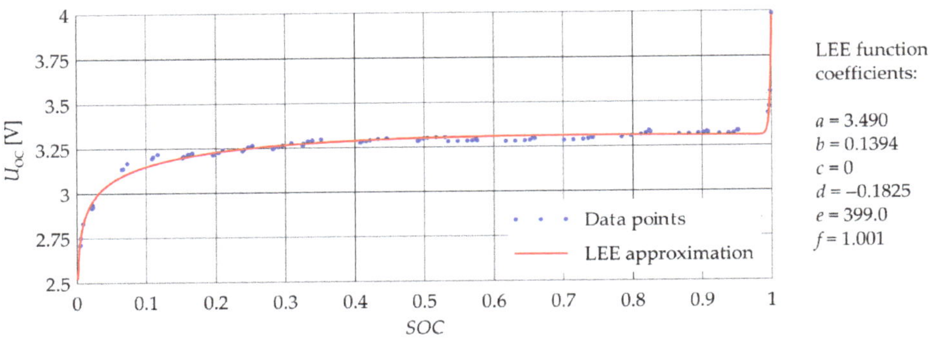

Figure 9. OCV characteristic approximated with LEE function.

3.3.3. Impulse Evaluation and Selection

Once identified and filtered (Section 3.3.1), the HPPC impulses need to be selected for further time constant identification. The problem is that not all recorded impulses are suitable for further analysis. They may contain defects resulting from the measurement method (CC/CV mechanism) or from the physical properties of the cell.

Figure 10 schematically shows the shape and interpretation of a healthy (suitable for further analysis) pulse [11,14] and two cases of faulty pulses. Examples of recorded healthy impulses are shown in Figures 6 and 8.

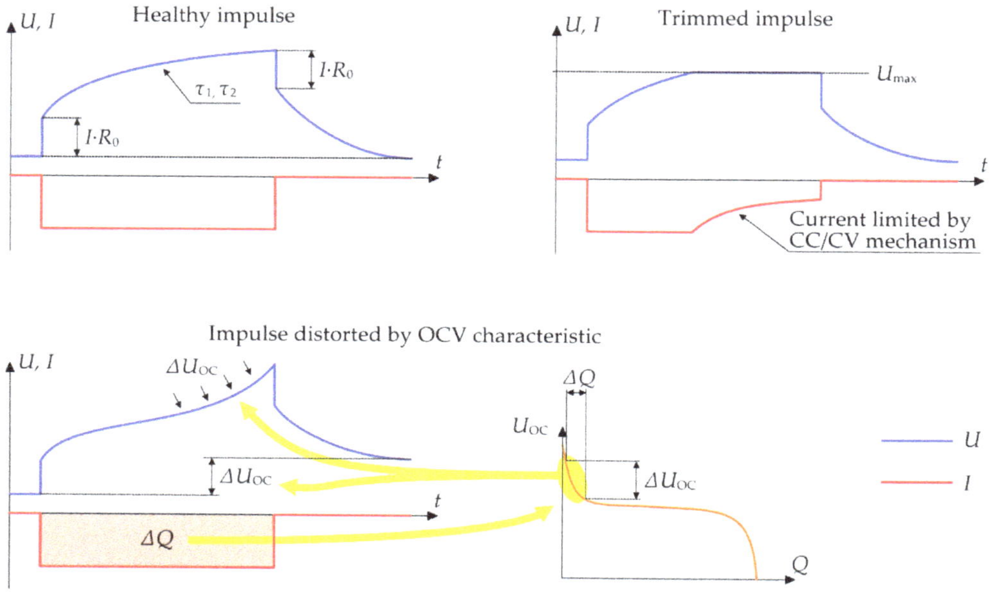

Figure 10. Shape of a single (charging) HPPC impulse and its potential flaws.

Pulse trimming (Figure 10) is easy to detect by analyzing the sequence of previously detected edges and the distances between them. A healthy pulse has two edges, falling and rising (which one is the first depends on whether the impulse is charging or discharging), separated in time by the assumed pulse length. Deviation from this pattern suggests that the pulse has been trimmed (Figure 7). Trimming occurs when the cell voltage reaches the limit during the duration of the pulse, and therefore, the measurement system switches from CC to CV mode.

In the performed tests, trimming always occurred together with the second defect, distortion by the OCV characteristic (Figure 10). Pulse distortion occurs for small SOC values, close to 0, and large ones, close to 1. This is because in these areas the OCV characteristic is the steepest (Figure 9). Therefore, even a slight change in SOC during the pulse duration causes a significant change in the voltage U_{OC} of the cell, which translates into the shape of the recorded waveform U (Figure 10). The shape of the waveform ceases to depend only on the time constants τ_1 and τ_2, which is a necessary assumption to make the identification of these constants possible. In the extreme case, the recorded waveform bends in a direction opposite (Figure 11) so that it results from (4), assuming that the time constants τ_1 and τ_2 are positive.

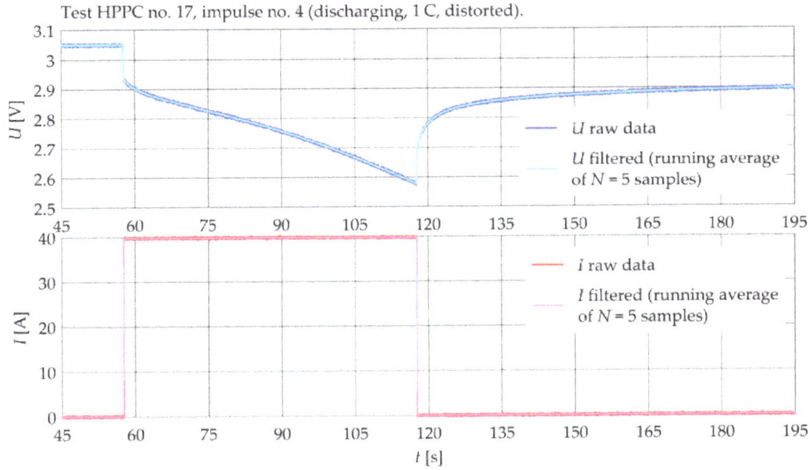

Figure 11. Exemplary HPPC impulse distorted by OCV characteristic.

In Table 3, summarizing the HPPC tests, trimmed pulses are marked in red and distorted pulses in yellow. Only healthy pulses marked in green were used in further analysis.

3.3.4. Impulse Waveform Approximation

The filtered and selected HPPC pulses were approximated by Function (4) to identify the time constants τ_1 and τ_2 and the resistances R_0, R_1, and R_2. Then, on the basis of Formula (1), capacities C_1 and C_2 were calculated. The approximation was carried out using the PSO optimization method. At this stage of the research, a configuration of the PSO algorithm was found that guaranteed high repeatability of the obtained results. The fully informed particle swarm cognition method and the 8th order ring lattice swarm topology were used. The cognition factor was 4.1, the swarm consisted of 64 particles, and the number of iterations of the algorithm was set to 180.

The optimization method used and the experiments performed with it were described in a separate article [63].

3.3.5. R and C vs. *SOC* Characteristics Approximation

The results of the approximation described in Section 3.3.4 are the values of the cell equivalent circuit parameters and the corresponding *SOC* values. These values are presented in the form of points on the graphs in Figure 12. These points are arranged in more or less regular bands, which should be approximated with continuous functions in order to create a mathematical model of the cell. A polynomial approximation was used, and several experiments were performed with polynomials of various orders. The best results were obtained for 3rd order polynomials:

$$f(SOC) = a + b\,SOC + c\,SOC^2 + d\,SOC^3. \quad (9)$$

The approximation was carried out using the Levenberg–Marquardt method. The resistance and capacitance characteristics were approximated and the polynomial coefficients obtained are summarized in Table 4. In Figure 12, in the graphs of R and C values, the blue lines are the waveforms of Function (9) with the parameters from Table 4. The blue lines in the graphs τ_1 and τ_2 are the product of the approximating functions, respectively R_1 and C_1 for τ_1, R_2 and C_2 for τ_2, according to (1).

3.4. Model Verification

The approximated OCV (Section 3.3.2), R_0, R_1, R_2, C_1, and C_2 (Section 3.3.5) characteristics fully describe the Thevenin equivalent circuit shown in Figure 3. This circuit, together with functions describing its parameters, was implemented in the MATLAB/Simulink environment by creating a simulation model of the cell, in a similar way as in [31]. The last parameter describing the model is the charge value Q corresponding to $SOC = 1$. Due to the problems with determining the actual capacity of the cell described in Section 3.2, this value was found by performing a series of simulations of the cell operating in model conditions and comparing their results with the transients recorded in the laboratory.

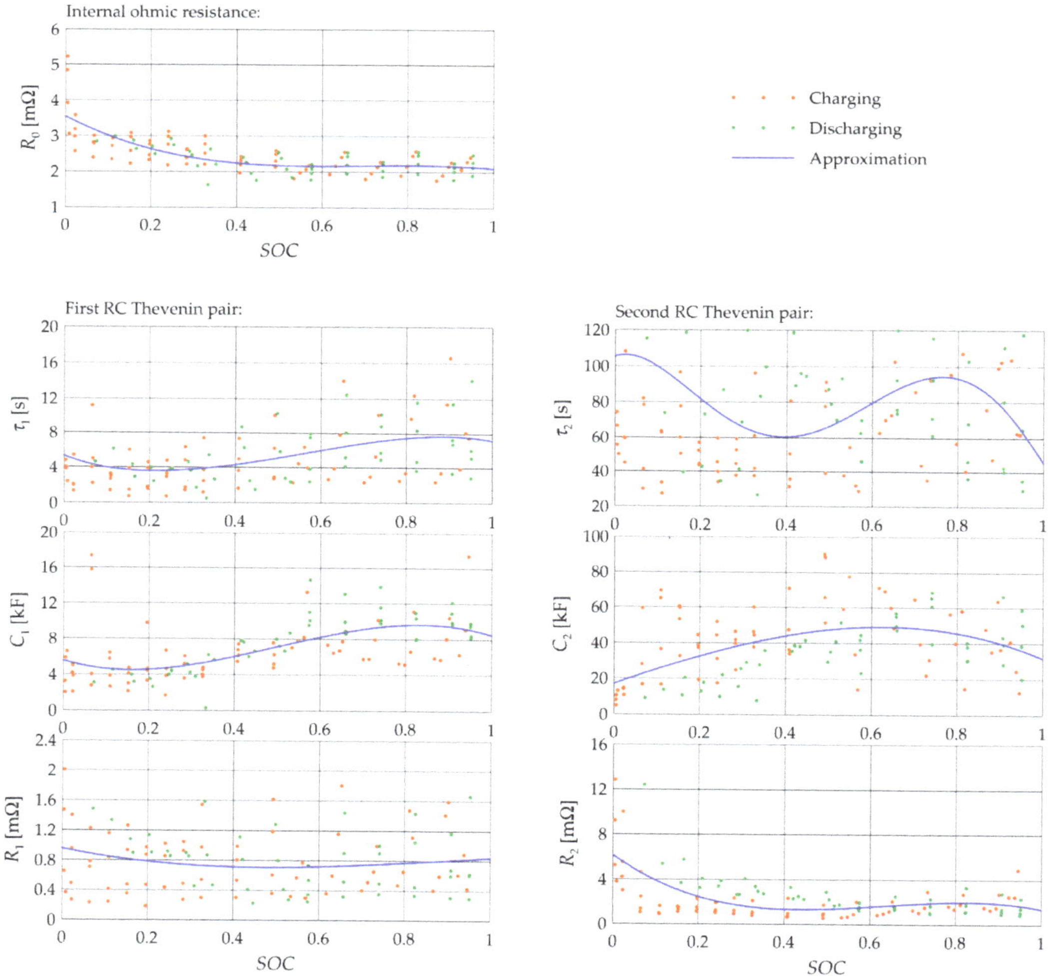

Figure 12. Thevenin equivalent circuit R and C parameter characteristics approximated with 3rd order polynomial.

As model operating conditions, a current load profile based on the CDC [64], used to test batteries of hybrid vehicles, was used. The application of cycle-based tests is a typical strategy for battery cell mathematical model verification. Other popular cycle-based tests are the dynamic stress test (DST) [46,51,65,68], ARTEMIS [57,77], and others [17,28,54,66,69,75,78,79]. The applied CDC cycle consists of a set of discharge pulses of different value (acceleration and driving at a constant speed) as well as charging ones (regenerative braking). A single cycle was about 7 min long and discharged the cell by 3.3% of Q_n, so it was repeated over 30 times until the cell was fully discharged.

Table 4. *R* and *C* parameters 3rd order polynomial approximation coefficients values.

	a	*b*	*c*	*d*
R_0	3.551×10^{-3}	-6.172×10^{-3}	8.993×10^{-3}	-4.267×10^{-3}
R_1	9.601×10^{-4}	-1.154×10^{-3}	1.611×10^{-3}	-5.716×10^{-4}
R_2	6.169×10^{-3}	-2.678×10^{-2}	4.690×10^{-2}	-2.485×10^{-2}
C_1	5549	-1.359×10^{4}	5.058×10^{4}	-3.397×10^{4}
C_2	1.712×10^{4}	8.510×10^{4}	-2.850×10^{4}	-4.243×10^{4}

Figure 13 shows the selected simulation results compared with the waveform recorded in the laboratory. The values of capacitance *Q* used in the simulation model are summarized in Table 5. The simulation accuracy has been evaluated with root-mean-square (RMS) error [49,65,68,69,74] given by the formula:

$$e_{\mathrm{rms}} = \sqrt{\frac{1}{K}\sum_{k=1}^{K}\left(U_{measurement}(t_k) - U_{simulation}(t_k)\right)^2}. \tag{10}$$

Table 5. Cell capacities used in simulations and resulting voltage error value.

Voltage RMS Error	Cell Capacity Q [Ah]	Comment
0.0432	45.7	Average for discharge characteristics, CC mode only
0.0487	47.7	Average for discharge characteristics, CC + CV
0.120	50.7	HPPC tests total discharge
0.167	40.0	Q_n—nominal cell capacity

The data in Table 5 are ordered from the lowest e_{rms} value (best result) to the highest. In Figure 13, the voltage relative error transients are shown, calculated as follows:

$$\delta U = \frac{U_{measurement} - U_{simulation}}{U_{measurement}} 100\%. \tag{11}$$

The δU error statistics for transients presented in Figure 13 are presented in Table 6. The order of the data in Table 6 is the same as in Table 5 and Figure 13.

Table 6. Simulation results—voltage relative error statistics.

Cell Capacity Q [Ah]	Average Error $\lvert\delta U\rvert$ [%]	Average Error for *t* from 5 min to 180 min $\lvert\delta U\rvert$ [%]	Peak Error $\lvert\delta U\rvert$ [%]	Peak Error for *t* from 5 min to 180 min $\lvert\delta U\rvert$ [%]
45.7	0.977	0.751	14.9	9.62
47.7	1.07	0.805	14.6	9.83
50.7	2.44	0.873	22.9	10.1
40	2.73	0.579	20.4	9.04

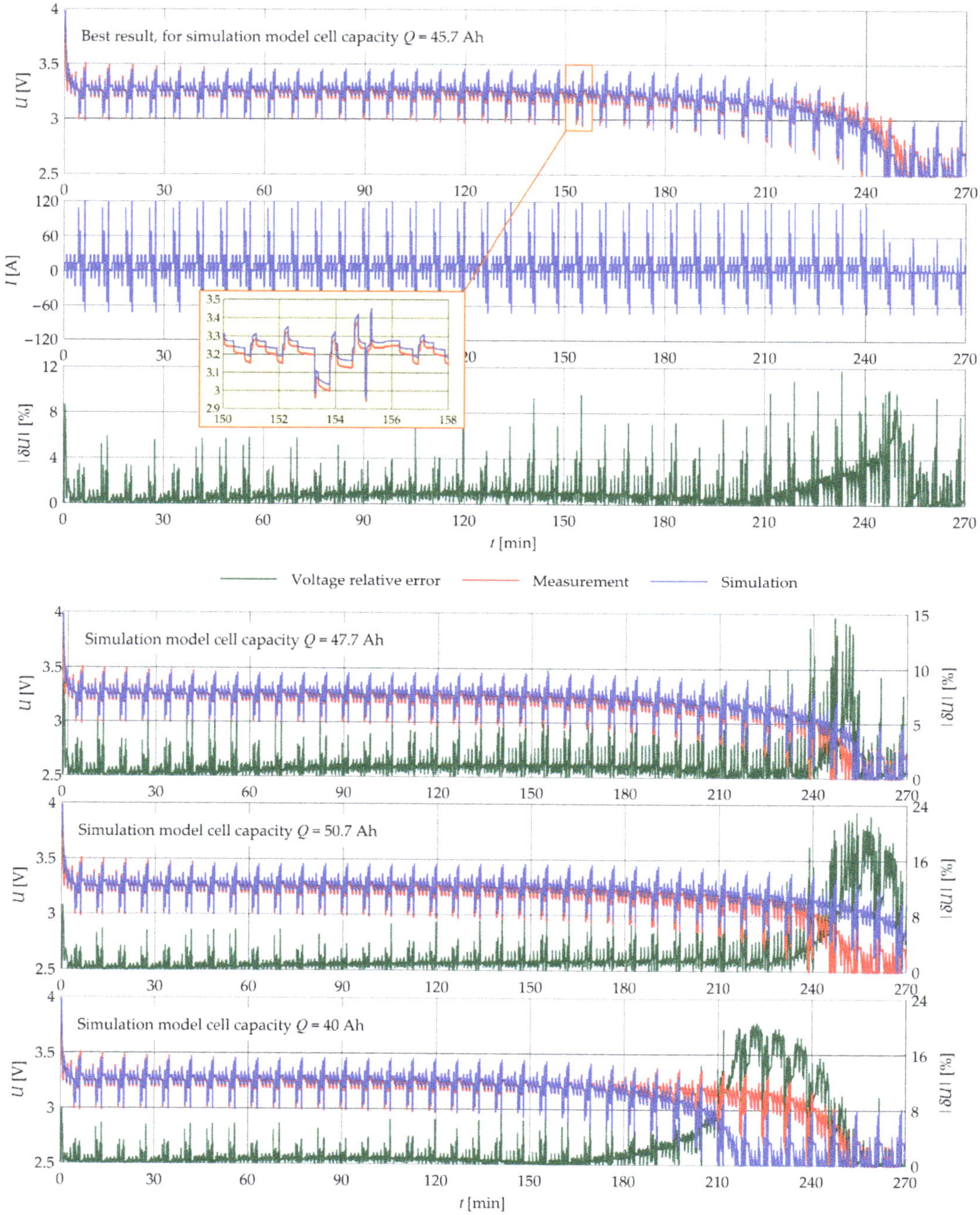

Figure 13. Comparison of simulation and measurement CDC test results. Simulations performed for various values of cell capacity.

4. Discussion

When identifying the parameters of the mathematical model of the cell, a major problem was determining its actual capacity, which comes from a comparison of the transients obtained for various Q values shown in Figure 13. The values obtained by the different methods (Section 3.2) varied considerably. They also differed from the nominal capacity Q_n. It should be noted that according to the discharge characteristics provided in the cell data sheet by the cell manufacturer, the cell capacity at normal temperature (i.e., the temperature at which the tests described herein were performed) varied with the discharge current from about 1.04 Q_n (3 C) to 1.15 Q_n (0.5 C).

These values correspond to Q_{CC} values in Table 2. Particularly significant here was the value of the discharge current. This is why the capacity determined from the HPPC tests was the largest. This was because, during these tests, the charge was taken from the cell in small increments separated by long relaxation times. Thus, the cell had a lot of time to regenerate and rebuild the voltage lowered by the discharge.

For these reasons, we decided to treat the result of the CDC test (Section 3.4) as an indication, because the working conditions during this test were close to the real working conditions of the battery in the vehicle. Cell capacitance identified using the method giving a result consistent with the CDC test will, therefore, have the highest value in use.

The applied edge detection algorithm was an effective method of extracting individual pulses from the entire recorded HPPC test transient (Section 3.3.1). It was also helpful in identifying trimmed pulses (Section 3.3.3). Nevertheless, the detection of slopes sometimes encountered problems resulting from the properties of the equipment used in the laboratory setup. For example, in Figure 7, the enlarged fragment of the Δ waveform shows disturbances in the form of short peaks. These disturbances often occurred just after the power supply switched from CC to CV mode. They probably resulted from the way the operation of the control system was implemented in the applied active power supply. In Figure 7, this disturbance caused the detection of an additional, non-existent edge. The result of the operation of the power supply control system is also visible in the current waveform in Figure 8, this being the cause of the "false impulse". This pulse, despite the high peak value, was very short, so it had no significant effect on the SOC of the cell. These types of pulses appeared in the no-current state when the change took place in the set value of the cut-off voltage (voltage at which the power supply switched from CC to CV mode). Such a change was performed before each change in the direction of the current flow: before charging, the value was set to 4 V, and before discharging, the value was set to 2.5 V. These pulses also sometimes resulted in the detection of a non-existent edge, which had to be taken into account in the analysis.

After identifying the time constants, the applied HPPC pulses, extended to 60 s, were too short to correctly identify the second time constant of Thevenin's model. The graphs in Figure 12 show that the points corresponding to the identified R_0 values are arranged in a narrow, regular band, which proves good quality of identification. In the case of the time constant τ_1, the obtained band is much wider and the dispersion of values is greater, but some regularity is still visible. In the case of the time constant τ_2, the dispersion of the results is very large, and their arrangement on the graph does not show any regularity. Note that the values of the time constant τ_2 in Figure 12 changed in the interval from 30 s to 120 s, i.e., by 400%. Probably, in individual cases, values greater than 120 s would have been obtained, if not for the fact that such a value was set as a limitation of the search space in the applied PSO algorithm. It should be noted that, as stated in Section 3.3, to ensure good quality identification of the exponential waveform time constants, the length of its recorded fragment should be several times greater than the length of its time constants. However, with the applied HPPC pulse length equal to 60 s, more than half of the identified τ_2 values were greater, even up to two times. Increasing the duration of the HPPC pulses would be undesirable, because it would cause changes too large in the SOC during the pulse duration. In the case of the tested LFP type cell, resignation from determining two time constants in favor of only one should be considered, as well as shortening the duration of the HPPC pulse. Let us also pay attention to the obtained resistance and capacitance values, given in Figure 12. The resistances are of the order of mΩ, which results in high short-circuit currents of lithium-ion cells. Capacitances are of the order of kF. Similar values were obtained, for example, in [32].

Reducing the HPPC pulse duration would reduce problems with distortion of their voltage response by the OCV characteristics (Section 3.3.3). It should be noted that the distortion effect in the form of a voltage waveform bent in the opposite direction shown in Figures 10 and 11 is an extreme case. When the distortion was small, the distorted impulse did not differ in shape from the healthy one, but the time constants identified on its basis had overestimated values. It is possible that this effect (at least partially) is responsible for the lack of regularity of the results presented in the τ_2 graph in Figure 12. This problem, however, requires confirmation and further analysis.

Among the functions known in the literature, the LEE function was selected to approximate the OCV characteristic. It has three *SOC*-dependent terms that are the most suitable for the specific shape of the LFP cell OCV characteristic. The logarithmic term describes the shape of the characteristic for *SOC* close to 0, linear describes the slope of the middle part of the characteristic, and exponential describes its shape for *SOC* close to 1. As the comparison of the measurement and simulation results showed, the OCV characteristic had the greatest impact on the accuracy of the simulation model. The enlarged fragment of the graph in Figure 13 shows that the simulated and measured voltage waveforms had a very similar shape, but there was a slowly varying offset between them. The Thevenin equivalent circuit (Figure 3) shows that the cell impedance, composed of the *R* and *C* elements, was responsible for the shape of the waveform, this being the response to current changes. The offset, on the other hand, is the result of differences in U_{OC} (OCV) voltages.

Table 6 summarizes the voltage RMS error statistics, corresponding to the waveforms in Figure 13. Error values averaged over time and peak values are presented. The data are presented for entire transients and for a limited time range, from 5 min to 180 min. In this range, the cell operates on the almost linear part of the OCV characteristic, i.e., in the most typical conditions from a practical point of view.

The data show that for the optimal cell capacity (Q = 45.7 Ah) the average voltage errors were less than 1%, which proves the very good fidelity of the obtained simulation model. Error peaks under typical operating conditions (5–180 min) are at an acceptable level of about 10%. The peak values correspond to the dynamic states (with rapid changes in the load current), and their values are influenced primarily by the quality of identifying the parameters of the RC pairs related to the time constants. Note, that for the reasons described in Section 3.3, only two time constants have been identified, which affects the precision of the model in dynamic states. It should also be noted that in the literature, models with only one time constant [24,25,30,31,39] are sufficiently considered to be accurate.

In conclusion, despite the previously described problems, the identification of time constants and *R* and *C* elements had a satisfactory effect, and the accuracy of the obtained simulation model can be improved by better methods of identification and approximation of the OCV characteristics.

5. Conclusions

The research showed the following:

- Among the various cell capacity values obtained as measurements, the best performance of the mathematical model was obtained for the averaged charge taken from the cell during discharge in the CC mode for different current values. Therefore, this method is recommended for determining the actual capacity of the cell.
- The OCV characteristics of the LFP cell are best approximated by the LEE function.
- Identification of the second time constant of the LFP cell is difficult, because of its large value, greater than a typical HPPC impulse duration.
- Suggestions for further research:

- It would be advisable to develop methods for automatic quality evaluation of HPPC impulses, based on the criteria given in Section 3.3.3, which would enable full automation of the HPPC test results processing.
- A method should be developed to detect the occurrence of distortion of HPPC pulses in cases where the distortion is small and does not significantly change the shape of the voltage waveform yet, but already overestimates the obtained values of time constants.
- Simulation model accuracy may be improved by better OCV characteristic approximation.

Author Contributions: Conceptualization, T.B. and R.N.; methodology, T.B. and R.N.; software, T.B.; validation, T.B., R.N., W.S. and W.K.; formal analysis, T.B.; investigation, T.B., R.N. and W.K.; resources, T.B., R.N., W.S. and W.K.; data curation, T.B.; writing—original draft preparation, T.B.; writing—review and editing, T.B., R.N., W.S. and W.K.; visualization, T.B.; supervision, W.S. and W.K.; project administration, W.K.; funding acquisition, W.S. and W.K. All authors have read and agreed to the published version of the manuscript.

Funding: This project was co-financed by the European Regional Development Fund in accordance with the contract POIR.01.01.01-00-1427/20-00.

Data Availability Statement: Not applicable.

Acknowledgments: The authors would like to acknowledge the help of Katarzyna Lota and Kamil Frączek (Łukasiewicz Research Network—Institute of Non-Ferrous Metals) in supplying cells for the research.

Conflicts of Interest: The authors declare no conflict of interest.

References

1. Diampovesa, S.; Hubert, A.; Yvars, P.A. Designing physical systems through a model-based synthesis approach. Example of a Li-ion battery for electrical vehicles. *Comput. Ind.* **2021**, *129*, 103440. [CrossRef]
2. Skarka, W. Model-Based Design and Optimization of Electric Vehicles. In Proceedings of the 25th ISPE International Conference on Transdisciplinary Engineering, Modena, Italy, 3–6 July 2018; Volume 7, pp. 566–575.
3. Niestrój, R.; Rogala, T.; Skarka, W. An Energy Consumption Model for Designing an AGV Energy Storage System with a PEMFC Stack. *Energies* **2020**, *13*, 3435. [CrossRef]
4. Mateja, K.; Skarka, W.; Peciak, M.; Niestrój, R.; Gude, M. Energy Autonomy Simulation Model of Solar Powered UAV. *Energies* **2023**, *16*, 479. [CrossRef]
5. Peciak, M.; Skarka, W.; Mateja, K.; Gude, M. Impact Analysis of Solar Cells on Vertical Take-Off and Landing (VTOL) Fixed-Wing UAV. *Aerospace* **2023**, *10*, 247. [CrossRef]
6. Giannelos, S.; Borozan, S.; Aunedi, M.; Zhang, X.; Ameli, H.; Pudjianto, D.; Konstantelos, I.; Strbac, G. Modelling Smart Grid Technologies in Optimisation Problems for Electricity Grids. *Energies* **2023**, *16*, 5088. [CrossRef]
7. Giannelos, S.; Djapic, P.; Pudjianto, D.; Strbac, G. Quantification of the Energy Storage Contribution to Security of Supply through the F-Factor Methodology. *Energies* **2020**, *13*, 826. [CrossRef]
8. Raventós, O.; Bartels, J. Evaluation of Temporal Complexity Reduction Techniques Applied to Storage Expansion Planning in Power System Models. *Energies* **2020**, *13*, 988. [CrossRef]
9. Tang, Z.; Song, A.; Wang, S.; Cheng, J.; Tao, C. Numerical Analysis of Heat Transfer Mechanism of Thermal Runaway Propagation for Cylindrical Lithium-ion Cells in Battery Module. *Energies* **2020**, *13*, 1010. [CrossRef]
10. Li, N.; Zhang, H.; Zhang, X.; Ma, X.; Guo, S. How to Select the Optimal Electrochemical Energy Storage Planning Program? A Hybrid MCDM Method. *Energies* **2020**, *13*, 931. [CrossRef]
11. Davis, K.; Hayes, J.G. Comparison of Lithium-Ion Battery Pack Models Based on Test Data from Idaho and Argonne National Laboratories. In Proceedings of the IEEE Energy Conversion Congress and Exposition (ECCE), Detroit, MI, USA, 11–15 October 2020; pp. 5626–5632. [CrossRef]
12. Rahmoun, A.; Biechl, H. Modelling of li-ion batteries using equivalent circuit diagrams. *Electr. Rev.* **2012**, *2*, 152–156.
13. Tremblay, O.; Dessaint, L.; Dekkiche, A. A Generic Battery Model for the Dynamic Simulation of Hybrid Electric Vehicles. In Proceedings of the IEEE Vehicle Power and Propulsion Conference, Arlington, TX, USA, 9–12 September 2007; pp. 274–279. [CrossRef]
14. Cipin, R.; Toman, M.; Prochazka, P.; Pazdera, I. Identification of Li-ion Battery Model Parameters. In Proceedings of the International Conference on Electrical Drives & Power Electronics (EDPE), The High Tatras, Slovakia, 24–26 September 2019; pp. 225–229. [CrossRef]
15. Chen, S.X.; Tseng, K.J.; Choi, S.S. Modeling of Lithium-Ion Battery for Energy Storage System Simulation. In Proceedings of the Asia-Pacific Power and Energy Engineering Conference, Wuhan, China, 28–30 March 2009; pp. 1–4. [CrossRef]

16. Huang, K.; Wang, Y.; Feng, J. Research on equivalent circuit Model of Lithium-ion battery for electric vehicles. In Proceedings of the 3rd World Conference on Mechanical Engineering and Intelligent Manufacturing (WCMEIM), Shanghai, China, 4–6 December 2020; pp. 492–496. [CrossRef]
17. He, H.; Xiong, R.; Fan, J. Evaluation of Lithium-Ion Battery Equivalent Circuit Models for State of Charge Estimation by an Experimental Approach. *Energies* **2011**, *4*, 582–598. [CrossRef]
18. Sibi Krishnan, K.; Pathiyil, P.; Sunitha, R. Generic Battery model covering self-discharge and internal resistance variation. In Proceedings of the IEEE 6th International Conference on Power Systems (ICPS), New Delhi, India, 4–6 March 2016; pp. 1–5. [CrossRef]
19. Wu, W.; Qin, L.; Wu, G. State of power estimation of power lithium-ion battery based on an equivalent circuit model. *J. Energy Storage* **2022**, *51*, 104538. [CrossRef]
20. Khattak, A.A.; Khan, A.N.; Safdar, M.; Basit, A.; Zaffar, N.A. A Hybrid Electric Circuit Battery Model Capturing Dynamic Battery Characteristics. In Proceedings of the IEEE Kansas Power and Energy Conference (KPEC), Manhattan, KS, USA, 13–14 July 2020; pp. 1–6. [CrossRef]
21. Mueller, K.; Schwiederik, E.; Tittel, D. Analysis of parameter identification methods for electrical Li-Ion battery modelling. In Proceedings of the World Electric Vehicle Symposium and Exhibition (EVS27), Barcelona, Spain, 17–20 November 2013; pp. 1–9. [CrossRef]
22. Meng, J.; Yue, M.; Diallo, D. Nonlinear extension of battery constrained predictive charging control with transmission of Jacobian matrix. *Int. J. Electr. Power Energy Syst.* **2023**, *146*, 108762. [CrossRef]
23. Komal, S.; Kamyar, M.; Zunaib, A. Online reduced complexity parameter estimation technique for equivalent circuit model of lithium-ion battery. *Electr. Power Syst. Res.* **2020**, *185*, 106356. [CrossRef]
24. Maletić, F.; Deur, J. Analysis of ECM-based Li-Ion Battery State and Parameter Estimation Accuracy in the Presence of OCV and Polarization Dynamics Modeling Errors. In Proceedings of the IEEE 29th International Symposium on Industrial Electronics (ISIE), Delft, The Netherlands, 17–19 June 2020; pp. 1318–1324. [CrossRef]
25. Simin, P.; Gang, S.; Yunfeng, C.; Xu, C. Control of different-rating battery energy storage system interface to a microgrid. *Przegląd Elektrotechniczny (Electr. Rev.)* **2011**, *87*, 256–262.
26. Meng., J.; Boukhnifer, M.; Diallo, D. Lihtium-ion battery monitoring and observability analysis with extended equivalent circuit model. In Proceedings of the IEEE Mediterranean Conference on Control and Automation (MED), Saint-Raphaël, France, 15–18 September 2020; pp. 764–769.
27. Westerhoff, U.; Kurbach, K.; Lienesch, F.; Kurrat, M. Analysis of lithium-ion battery models based on electrochemical impedance spectroscopy. *Energy Technol.* **2016**, *4*, 1620–1630. [CrossRef]
28. Stroe, D.I.; Swierczynski, M.; Stroe, A.I.; Knudsen Kær, S. Generalized Characterization Methodology for Performance Modelling of Lithium-Ion Batteries. *Batteries* **2016**, *2*, 37. [CrossRef]
29. Cordoba-Arenas, A.; Onori, S.; Rizzoni, G.A. Control-oriented lithium-ion battery pack model for plug-in hybrid electric vehicle cycle-life studies and systems design with consideration of health management. *J. Power Sources* **2015**, *279*, 791–808. [CrossRef]
30. Liaw, B.Y.; Nagasubramanian, G.; Jungst, R.G.; Doughty, D.H. Modeling of lithium ion cells—A simple equivalent-circuit model approach. *Solid State Ion.* **2004**, *175*, 835–839. [CrossRef]
31. Huria, T.; Ceraolo, M.; Gazzarri, J.; Jackey, R. High fidelity electrical model with thermal dependence for characterization and simulation of high power lithium battery cells. In Proceedings of the IEEE International Electric Vehicle Conference, Greenville, SC, USA, 4–8 March 2012; pp. 1–8. [CrossRef]
32. Sockeel, N.; Shahverdi, M.; Mazzola, M.; Meadows, W. High-Fidelity Battery Model for Model Predictive Control Implemented into a Plug-In Hybrid Electric Vehicle. *Batteries* **2017**, *3*, 13. [CrossRef]
33. Li, K.; Soong, B.H.; Tseng, K.J. A high-fidelity hybrid lithium-ion battery model for SOE and runtime prediction. In Proceedings of the IEEE Applied Power Electronics Conference and Exposition (APEC), Tampa, FL, USA, 26–30 March 2017; pp. 2374–2381. [CrossRef]
34. Hemi, H.; M'Sirdi, N.K.; Naamane, A.; Ikken, B. Open Circuit Voltage of a Lithium ion Battery Model Adjusted by Data Fitting. In Proceedings of the 6th International Renewable and Sustainable Energy Conference (IRSEC), Rabat, Morocco, 5–8 December 2018; pp. 1–5.
35. Zhang, Q.; Shang, Y.; Li, Y.; Cui, N.; Duan, B.; Zhang, C. A novel fractional variable-order equivalent circuit model and parameter identification of electric vehicle Li-ion batteries. *ISA Trans.* **2020**, *97*, 448–457. [CrossRef]
36. Baczyńska, A.; Niewiadomski, W.; Gonçalves, A.; Almeida, P.; Luís, R. Li-NMC Batteries Model Evaluation with Experimental Data for Electric Vehicle Application. *Batteries* **2018**, *4*, 11. [CrossRef]
37. Somakettarin, N.; Funaki, T. Study on Factors for Accurate Open Circuit Voltage Characterizations in Mn-Type Li-Ion Batteries. *Batteries* **2017**, *3*, 8. [CrossRef]
38. Gao, Y.; Ji, W.; Zhao, X. SOC Estimation of E-Cell Combining BP Neural Network and EKF Algorithm. *Processes* **2022**, *10*, 1721. [CrossRef]
39. Rothenberger, M.J.; Docimo, D.J.; Ghanaatpishe, M.; Fathy, H.K. Genetic optimization and experimental validation of a test cycle that maximizes parameter identifiability for a Li-ion equivalent-circuit battery model. *J. Energy Storage* **2015**, *4*, 156–166. [CrossRef]

40. Nemes, R.; Ciornei, S.; Ruba, M.; Hedesiu, H.; Martis, C. Modeling and simulation of first-order Li-Ion battery cell with experimental validation. In Proceedings of the 8th International Conference on Modern Power Systems (MPS), Cluj-Napoca, Romania, 21–23 May 2019; pp. 1–6. [CrossRef]
41. Nemes, R.O.; Ciornei, S.M.; Ruba, M.; Martis, C. Parameters identification using experimental measurements for equivalent circuit Lithium-Ion cell models. In Proceedings of the 11th International Symposium on Advanced Topics in Electrical Engineering (ATEE), Bucharest, Romania, 28–30 March 2019; pp. 1–6. [CrossRef]
42. Hua, X.; Zhang, C.; Offer, G. Finding a better fit for lithium ion batteries: A simple, novel, load dependent, modified equivalent circuit model and parameterization method. *J. Power Sources* **2021**, *484*, 229117. [CrossRef]
43. Li, Z.; Shi, X.; Shi, M.; Wei, C.; Di, F.; Sun, H. Investigation on the Impact of the HPPC Profile on the Battery ECM Parameters' Offline Identification. In Proceedings of the Asia Energy and Electrical Engineering Symposium (AEEES), Chengdu, China, 28–31 May 2020; pp. 753–757. [CrossRef]
44. Haghjoo, Y.; Khaburi, D.A. Modeling, simulation, and parameters identification of a lithium-ion battery used in electric vehicles. In Proceedings of the 9th Iranian Conference on Renewable Energy & Distributed Generation (ICREDG), Mashhad, Iran, 23–24 February 2022; pp. 1–7. [CrossRef]
45. Tran, M.K.; Mathew, M.; Janhunen, S.; Panchal, S.; Raahemifar, K.; Fraser, R.; Fowler, M. A comprehensive equivalent circuit model for lithium-ion batteries, incorporating the effects of state of health, state of charge, and temperature on model parameters. *J. Energy Storage* **2021**, *43*, 103252. [CrossRef]
46. Deng, S.D.; Liu, S.Y.; Wang, L.; Xia, L.L.; Chen, L. An improved second-order electrical equivalent modeling method for the online high power Li-ion battery state of charge estimation. In Proceedings of the IEEE 12th Energy Conversion Congress & Exposition—Asia (ECCE-Asia), Singapore, 24–27 May 2021; pp. 1725–1729. [CrossRef]
47. Parthasarathy, C.; Laaksonen, H.; Halagi, P. Characterisation and Modelling Lithium Titanate Oxide Battery Cell by Equivalent Circuit Modelling Technique. In Proceedings of the IEEE PES Innovative Smart Grid Technologies—Asia (ISGT Asia), Brisbane, Australia, 5–8 December 2021; pp. 1–5. [CrossRef]
48. Navas, S.J.; Cabello González, G.M.; Pino, F.J.; Guerra, J.J. Modelling Li-ion batteries using equivalent circuits for renewable energy applications. *Energy Rep.* **2023**, *9*, 4456–4465. [CrossRef]
49. Wang, J.; Jia, Y.; Yang, N.; Lu, Y.; Shi, M.; Ren, X.; Lu, D. Precise equivalent circuit model for Li-ion battery by experimental improvement and parameter optimization. *J. Energy Storage* **2022**, *52*, 104980. [CrossRef]
50. Sörés, M.A.; Hartmann, B. Overview of possible methods of determining self-discharge. In Proceedings of the IEEE International Conference on Environment and Electrical Engineering and IEEE Industrial and Commercial Power Systems Europe (EEEIC/I&CPS Europe), Madrid, Spain, 9–12 June 2020; pp. 1–5. [CrossRef]
51. Tang, A.; Gong, P.; Li, J.; Zhang, K.; Zhou, Y.; Zhang, Z. A State-of-Charge Estimation Method Based on Multi-Algorithm Fusion. *World Electr. Veh. J.* **2022**, *13*, 70. [CrossRef]
52. Jarrraya, I.; Degaa, L.; Rizoug, N.; Chabchoub, M.H.; Trabelsi, H. Comparison study between hybrid Nelder-Mead particle swarm optimization and open circuit voltage—Recursive least square for the battery parameters estimation. *J. Energy Storage* **2022**, *50*, 104424. [CrossRef]
53. Castanho, D.; Guerreiro, M.; Silva, L.; Eckert, J.; Antonini Alves, T.; Tadano, Y.d.S.; Stevan, S.L., Jr.; Siqueira, H.V.; Corrêa, F.C. Method for SoC Estimation in Lithium-Ion Batteries Based on Multiple Linear Regression and Particle Swarm Optimization. *Energies* **2022**, *15*, 6881. [CrossRef]
54. Pizarro-Carmona, V.; Castano-Solís, S.; Cortés-Carmona, M.; Fraile-Ardanuy, J.; Jimenez-Bermejo, G. GA-based approach to optimize an equivalent electric circuit model of a Li-ion battery-pack. *Expert Syst. Appl.* **2021**, *172*, 114647. [CrossRef]
55. Huang, Y.; Li, Y.; Jiang, L.; Qiao, X.; Cao, Y.; Yu, J. Research on Fitting Strategy in HPPC Test for Li-ion battery. In Proceedings of the IEEE Sustainable Power and Energy Conference (iSPEC), Beijing, China, 21–23 November 2019; pp. 1776–1780. [CrossRef]
56. Wang, C.; Xu, M.; Zhang, Q.; Feng, J.; Jiang, R.; Wei, Y.; Liu, Y. Parameters identification of Thevenin model for lithium-ion batteries using self-adaptive Particle Swarm Optimization Differential Evolution algorithm to estimate state of charge. *J. Energy Storage* **2021**, *44*, 103244. [CrossRef]
57. Hamida, M.A.; El-Sehiemy, R.A.; Ginidi, A.R.; Elattar, E.; Shaheen, A.M. Parameter identification and state of charge estimation of Li-Ion batteries used in electric vehicles using artificial hummingbird optimizer. *J. Energy Storage* **2022**, *51*, 104535. [CrossRef]
58. Szewczyk, P.; Łebkowski, A. Comparative Studies on Batteries for the Electrochemical Energy Storage in the Delivery Vehicle. *Energies* **2022**, *15*, 9613. [CrossRef]
59. Łebkowski, Ł. Temperature, Overcharge and Short-Circuit Studies of Batteries used in Electric Vehicles. *Prz. Elektrotechniczny* **2017**, *93*, 67–73. [CrossRef]
60. Ohneseit, S.; Finster, P.; Floras, C.; Lubenau, N.; Uhlmann, N.; Seifert, H.J.; Ziebert, C. Thermal and Mechanical Safety Assessment of Type 21700 Lithium-Ion Batteries with NMC, NCA and LFP Cathodes–Investigation of Cell Abuse by Means of Accelerating Rate Calorimetry (ARC). *Batteries* **2023**, *9*, 237. [CrossRef]
61. Kiemel, S.; Glöser-Chahoud, S.; Waltersmann, L.; Schutzbach, M.; Sauer, A.; Miehe, R. Assessing the Application-Specific Substitutability of Lithium-Ion Battery Cathode Chemistries Based on Material Criticality, Performance, and Price. *Resources* **2021**, *10*, 87. [CrossRef]
62. Forte, F.; Pietrantonio, M.; Pucciarmati, S.; Puzone, M.; Fontana, D. Lithium Iron Phosphate Batteries Recycling: An Assessment of Current Status. *Crit. Rev. Environ. Sci. Technol.* **2021**, *51*, 2232–2259.

63. Białoń, T.; Niestrój, R.; Korski, W. PSO-Based Identification of the Li-Ion Battery Cell Parameters. *Energies* **2023**, *16*, 3995. [CrossRef]
64. Belt, J.R. *Battery Test Manual for Plug-In Hybrid Electric Vehicles*, 2nd ed.; U.S. Department of Energy Vehicle Technologies Program: Idaho Falls, ID, USA, 2010. [CrossRef]
65. Yang, Z.; Wang, X. An improved parameter identification method considering multi-timescale characteristics of lithium-ion batteries. *J. Energy Storage* **2023**, *59*, 106462. [CrossRef]
66. Karimi, D.; Behi, H.; Van Mierlo, J.; Berecibar, M. Equivalent Circuit Model for High-Power Lithium-Ion Batteries under High Current Rates, Wide Temperature Range, and Various State of Charges. *Batteries* **2023**, *9*, 101. [CrossRef]
67. Guenther, C.; Barillas, J.K.; Stumpp, S.; Danzer, M.A. A dynamic battery model for simulation of battery-to-grid applications. In Proceedings of the 3rd IEEE PES Innovative Smart Grid Technologies Europe (ISGT Europe), Berlin, Germany, 14–17 October 2012; pp. 1–7. [CrossRef]
68. Shi, J.; Guo, H.; Chen, D. Parameter identification method for lithium-ion batteries based on recursive least square with sliding window difference forgetting factor. *J. Energy Storage* **2021**, *44*, 103485. [CrossRef]
69. Tran, M.-K.; DaCosta, A.; Mevawalla, A.; Panchal, S.; Fowler, M. Comparative Study of Equivalent Circuit Models Performance in Four Common Lithium-Ion Batteries: LFP, NMC, LMO, NCA. *Batteries* **2021**, *7*, 51. [CrossRef]
70. Feng, D.; Huang, J.; Jin, P.; Chen, H.; Wang, A.; Zheng, M. Parameter Identification and Dynamic Simulation of Lithium-Ion Power Battery Based on DP Model. In Proceedings of the 14th IEEE Conference on Industrial Electronics and Applications (ICIEA), Xi'an, China, 19–21 June 2019; pp. 1275–1279. [CrossRef]
71. Einhorn, M.; Conte, V.F.; Kral, C.; Fleig, J.; Permann, R. Parameterization of an electrical battery model for dynamic system simulation in electric vehicles. In Proceedings of the IEEE Vehicle Power and Propulsion Conference, Lille, France, 1–3 September 2010; pp. 1–7. [CrossRef]
72. Yu, Q.; Wan, C.; Li, J.; E, L.; Zhang, X.; Huang, Y.; Liu, T. An Open Circuit Voltage Model Fusion Method for State of Charge Estimation of Lithium-Ion Batteries. *Energies* **2021**, *14*, 1797. [CrossRef]
73. Gao, L.; Liu, S.; Dougal, R.A. Dynamic lithium-ion battery model for system simulation. *IEEE Trans. Compon. Packag. Technol.* **2002**, *25*, 495–505. [CrossRef]
74. Wen, F.; Duan, B.; Zhang, C.; Zhu, R.; Shang, Y.; Zhang, J. High-Accuracy Parameter Identification Method for Equivalent-Circuit Models of Lithium-Ion Batteries Based on the Stochastic Theory Response Reconstruction. *Electronics* **2019**, *8*, 834. [CrossRef]
75. Baccouche, I.; Jemmali, S.; Manai, B.; Omar, N.; Amara, N.E.B. Improved OCV Model of a Li-Ion NMC Battery for Online SOC Estimation Using the Extended Kalman Filter. *Energies* **2017**, *10*, 764. [CrossRef]
76. Pillai, P.; Sundaresan, S.; Kumar, P.; Pattipati, K.R.; Balasingam, B. Open-Circuit Voltage Models for Battery Management Systems: A Review. *Energies* **2022**, *15*, 6803. [CrossRef]
77. Shaheen, A.M.; Hamida, M.A.; El-Sehiemy, R.A.; Elattar, E.E. Optimal parameter identification of linear and non-linear models for Li-Ion Battery Cells. *Energy Rep.* **2021**, *7*, 7170–7185. [CrossRef]
78. Plett, G.L. High-performance battery-pack power estimation using a dynamic cell model. *IEEE Trans. Veh. Technol.* **2004**, *53*, 1586–1593. [CrossRef]
79. Marušić, D.; Vašak, M. Efficient Method of Identifying a Li-Ion Battery Model for an Electric Vehicle. In Proceedings of the IEEE 20th International Power Electronics and Motion Control Conference (PEMC), Brasov, Romania, 25–28 September 2022; pp. 421–426. [CrossRef]

energies

MDPI

Article

A Model-Aware Comprehensive Tool for Battery Energy Storage System Sizing

Matteo Spiller [1], Giuliano Rancilio [1], Filippo Bovera [1], Giacomo Gorni [2], Stefano Mandelli [3], Federico Bresciani [2] and Marco Merlo [1,*]

1 Politecnico di Milano—Department of Energy, Via Lambruschini 4a, 20156 Milano, Italy; matteo.spiller@polimi.it (M.S.); giuliano.rancilio@polimi.it (G.R.); filippo.bovera@polimi.it (F.B.)
2 Eni S.p.A., Renewable, New Energies and Material Science Research Center, Via Fauser 4, 28100 Novara, Italy; giacomo.gorni@eni.com (G.G.); federico.bresciani@eni.com (F.B.)
3 Plenitude, Via Giuseppe Ripamonti 85, 20141 Milano, Italy; stefano.mandelli@plenitude.com
* Correspondence: marco.merlo@polimi.it

Abstract: This paper presents a parametric procedure to size a hybrid system consisting of renewable generation (wind turbines and photovoltaic panels) and Battery Energy Storage Systems (BESS). To cope with the increasing installation of grid-scale BESS, an innovative, fast and flexible procedure for evaluating an efficient size for this asset has been developed. The tool exploits a high-fidelity empirical model to assess stand-alone BESS or hybrid power plants under different service stacking configurations. The economic performance has been evaluated considering the revenue stacking that occurs when participating in up to four distinct energy markets and the degradation of the BESS performances due to both cycle- and calendar-aging. The parametric nature of the tool enables the investigation of a wide range of system parameters, including novel BESS control logic, market prices, and energy production. The presented outcomes detail the techno-economic performances of a hybrid system over a 20-year scenario, proposing a sensitivity analysis of both technical and economic parameters. The case study results highlight the necessity of steering BESS investment towards the coupling of RES and accurate planning of the service stacking. Indeed, the implementation of a storage system in an energy district improves the internal rate of return of the project by up to 10% in the best-case scenario. Moreover, accurate service stacking has shown a boost in revenues by up to 44% with the same degradation.

Keywords: battery energy storage system; renewables; market service stacking

Citation: Spiller, M.; Rancilio, G.; Bovera, F.; Gorni, G.; Mandelli, S.; Bresciani, F.; Merlo, M. A Model-Aware Comprehensive Tool for Battery Energy Storage System Sizing. *Energies* **2023**, *16*, 6546. https://doi.org/10.3390/en16186546

Academic Editor: Quanqing Yu

Received: 3 July 2023
Revised: 16 August 2023
Accepted: 25 August 2023
Published: 12 September 2023

1. Introduction

The energy sector is responsible for a large share of anthropogenic carbon emissions that lead to climate change and global warming. The 2015 United Nations Climate Change Conference in Paris set the milestone of limiting the average temperature increase to below 1.5 °C [1]. A total of 160 countries around the world have agreed to combat global heating through the installation of renewable energy sources (RES) for a more sustainable energy scenario. Consequently, the overall energy production by RES has increased from 20 to 28% since 2010, and the target is for production to reach at least 43% by 2030 [2]. Focusing on the European Union (EU), the EU Green Deal and the "Fit-for-55" package set the goal is of reducing greenhouse gas emissions by 55% with respect to 1990 by 2030 [3].

RES have a low carbon footprint and are, therefore, among the main candidates for energy sources that could be used to reach these decarbonization targets. In any case, their integration into the energy system is not straightforward. These sources are characterized by the high intermittency and non-programmability of the energy output, which complicates the balance between supply and demand in the power system [4]. Grid-connected battery energy storage systems (BESS) represent a viable resource to cope with those issues and guarantee the balance, stability, and adequacy of a decarbonizing power system. Compared to other storage systems, BESS are close to market maturity, with prices dropping by 87% from 2010 to 2019 [5]. The installed grid-scale battery storage capacity will expand 44-fold between 2021 and 2030 to 680 GW according to IEA [6]. In addition, lithium-ion batteries outperform other storage technologies in terms of energy density, power density, and round-trip efficiency. Moreover, their operational reliability can last up to 20 years with suitable management strategies, such as temperature regulation and capacity augmentation [7]. The success of this technology is attributed to the flexibility and scalability of these assets, coupled with their ability to behave as bulk energy systems. These characteristics enable the provision of a wide set of services to system operators and active users. Such services could reduce RES uncertainty and make BESS attractive investments.

Despite the possible enhancement that BESS could provide to electric power system operation, this technology still has shortcomings. Although battery prices are decreasing yearly, the overall cost per kWh is still high, representing the most common challenge. To cope with high capital costs, it is crucial to capture multiple cash flows to increase the financial viability of the project. The dynamic stacking of BESS services ensures higher profits for the asset [8]. Therefore, it is of paramount importance to evaluate an efficient system operation that creates multiple streams of revenues to increase the economic benefits. To properly assess and optimize the cash flow, the owner of the system must optimally size the BESS, accounting for several economic aspects such as cell technology, installation cost, and maintenance cost. These expenditures are mainly affected by lifetime, battery capacity, and overall performance, which interact in a non-linear manner. The proper evaluation of these aspects allows for more accurate modeling and, consequently, a more precise economic analysis.

High-fidelity BESS modelization is mandatory to ensure accurate economic evaluation. This paper proposes a model-aware BESS-sizing procedure that accurately represents the performance of BESS in different energy markets during their lifetime, accounting for the main non-linearities. In general, the stacking of the services is mostly addressed by constant BESS models that do not consider the non-linearities of this technology and the presence of auxiliaries. Furthermore, service stacking with grid-scale storage is mainly investigated in a stand-alone configuration, without evaluating the possible services that the system can provide to an RES power plant. Lastly, services such as the capacity market have not been assessed to date. Based on the state-of-art described in the next section, the novelties of this work are as follows:

- A sizing procedure is developed that investigates a 20-year BESS investment with a high-fidelity empirical model developed in [9] and updated with equations capable of emulating the capacity degradation of the system. The latter structure has been exploited to create a fast and flexible tool that is able to evaluate the most cost-effective storage investment, ensuring an efficient trade-off between computational effort and accuracy.
- Innovative algorithms are developed that are capable of stacking multiple services with a sequential approach. The implemented solutions produce results seamlessly, with two distinct configurations: stand-alone and hybrid-renewable power plants.

The work is structured as follows. Section 2 reviews the modelization and the algorithms exploited for sizing BESS in the literature. Section 3 describes the proposed empirical model, the methodology of the sizing procedure, and the novel algorithm proposed for stacking the energy markets. Section 4 introduces the study cases. Section 5 discusses the main results. Lastly, Section 6 summarizes the activities and lists future works.

2. Literature Review

The BESS sizing procedure consists of identifying the most cost-effective configuration for the stakeholders. The application is complex and non-linear. This section aims to describe two different fundamental aspects of the procedure: the modeling, and solution methods [10].

2.1. Modeling

Modeling focuses on the mathematical representation of the key components of BESS. A wide number of approaches have been developed, with different levels of complexity and computational effort. According to the degree of physical insight, battery models can be divided into three different levels: electrochemical model, circuit-oriented (or electrical) model, and black box models (empirical or stochastic) [11]. The selection of a model is bounded to the application area. Namely, a specific model ensures a different degree of accuracy and computational effort, and depending on the details required by the implementation, an efficient balance between the two characteristics improves the quality of the analysis. For instance, electrochemical models are the most accurate approach to battery representation. They describe the chemical reactions that take place in the electrodes and the electrolytes using a set of non-linear differential equations [12]. The expressions detail the effect of the electrochemical reactions, such as the diffusion, migration, kinetic phenomena, and lithium concentration, to truly represent the state variables of the system [13]. A description of the batteries at a microscopic scale accurately represents the key behaviors of the nonlinear system [14]. Although there is no doubt regarding the accuracy of electrochemical models, they are used in low-speed applications, such as online capacity estimation [15,16], predictive maintenance [17] or the validation of complementary models [18]. Circuit-oriented models are electrical equivalent models that are able to represent the state variables of the batteries [11]. Electrical models consist of electrical circuits made by capacitors and resistances, whose proper connection emulates the behavior of the batteries. The basic electrical model, known as Rint, has a big capacitor that can describe the open-circuit voltage of the cell, and a series resistance that can simulate the battery's internal resistance [19]. Moreover, RC circuits can be connected in series to the Rint model to increase the order of the system and emulate relaxation and polarization effects [20]. Despite the empirical nature of those elements, these modelizations are widely adopted thanks to their computational efficiency in capturing the dynamic response of the system [21]. The ability to estimate the state of batteries in real-time makes these models suitable for applications such as EV state estimation [20], and grid stability [22,23]. Furthermore, electrical models ensure a good performance when evaluating the terminal voltage and SOC of the batteries. Articulated modelization, such as a three-order model, allows for these state variables to be described with errors lower than 1% [21,24]. However, if the accuracy of the model increases, the same occurs regarding the computational effort [25]. In sizing applications, computational efficiency is a priority since it requires the evaluation of steady-state performances over a long time-window (e.g., 15–20 years). Therefore, the complexity of the electrical model could not produce results that are valuable for the procedure in a reasonable time. Empirical models ensure the optimal computational effort for this kind of analysis. The modelization is characterized by mathematically constructed models that utilize the observed data and measurements to represent the behavior and performance of BESS systems. These models are developed by an analysis of real-world operational data and the characteristics gathered from datasheets or experiments, allowing for them to capture the key relationships and patterns between various parameters [26,27]. By leveraging statistical techniques, regression analyses, or other mathematical approaches, empirical models provide a parametric representation of how different factors, such as battery lifetime, efficiency, and capacity, interact and impact the overall performance of BESS systems [9].

A fair share of BESS sizing procedures are based on scalar linear empirical systems that resemble the BESS performances. In [28], a constant battery and inverter efficiency model has been exploited to size a PV residential system by evaluating the economic criteria of the annuity method. Study [29] investigates the optimal sizing of BESS through a life-cycle cost model. This model inspects the different phases of the system, also accounting for the final decommissioning, maintenance and recycling, and disposal as a reduction in the economics of the system. Ref. [30] proposes a constant-efficiency empirical model to size an energy storage system, accounting for hosting capacity and reductions in wind curtailment. Although these simplified empirical models represent the easiest approach to configuring and producing results for BESS, they are generally the least accurate [31]. Nevertheless, high-fidelity empirical models that represent varying efficiencies and power capabilities depending on SoC and C-rate improve the accuracy of the analysis [32]. These approaches rely on experimental campaigns to create a computationally efficient BESS model that contains an error. In [33], a regression technique has been exploited to build a non-linear BESS with an average SOC root mean square error of 3%. In [34], a detailed non-linear power losses model has been implemented to more precisely capture the low-efficiency working region of the storage system. In [9], a grid-scale BESS has been modeled with lookup tables to represent the non-linear efficiency and the auxiliary consumption of the system, ensuring an average SOC error of 0.168%. Despite the high fidelity of the modelization, the latter applications are limited to the operation perspective, and no one has used the accuracy of the model to size the BESS.

2.2. Solution Method

The solution method consists of the approach used to investigate the optimal size of the BESS, i.e., the mathematical procedure adopted to set the schedule and the dispatching of different services. The goal of the algorithm selected in the sizing procedure is to identify the best power setpoint to efficiently maximize the benefits generated by the BESS. The algorithm needs to emulate the provision of diverse and multiple services by the BESS. Namely, at each time step, given the actual SOC, the storage system is charged or discharged according to a logic that improves the welfare of the BESS owner. Most existing studies are based on mathematical programming due to their ability to identify a global optimum for the objective function under analysis. Stochastic Dynamic Programming (DP), and stochastic Mixed-Integer Linear Programming (MILP) or techniques derived from these two approaches are the most-adopted solutions to the sizing problem [35]. For instance, studies [36,37] propose a stochastic MILP and DP, respectively, to optimize the sizing of a grid-scale storage system with constant efficiency. However, these methodologies show important limitations in terms of the modeling and objectivity of the solution [38]. The optimal sizing problem is a non-convex and non-linear combinatorial optimization problem [39]. Therefore, MILP-sizing procedures need to rely on techniques such as relaxation, piecewise linear approximation, or the implementation of a constant efficiency to generate a problem that can be treated by the standard solvers [32,40,41]. However, DP is afflicted by dimensionality problems that need to be solved with the so-called approximated dynamic programming [42,43]. Although these techniques solve the main issues related to computational effort, their implementation comes at the expense of accuracy. Moreover, sizing problems are computationally intensive, and many works consider a timeframe that does not evaluate the whole investment lifespan. Lastly, to properly manage technical and economic targets, multi-objective functions are typically required [44]. These lead to a lack of objectivity in the solutions, which needs to be properly evaluated with Pareto analysis [45]. It follows that mathematical programming cannot provide diversified system operation. However, without the support provided by these techniques, it is necessary to develop algorithms that are capable of efficiently scheduling the assets. Furthermore, considering the actual cost of BESS, the profits from multiple applications are fundamental to generating a positive investment. Therefore, algorithms capable of effectively stacking the services are necessary to improve the economics of a storage system. Implementing those algorithms is not straightforward, since BESS are limited in power and energy. The desired output of the service stacking problem is a strategy for optimal capacity allocation during a given period, accounting for market prices and system dynamics. Three distinct types of service stacking are defined in the literature: sequential, parallel, and dynamic [46]. In parallel stacking, a constant allocation of storage capacity is given to the services whilst the sequential multi-use provides these services in turn. The dynamic multi-use aims to increase profit as it combines the advantages of the two predecessors. It follows that service stacking in sizing applications is a complex topic and aspects such as the type of stacking or the services' optimum sizing represent important challenges for the stakeholder [47]. Furthermore, power plant configuration has a relevant role in the BESS-sizing procedure due to the different services that this technology can provide. For instance, a BESS coupled with an energy district could perform tasks such as load peak reductions and smooth power injections. Vice versa, in a stand-alone setup, the main activities are exclusively grid- and market-oriented. BESS-sizing approaches have been categorized into four main categories to deal with different configurations: microgrids, distributed renewable energy systems, standalone hybrid renewable energy systems, and renewable energy power plants [48]. Despite the wide interest in BESS-sizing in the literature, the proposed approaches typically focus on a single configuration. Conversely, industrial and utility-scale BESS stakeholders are interested in flexible tools that are able to evaluate heterogeneous configurations of power plants, storage systems, markets, and services.

Table 1 lists the studies that have addressed the stacking of services for a grid-scale BESS.

Table 1. List of studies that have inspected the stacking of services for a grid-scale BESS.

Reference	Services	Performance Assumptions	Configuration	Scope
[49]	Fast Frequency—Balancing market	Efficiency as a function of SOC and power	Stand-alone	Operation
[50]	Arbitrage—Frequency regulation	Constant efficiency	Stand-alone	Operation
[51]	Arbitrage—Frequency regulation	Constant efficiency	Stand-alone	Operation
[52]	Arbitrage—Frequency regulation-	Constant efficiency	Stand-alone	Sizing
[53]	Power shifting—aFRR	Constant efficiency	Wind farm coupling	Operation
[54]	Ancillary services market	Constant efficiency	Stand-alone	Operation
[55]	Arbitrage—distribution investment deferral—frequency regulation	Constant efficiency	Stand-alone	Operation
[34]	Arbitrage—Frequency regulation	Non-linear power losses depending on C-rate and SOC	Stand-alone	Operation
[56]	Frequency regulation—power shifting	Constant efficiency	Microgrid	Sizing
This work	Arbitrage—Frequency regulation—mFRR—capacity market	Efficiency as a function of SOC and power	Stand-alone and PV–wind coupled	Sizing

As stated in the Introduction, this work improves on the state-of-the-art by developing a high-fidelity model of BESS and novel stacking algorithms, considering both stand-alone and RES-coupled operations.

3. Methodology

This work proposes a novel flexible tool evaluating the size of a BESS and computing its cost-effectiveness while providing multiple services. The exploitation of the tool has been eased thanks to a graphic user interface (GUI) developed in $MATLAB^{TM}$. The structure allows for the easy evaluation of candidate groups of BESS and identifies the configuration that ensures the best economic return. The GUI compares different specific BESS sizes to accurately assess the trade-off between the size and the cost. The procedure is designed to be compatible with the generic EU market (e.g., it implements standard balancing products, and considers the day-ahead market according to the EU framework), meaning that it can be easily adapted to different requirements. However, the focus of the proposed algorithm in this paper is specifically related to the Italian market (e.g., the ancillary services market price scenarios and award rates, a well as the rules and prizes of the capacity remuneration mechanism, suit the Italian case).

This procedure may investigate both stand-alone and BESS hybrid-renewable energy power plants with wind and photovoltaic services connected to the national grid. Furthermore, the power managed at the point of delivery (POD) with the grid can be limited to a specified value to address study cases where there is an existing contractual connection power, and the curtailment of the overproduction is foreseen. The proxy model adopted in the procedure is a high-fidelity (cf. realistic) empirical battery model based on an experimental campaign at the Joint Research Centre (JRC) of Ispra (VA-Italy) on nickel–manganese cobalt (NMC) BESS [9]. The main features are the efficiency of the overall system, including transformer and power-conversion systems, and the auxiliary consumption. The nonlinear round-trip efficiency is expressed as a function of power and SOC through look-up tables. Instead, the auxiliaries' consumption relies on the power flown in the BESS and the ambient temperature. Although the model emulates the performances of NMC technology, its structures easily allow for the fitting of the data of different electrochemical technologies (e.g., lithium iron phosphate or even non-lithium-based batteries). Figure 1 depicts the block diagram of the high-fidelity empirical model adopted in this work using $SIMULINK^{TM}$. Each block has a specific purpose that ensures the emulation of a large-scale BESS. The overall efficiency block takes the AC power, and the SOC as input and converts them into DC power using the previously mentioned look-up table. After this, the DC input is processed by the capability curve that limits the C-rate depending on the actual SOC. Lastly, the actual C-rate is used to update the SOC of the system. Inside the latter block, the energy content of the system is evaluated at each time sample.

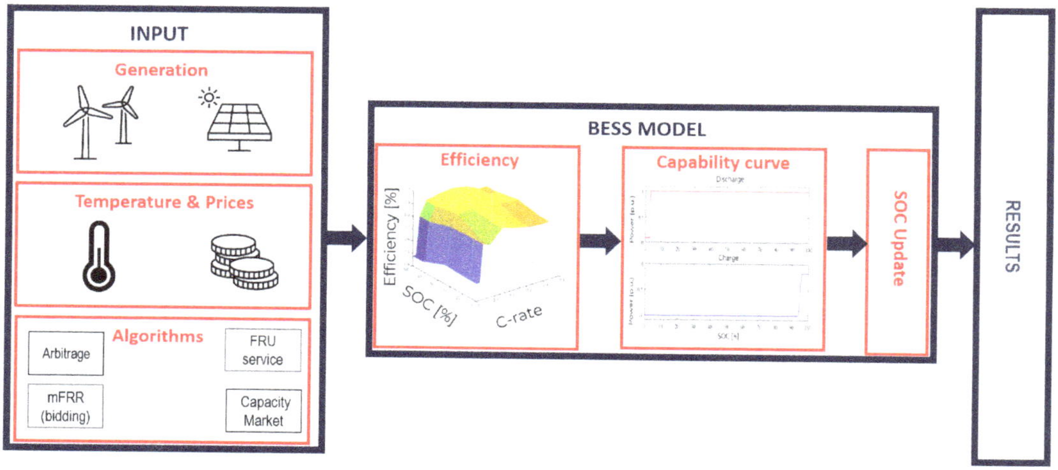

Figure 1. Block diagram of the sizing tool developed in this work.

To ensure a realistic representation of BESS performances, an aging model from the literature was inserted into this framework. It is well-known that capacity degradation is affected by various chemical reactions, which can typically be classified into two major phenomena: calendar and cycle aging [57]. To account for these processes, two equations that describe the cycle and the calendar aging of the system have been inserted into the SOC update block to reduce the nominal capacity throughout the simulation. Both expressions were obtained from the literature. The cycle aging is related to the use of the BESS and its C-rate. The complete expression implemented in the model was obtained in [58].

$$\text{Cycledegrad.}[\%] = -[\text{SOC}(t) - \text{SOC}(t-1)] \times 3.57 \times 10^{-5} \times e^{0.465 \times \text{C-rate}} \tag{1}$$

Equation (1) describes the system degradation due to cycling by evaluating the depth of discharge between each time sample $[\text{SOC}(t) - \text{SOC}(t-1)]$ and the specific C-rate at which the BESS is cycled. The proposed tool investigates the BESS operation with a 15-min granularity, over a very long time window of up to several years; consequently, the equation computes the percentage of capacity reduction generated in 15 min by a specific constant C-rate.

The second degradation mechanism has been modeled as a constant-capacity degradation in the square root of time function. This trend has been widely adopted to describe the chemical reaction inside the system that occurs due to calendar aging. The expression adopted in this work is based on [59], assuming an average SOC of 50% for the system.

$$\text{Calendardegrad.} = -99.43 \times 10^3 \times e^{-\frac{42577}{RT}} \times \sqrt{t} \tag{2}$$

Equation (2) reports the capacity calendar degradation depending on the age of the storage system, considering the gas constant R and the temperature T at which the system is kept by the auxiliaries. It has been assumed that the two effects superimpose to continuously update the SOH of the system. Therefore, at each time sample the BESS' nominal energy is reduced depending on the C-rate and the time that has elapsed.

The left side of Figure 1 reports the possible power plant setup evaluated by the tool. The inputs for the analysis are time series that can be ascribed into three different categories: energy, market, and ambient inputs. An additional block considering the optional presence of RES plants was also included to investigate hybrid configurations. Energy input returns the MW production of a given RES plant, if present. Market input contains fundamental information about the structure of each service being addressed. Lastly, temperature inputs are necessary for the auxiliary consumption of the BESS model. These parameters can be easily changed thanks to the GUI, which allows for the selection of the Excel file that is necessary for the analysis.

The evaluation of the nominal energy of the BESS depending on a given set of inputs is evaluated by the tool with iterations of different sizes through the definition of two distinct sets of parameters: nominal power and energy-to-power ratio (EPR). Moreover, if the storage system is coupled with RES, the tool can evaluate variable power plant sizes.

The cost-effectiveness of BESS relies on the profits that the asset can generate by participating in different services. Therefore, different algorithms were developed to emulate the participation of the BESS in various energy markets. Each algorithm elaborates the inputs and returns a power setpoint that simulates the dispatching of the BESS power flows or the possibility of charging using the assets in the energy district.

The algorithms proposed in this paper are analytical adaptive algorithms. Every day, the input data are fed to the algorithms that compute the operation of the BESS as a combination of charge and discharge signals. The simulation calculates the cash flow for twenty years for each size. The outcomes of each analysis are exploited to compute the performance indicators fundamentals to determine the optimal size of the system. The tool evaluates the internal rate of return (IRR) and the CAPEX that is covered to quantify the cost-effectiveness of each investment. The IRR was computed with the financial toolbox implemented in $MATLAB^{TM}$. Instead, the CAPEX covering 20 years was obtained following Equation (3), using the constant interest rate r.

$$\text{CAPEX}_{\text{covered}}[\%] = \frac{100}{\text{CAPEX}} \times \sum_{t=1}^{20} \frac{\text{Revenues}(t) - \text{OPEX}(t) - \text{Marketpenalties}}{(1+r)^t} \tag{3}$$

CAPEX and OPEX are properly addressed in the analysis. Equation (4) describes the capital cost as a product of the nominal energy and an exponential function of the duration. This expression was obtained through an interpolation of the values reported in [60].

$$\text{CAPEX}[€] = E^{nom}\left(220 \times \text{duration}^{-0.9795} + 287.1\right) \tag{4}$$

This condition is justified by the fact that equipment cost is strictly correlated with the power being handled and not only with the nominal energy. The hypothesis adopted for this work is that yearly OPEX corresponds to 2.5% of the CAPEX [61].

Lastly, due to the novelties of the application, a validation procedure based on previous works has not been added. Nevertheless, the high-fidelity empirical model represents an updated version of a previously developed structure. In [62], the adopted model has been compared with state-of-the-art models: the accuracy of performance representation increases when using a variable BESS efficiency and considering auxiliary system demand.

The algorithms implemented in the tools aim to emulate participation in different services. The next paragraphs briefly describe the logic adopted for their implementation. Four services typically present in EU markets have been modeled: energy arbitrage, capacity market participation, the provision of manual Frequency Restoration Reserve (mFRR), and the provision of fast frequency regulation within the Fast Reserve (FR) project. Each algorithm generates a signal that simulates the charge and discharge requirement for the BESS that participates in the specific service.

3.1. Energy Arbitrage

The term energy arbitrage refers to the possibility of storage systems exploiting day-ahead market spread to generate profits. To achieve this control, the algorithm identifies the minimum and maximum prices of the market each day to set up the proper control logic. The algorithm was limited to a cycle per day to cope with the uncertainty of the prices that are always present in a multi-year analysis. To grant an economic profit, the procedure has a feedback control that checks whether the cycling cost of the system cancels the profits per cycle. Namely, this cost can be described as the total lifetime cost of the investment in electricity storage, divided by the cumulative delivered electricity. This parameter is customizable from the GUI and impacts the arbitrage provision. Furthermore, in case of BESS coupled with RES, if the RES power plant produces a power higher than the POD limit (i.e., the contractual connection power), the algorithm directs the exceeding generation toward the BESS if it is not fully charged. The latter charging process is considered free of charge since that energy will be curtailed if not stored. To properly address this aspect, an SOC control is implemented inside the algorithm to check the boundary condition of the storage system. Lastly, it has been hypothesized that the BESS acts as a price-taker in the day-ahead market. Algorithm 1 describes the structure of the energy arbitrage algorithm in detail.

3.2. Capacity Market

The electricity capacity market is coupled with the energy market. The scope of this service is to ensure power plant owners have sufficient capacity to meet the system demands and reliability. Participation in the market is defined through a tender procedure. The winners receive the capacity payment if they correctly provide power during the scarcity hours defined by the authority. In Italy, the assets that won the tender are entitled to the capacity payment if they participate in the day-ahead market in 1000 mandatory hours. The capacity payment differs from each power plant and is based on the probability of derating a given technology. Namely, the Italian authority associated a parameter with each type of power plant, related to their reliability in producing the nominal power [63]. This value reduces the capacity payment depending on the characteristics of each technology. For instance, BESS derating is shown in Table 2 and is proportional to the duration of the storage system, which means that a larger duration ensures better reliability in the provision of the capacity. Thus, the remunerated power of a BESS is its qualified power times (1—derating).

Algorithm 1: Energy Arbitrage

Input: $P_i^{prod}, P^{nom}, E^{nom}, POD_{limit}, P_i^{DAM}, SOC_0^{initial}$
Output: P_{cha}, P_{dis}
<u>**CHARGE PHASE**</u>
for I in 96 (quarter of hours in a day) **do**
  if $P_i^{prod} > POD_{limit}$
    if $P_i^{prod} - POD_{limit} < P^{nom}$
      $P_i^{cha} = P_i^{prod} - POD_{limit}$
      Else
        $P_i^{cha} = P_{nom}$
      $E^{cha} = E^{cha} + \frac{P_i^{cha}}{4}$
If $sum(E_i^{cha}) < E^{nom}$
$j = find\left(min\left(P_i^{MGP}\right)\right)$
  for i in 96 (quarter of hours in a day) **do**
    if i == j
    if $P_i^{prod} < P^{nom}$
      $P_i^{cha} = P_i^{prod}$
    Else
      $P_i^{cha} = P^{nom}$
    $E^{cha} = E^{cha} + \frac{P_i^{cha}}{4}$
<u>**DISCHARGE PHASE**</u>
$E^{dis} = E^{cha}$
$j = find\left(max\left(P_i^{MGP}\right)\right)$
for i in 96 (quarter of hours in a day) **do**
  while $E^{dis} > 0$
  if i == j
    if $P_i^{nom} + P_i^{prod} > POD_{limit}$
      $P_i^{dis} = POD_{limit} - P_i^{prod}$
      Else
      $P_i^{dis} = P^{nom}$
  $E^{dis} = E^{dis} - \frac{P_i^{dis}}{4}$
<u>**SOC CONTROL**</u>
for i in 96 (quarter of hours in a day) **do**
  $SOC_i = SOC_{i-1} - \frac{P_i^{dis}}{4} + \frac{P_i^{cha}}{4}$
  If $SOC_i > 100$
    $SOC_i = 100$
    $P_i^{cha} = \frac{100 - SOC_{i-1}}{4 * E^{nom}}$
  Else if $SOC_i < 0$
    $SOC_i = 0$
    $P_i^{cha} = \frac{SOC_{i-1}}{4 * E^{nom}}$
<u>**PROFITS PER CYCLE CONTROL**</u>
for i in 96 (quarter of hours in a day) **do**
  If $P_i^{prod} > POD^{limit}$
    $P^{chargefree} = min\left(P^{nom}, P_i^{prod} - POD^{limit}\right)$
  $P^{chargepaid} = P_i^{charge} - P^{chargefree}$
  $profit(i) = \left(0.9 \times P_i^{discharge} - P^{chargepaid}\right) \times \frac{DAM_i}{4}$
If $sum\left(profit - Profits^{Cycle}/E^{nom}\right) \leq 0$
  $P_{1:96}^{discharge} = 0$
  $P_{1:96}^{charge} = 0$

Table 2. Storage derating depending on the duration of the capacity payment in the capacity market defined by the Italian authority.

Duration [h]	Derating [%]
1	76
2	66
4	33
6	19
8	10

Furthermore, the non-programmable RES, due to their reliance on natural phenomena, face difficulties in providing a constant power setpoint. Hereby, the Italian authority has defined a derating factor for the capacity payment that these power plants can receive. Table 3 details the derating factor for the capacity payment of wind and PV power plants.

Table 3. Power plant derating factor defined by the Italian authority for the capacity payment.

Technology	Derating [%]
PV	84
Wind	88

In this work, when the capacity algorithm is enabled, it is assumed that the hybrid power plant has won the tender for, and participates in, the capacity market. The capacity payment can be extended to account for a hybrid system with multiple energy sources following Equation (5), where the overall remunerated capacity CP is equal to the summation of the product between the peak power $P^{peak}_{powerplant}$ and the derating $Derating_{powerplant}$ of each technology multiplied by the capacity payment $Capacity^{MW}_{payment}$ specified by the user. Algorithm 2 describes the logic used to emulate the capacity market inside the model.

$$CP[€] = \sum_{powerplant=1}^{N} P^{peak}_{powerplant} \times \left(1 - Derating_{powerplant}\right) \times Capacity^{MW}_{payment} \quad (5)$$

Algorithm 2: Capacity Market Algorithm

Input: P^{prod}_i, P^{nom}, POD_{limit}, CDP, CMS^{input}_i, CMS^{charge}_i
Output: P_{cha}, P_{dis}
If sum(MDC^{input}_i) > 0
    CHARGE PHASE
    for i in 96 (quarter of hours in a day) **do**
        if $CMS^{charge}_i == 1$
            if $P^{prod}_i > 0$ **and** $P^{prod}_i < P_{nom}$
                $P^{cha}_i = P^{prod}_i$
            Else if $P^{prod}_i > 0$ **and** $P^{prod}_i > P_{nom}$
                $P^{cha}_i = P_{nom}$
    DISCHARGE PHASE
    for i in 96 (quarter of hours in a day) **do**
        if $CMS^{input}_i == 1$
            if $P^{prod}_i < CDP$
                $P^{dis}_i = CDP - P^{prod}_i$
Else
    Energy Arbitrage Algorithm (Algorithm 1)

The derating parameters adopted in this work are those proposed by the Italian authority and reported in Tables 2 and 3. Nevertheless, the tool allows for those values to be adapted to a different regulatory scenario.

Lastly, the inadequacy of an asset to properly participate in the capacity market is computed ex-post. This calculation is mandatory to verify if the system correctly provides the requested capacity and is entitled to the capacity payment. The criteria established by the Italian authority foresee a neglection of the payment when the power plant does not provide more than 80% of the capacity for at least three months.

3.3. Manual Frequency Restoration Reserve (mFRR)

mFRR is the manual activation of frequency reserve that has the purpose of restoring the power balance of the electric power system. In Italy, this service is traded in the ancillary services market with a pay-as-bid approach. In [64], it has been pointed out that distrusted energy sources such as hybrid RES plants have a very low share of acceptance in the market. Despite the small liquidity of the ancillary services market, the participation of the asset in these services could be profitable. Indeed, as an inherent structure of the process, the mFRR prices tend to be less competitive than those in the day-ahead market and consequently generate more profit. To properly model the opportunity generated by participation in the mFRR, an acceptance criterion has been developed in [65]. The procedure, based on historical results on mFRR, develops around two important pieces of information. Firstly, among all the bids that were submitted, only 13% were accepted. Secondly, a correlation between bid acceptance and the submission price is present. In light of these statements, the tool was programmed to create a binary input vector that expresses the acceptance of the bids in the mFRR. The acceptance array changes depending on the average price that the user foresees being submitted in the market. Lastly, the accepted bids are remunerated at a value that is equal to the constant price times a gain that is proportional to the hour at which the bid is accepted. This solution was implemented to grasp the hourly price fluctuations in the bids in the ancillary services market.

Algorithm 3 describes the logic behind the control of the system participating in the mFRR. In particular, the approach evaluates whether the mFRR bids are more profitable than those in the day-ahead market to ensure higher revenues.

Algorithm 3: mFRR Algorithm

Input: P_i^{prod}, P^{nom}, E^{nom}, POD_{limit}, P_i^{MGP}, P_i^{mFRR}
Output: P_{cha}, P_{dis}
DISCHARGE PHASE
for I in 96 (quarter of hours in a day) **do**
    if $sum(P_i^{mFRR}) > 0$
        if $j = find\left(max\left(P_i^{mFRR}\right)\right)$
            if $P_j^{MSD} > P_j^{MGP}$
                if $P_j^{prod} > POD_{limit}$
                    $P_j^{dis} = 0$
                Else if $POD_{limit} - P_j^{prod} > P^{nom}$
                    $P_j^{dis} = P^{nom}$
            Else
                $P_j^{dis} = POD_{limit} - P_j^{prod}$
        $E^{dis} = E^{dis} - \frac{P_j^{dis}}{4}$

3.4. Fast Frequency Reserve

Fast frequency reserve is a service enabled by the Italian authority with the resolution 200/2020/R/eel [66]. It consists of the rapid provision of power from BESS to counteract the more usual frequency swings in the electric power system. The service is exclusive to

BESS selected through a tender procedure. In detail, the pilot project launched in 2021 in Italy mandated the provision of a fast frequency reserve for 1000 specified hours for the storage systems entitled to participate in this market. In those periods, the assets must ensure the provision of the power submitted during the tender to the service and grant it in all the hours specified by the authority. The correct interaction of the BESS with the grid is remunerated with a capacity payment from the TSO that is proportional to the power made available to the service. Fast frequency reserve requirements limit both the power and energy of the storage system. Algorithm 4 describes the fast frequency reserve algorithm as a limitation of the available power and energy during the hour mandated by the authority for the service.

Algorithm 4: Fast Frequency Reserve

Input: P_i^{prod}, P^{nom}, E^{nom}, POD_{limit}, P_i^{MGP}, P_i^{mFRR}, $P^{qualified}$, P^{FRU}, FRU^{signal}
Output: P_{cha}, P_{dis}
SOC Control
for i in 96 (quarter of hours in a day) **do**
  $SOC_i^{max} = 100 - \left(\frac{P^{qualified}}{4 \times E^{nom}} \times 100\right) \times FRU^{signal}$
  $SOC_i^{min} = \left(\frac{P^{qualified}}{4 \times E^{nom}} \times 100\right) \times FRU^{signal}$
  If $P_i^{cha} > P^{nom} - P^{FRU}$ **and** $FRU^{signal} = 1$
    $P_i^{cha} = P^{nom} - P^{FRU}$
  Else if $P_i^{dis} > P^{nom} - P^{FRU}$
    $P_i^{dis} = P^{nom} - P^{FRU}$
  If $SOC_i > SOC_i^{max}$
    $SOC_i = SOC_i^{max}$
    $P_i^{dis} = \frac{SOC_i^{max} - SOC_{i-1}}{4 \times E^{nom}}$
  Else if $SOC_i < SOC_i^{min}$
    $SOC_i = SOC_i^{min}$
    $P_i^{cha} = \frac{SOC_{i-1} - SOC_i^{min}}{4 \times E^{nom}}$
$SOC_i = SOC_{i-1} - \frac{P_i^{dis}}{4} + \frac{P_i^{cha}}{4}$

4. Study Cases

The sizing procedure was divided into two different study cases, labeled as stand-alone and RES-coupled. The purpose of this classification is to highlight the different benefits and shortcomings of the two configurations and the tool's ability to operate in different configurations. The stand-alone case analyzes the performance of the sole BESS interacting with the grid. The second one foresees a hybrid RES power plant coupled with a BESS that is capable of exchanging energy with the grid and the RES. In each study case, four different sets of simulations were carried out with different combinations of the algorithms, as described in Section 3. Each set of simulations aims to properly allocate different combinations of services to investigate the impact that a multi-use BESS has on its economics. The first simulation exclusively addresses the provision of energy arbitrage by the storage system. The second set couples the first service that was investigated with the capacity market through sequential stacking. In detail, the capacity algorithm has the dispatch priority due to the necessity of satisfying the energy provision mandated by the service regulations. The third simulation package couples the mFRR with the arbitrage and capacity algorithms. The ancillary service structure interacts with the arbitrage to identify the most profitable trading strategy for each day. It follows that the most remunerative dispatch is selected depending on the day-ahead market (DAM) prices and the acceptance in the mFRR. Lastly, the fast frequency reserve is evaluated in the fourth set of simulations stacked with the other services. The provision of this application is superimposed with the previous markets that were modeled. As a consequence, the fast reserve and a further service can be provided in parallel, and the latter is derated by the power and capacity required for the frequency regulation. The list of simulations described has allowed for an investigation of the impact that service stacking has on the revenues of the BESS. The tool iterates the power and EPR of the system and identifies the optimal size of different configurations using the CAPEX being covered and IRR. Figure 2 summarizes the sets selected for each study case. Furthermore, the figure specifies the stacking classification of the investigated service. The number of services included in the stacking configuration is incremental. The first one investigates sole arbitrage. Configuration 2 adds the capacity market to the arbitrage. The third one involves the mFRR in the previous services. Finally, configuration 4 investigates the stacking of all the services.

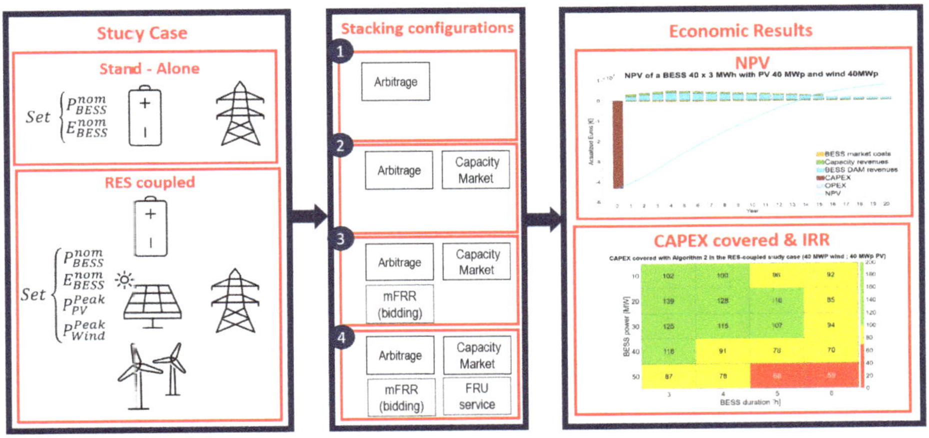

Figure 2. Summary of the study case and the service stacking proposed for this study.

It is worth stating that the parameters reported in Figure 2 are completely customizable through the GUI to cover different setups. The tool evaluates the economic performance of the BESS, allowing for the charge from the grid or the power plants.

Moreover, a POD limit was established to limit the power injection into the grid. Table 4 details the parameters adopted for the study cases. The capacity payment duration complies with the current Italian framework [67]. Furthermore, the fast reserve being implemented reflects the actual one in place in Italy [68]. Both capacity payments refer to the outcomes of the auction for the Sardinia market zone. To correctly compare the latter service with BESS of various sizes, the power enslaved to the fast reserve is defined as a percentage of the nominal power of the system.

Table 4. Parameters adopted in the study case presented in this work.

Parameter	Stand-Alone	RES-Coupled
Power [MW]	[10, 20, 30, 40, 50]	[10, 20, 30, 40, 50]
Duration [h]	[3, 4, 5, 6]	[3, 4, 5, 6]
Wind peak power [MWp]	none	[20, 30, 40]
PV peak power [MWp]	none	[20, 30, 40]
POD limit [MW]	none	[40]
Temperature	Sardinia Italy, 2021	Sardinia Italy, 2021
Prices [years]	2019–2022	2019–2022
Capacity market duration [years]	15	15
Fast reserve duration [years]	3	3
Capacity market payment [€ × year/MW]	51,012	51,012
Fast Reserve payment [€ × year/MW]	64,890	64,890
Fast Reserve Power [%]	20% P_{nom}	20% P_{nom}
Interest rate [%]	6	6
Profits per cycle [€/MWh]	25	25

The input vectors describe a twenty-year time series with a time sample of 15 min. The main input vectors are reported in Figure 3. The wind and PV profiles were generated by the open-access software AtlanteEolico-RSE [69] and PVGIS, respectively, for Southern Italy. The lack of seasonality of the wind was modeled by randomly sampling the weeks of the three years wind profile used as input. The PV power profile was reduced every year by a degradation equal to 1%/year. The historical temperatures were downloaded from the ARPA website for Southern Italy and linearly increased up to 2 °C to account for global warming [70]. Lastly, the day-ahead market prices of the year 2019 were exploited for the study case [71]. The profile was chosen because it represents the last year of business as usual before the pandemic and the gas shortages. Furthermore, due to the high prices recorded in 2021 and 2022, the 2019 price profile can be adopted as a conservative solution for the investment cash flow. A one-year simulation exclusively involving the arbitrage with the 2019-2020-2021 prices profile was developed, resulting in different realized revenues, as depicted in Table 5.

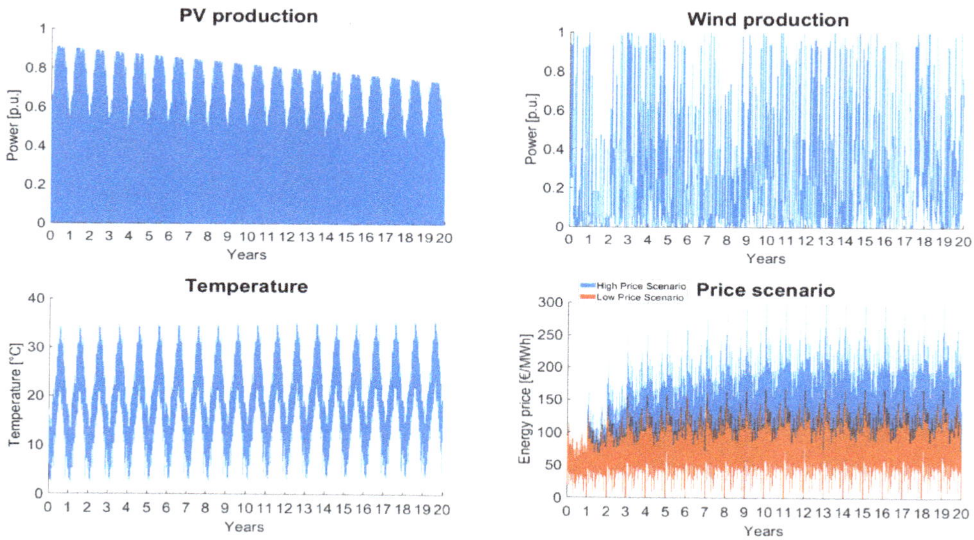

Figure 3. Input arrays for the BESS sizing tool.

Table 5. Revenues for one year of arbitrage made by a 20 MW/60 MWh BESS with three different price scenarios.

Price Scenario	Revenues [k€]	CAPEX Covered in the First Year [%]
DAM 2019	388.77	1.82
DAM 2021	762.22	3.58
DAM 2022	2125.54	9.98

The scope of this brief estimation has been the identification of the price profile that grants a lower economic return for the investment and can be used as input for the main analysis. The considered trend was extended to the period under evaluation and coupled with two different yearly gains. The scope of the gains is to create two distinct price scenarios to cope with the possible evolution of the prices in the future and provide a wider economic analysis of the investment. The yearly gains, defined as low- and high-price scenarios, were extrapolated from the results reported in [72]. The profits per cycle were kept low to understand the main differences that arise in the BESS cycling due to this parameter.

5. Results

This section presents the results obtained from the different algorithms implemented in the sizing tool. The results focus strictly on the economic indexes that are fundamental for evaluating the profitability of an investment. For the sake of completeness, both the covered CAPEX and IRR are reported. Furthermore, technical pieces of information are provided through the full equivalent cycle (FEC) and the capacity fades of the storage system under analysis.

Firstly, it is worth stating that the tool takes from 4.3 to 12.1 min (Intel i7 1260 CPU—16 GB) to simulate 20 years of service of a sole BESS with the lightest and heaviest algorithm; such a limited computational effort validates the approach's viability in tecno-economical studies. In the specified period, the procedure evaluates the power flow, degradation, and revenues of the BESS to provide valuable information on the investment.

The first reported outcomes concern the identification of a price profile that can act as a conservative solution. Table 5 details the revenues generated in one year by a 20 MW/60 MWh BESS, which performs energy arbitrage with three different price scenarios: the day-ahead market prices of 2019, 2021, and 2022.

It is possible to note that the larger volatility of the prices during the gas shortages returns better economics for the storage system. In particular, almost 10% of the 21.3 M€ of BESS CAPEX is covered in a single year in the most profitable price scenario. Nevertheless, these profits were generated by the abnormal market conditions of 2022 that, thanks to the more widespread adoption of RES, will not recur in the coming years. It follows that DAM prices for 2019 were exploited as a base to develop high and low price scenarios for the next 20 years.

The algorithms have different impacts on the economics of the system. The aim of the following paragraphs is to detail the benefits of each approach and their limitations in both stand-alone and RES coupled configurations to understand the outcomes of the tool in more depth.

5.1. Study Case 1: Stand-Alone BESS

5.1.1. Stacking Configuration 1: Energy Arbitrage

The arbitrage algorithm investigated in stacking configuration 1 is by far the less remunerative scheme for a stand-alone system. Firstly, the profit is strictly related to the spot market outcomes and, generally, a certain degree of uncertainty characterizes these variables. Furthermore, the operation of the BESS is connected to the market price spread. The high-price scenario has a larger spread than the low-price profile; consequently, it represents the most remunerative profile. The IRR difference between the same BESS with the two price trends corresponds to 4.5% on average. This value corresponds to a difference in revenues between the two scenario prices equal to 12.0% of the CAPEX for the whole investment lifespan. Furthermore, it is important to correctly estimate the profits per cycle to seize the correct trade-off between the operation of the system and the covered CAPEX. A reduction in the capital cost will directly impact the cycling cost of the system and increase the possibility of profits for arbitrage. The degradation is strictly related to these parameters, since a larger cycling will correspond to a higher cycle aging. The final capacity after 20 years for the arbitrage service is, on average, equal to 81.3%, considering a periodic capacity augmentation. Revenues are linearly impacted by the degradation. The phenomenon leads to a reduction in the profit at the end of the investment equal to 12.2%. A possible solution to increase the revenues from the arbitrage algorithm consists of increasing the cycling limits of the BESS set to one cycle per day. This condition may seem quite conservative for a system with a small EPR. Indeed, it is possible to increase profits if the market price spread allows for multiple cycles. However, profits larger than the profits per cycle must be always granted to perform the arbitrage. This condition is difficult to ensure with a higher number of cycles because the spread between the charge and discharge must be higher than the cycling cost. Furthermore, for a large EPR (>5 h), the benefits of multiple cycles are null due to the inherent structure of the market outcomes, which, in general, has two peaks lasting 2–3 h each [73]. It follows that BESS with durations over 5 h are less-suited to the daily energy arbitrage. As a consequence, the best configuration for the sole arbitrage is a 3 h BESS that correctly grasps the variation in the day-ahead market. Lastly, the revenues for BESS with the same EPR are linear with power because the efficiency of the service is exclusively related to the duration of the storage system. Therefore, BESS with the same duration, which mostly provide the arbitrage at the nominal power, have a similar IRR.

5.1.2. Stacking Configuration 2: Energy Arbitrage + Capacity Market

The capacity market represents an important source of income for the storage system as evaluated in configuration 2. The fixed revenues ensured by the provision of the services increase the economic return of the asset. However, in the best scenario, the profits cover only 55.8% of the CAPEX. Nevertheless, a stand-alone system must participate in this service to cover the missing money due to the excessive capital cost. Moreover, the service does not increase the degradation compared to the sole arbitrage strategy: indeed, in Italy, capacity remuneration is due if the obligation of bidding the remunerated capacity on the market is respected, as BESS would do in the case of energy arbitrage. However, capacity fade is slightly lower since the algorithms limit the BESS cycling to ensure the provision of the energy required by the regulation. Furthermore, due to the structure of the Italian capacity market, a longer duration of BESS benefits more from the capacity market. Indeed, as reported in Table 2 a lower derating factor is mandated by the authority of storage systems with a longer duration. This condition coupled with the economy of scale granted by a larger system EPR, which identifies BESS with a 5 h duration as the best candidate for the provision of arbitrage and capacity services.

5.1.3. Stacking Configuration 3: Energy Arbitrage + Capacity Market + mFRR

The addition of the mFRR to the services provided in stacking configuration 3 grants slightly better revenues thanks to the higher profits generated by the bids in the ancillary services market. On average, the increase in profits attests to around 1–2% of the CAPEX for the whole investment lifespan. In this case, the limits are mainly driven by the low award rate of assets on the Italian ancillary services market. However, with the increased penetration of RES, the liquidity of these assets inside the ancillary services market could increase. Therefore, future updates based on new market analysis may lead to an important increase in the cash flows. In the upper part of Figure 4, it is possible to observe how the BESS is dispatched by Algorithm 3, which selects the most profitable solution between mFRR and DAM. Furthermore, an SOC control, as depicted in the bottom part of the image, ensures that the BESS can efficiently participate in multiple services.

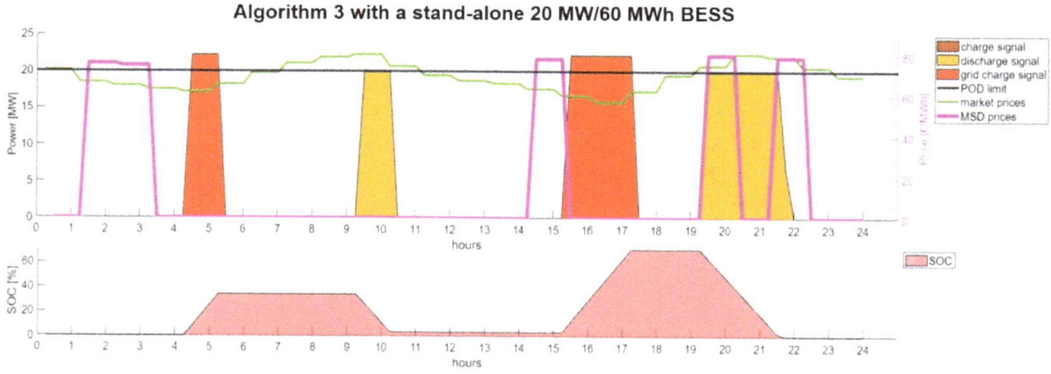

Figure 4. Algorithm 3 output for a stand-alone 20 MW/60 MWh BESS.

5.1.4. Stacking Configuration 4: Energy Arbitrage + Capacity Market + mFRR + FR

Lastly, the fast frequency reserve algorithm has been coupled with the other three in the stacking configuration 4. It has been noted that the impact of these initiatives on the economics of the system is inversely proportional to the nominal energy of the asset. Indeed, the regulation foresees a capacity payment based on the percentage of the nominal power enslaved to the service. However, the CAPEX of the asset increases with the duration of the BESS. Therefore, the service leads to a higher improvement in the economics of systems with a low duration. The covered CAPEX, thanks to three years of fast reserve service, goes from 2.0% of the 6 h BESS to 3.5% of the 3 h system. It is worth stating that the revenues generated by the capacity market maintain the 5 h duration system as the best investment for the stand-alone configuration. Nevertheless, the hypothesis adopted for the fast reserve is conservative. Indeed, after the three years mandated by the authority, no further service is foreseen for fast reserve provision inside the tool. However, the fast reserve provided by BESS may be standardized in the future and remunerated with regulations that are not yet specified. Table 6 lists the best results for the stand-alone study case for each stacking configuration. It is possible to observe that, despite being negative in all stacking configurations under our assumptions, the best IRR is granted by configuration 4, which is the solution that couples the higher number of services. Nevertheless, it is worth noting that the adopted price scenarios are conservative and return the minimum profit that BESS can generate. Furthermore, better economics are not strictly related to the asset cycling if capacity payments are granted to the storage system. Indeed, the same number of cycles allows for the revenues of the storage system to be tripled thanks to its participation in the capacity market and fast reserve provision. The general outcome of the stand-alone configuration is that, for a one-cycle-per-day system, the 20-year investment is not remunerative, no matter the adopted revenue stacking strategy: this unprofitable result is mainly due to the conservative assumption of using the 2019 DAM price scenario. However, the coupling of various services boosts the CAPEX covered by the profits from 49% to 60%. Therefore, the asset operator must carefully evaluate which services the system can provide to increase profit and ensure capacity payment. This is fundamental, especially for those services where participation is granted through an auction mechanism. Indeed, thanks to the capacity payment, the best investment is represented by 10 MW/50 MWh BESS. In general, systems with a duration of 5 h have better economics because they represent the optimal trade-off between capacity payment and CAPEX.

Table 6. Main results for the stand-alone study case (study case 1).

Stacking Configuration	Best BESS Sizing	IRR [%]	CAPEX Covered [%]	FEC	Energy Fade [%]
1	10 MW/30 MWh	−1.72	49	3183.3	18.68
2	10 MW/50 MWh	−1.03	55	3104.7	18.05
3	10 MW/50 MWh	−1.00	57	3141.1	18.40
4	10 MW/50 MWh	−0.95	60	3141.1	18.51

5.2. Study Case 2: RES-Coupled BESS

5.2.1. Stacking Configuration 1: Energy Arbitrage

The second study case exploits the presence of RES to increase profits. Indeed, the energy curtailment mandated by the POD limit is equivalent to free energy that will be lost without the storage system. This parameter, together with the PV and wind size, show a large sensitivity regarding the free charging energy for the BESS and the energy lost by the power plant. Although the overproduction of RES provides advantages to the BESS, it is fundamental to highlight that the configurations with the lowest peak power production barely improve the economics of the BESS. In the worst case, namely, a covered CAPEX of 50 MW/300 MWh BESS this increases by only 3% following the addition of 20 MWp of both PV and wind compared to the stand-alone case. A further observation related to the coupling of the asset to the RES is the higher number of cycles. The availability of cheap energy for charging the system due to the presence of RES increases the use of the BESS and its degradation, which reaches over 30% for the system with a smaller duration. However, increased use of the storage system corresponds to better economics; as augmentation of the asset is usually included in the OPEX, it is always better to cycle the system and boost the profits.

The presence of a power plant allows for the BESS to be charged for free and save energy that, without the storage system, will be curtailed. This condition leads to benefits proportional to the energy being stored. In the best scenarios, namely high prices and high-power plant peak power, the investment of a BESS generates value with the sole arbitrage. The revenues from the sole arbitrage are almost doubled in the best-case scenario. The most cost-effective BESS size with the provision of the sole arbitrage is 10 MW/30 MWh. This system represents an optimal trade-off between the CAPEX and the ability to collect the byproducts caused by RES plant overproduction. Higher powers reduce the curtailment but not linearly. Therefore, the expenditure increase is higher than the revenue generated for the increment in energy savings. Instead, the EPR increases only the expenditures without further energy savings, leading to a higher cost and similar profit.

5.2.2. Stacking Configuration 2: Energy Arbitrage + Capacity Market

Stacking configuration 2, which couples the capacity market with the arbitrage, improves the economic results. Following Equation (5), coupling with RES led to a higher-capacity payment compared to the stand-alone case. This condition drastically improves the economics of the system. However, this increase in capacity may lead to a mandatory power provision value that is larger than the nominal power of the BESS. Therefore, if the unpredictable production of RES goes to zero, in some cases, the BESS does not have the necessary characteristics to fulfill the constraints of the capacity market. It follows that, for a small BESS and large RES-size capacity, payment is not received due to the power limits of the storage system. However, larger BESS benefit from this condition, with an improvement of up to 30% in the CAPEX covered compared to the sole arbitrage provision. This aspect is fundamental to justify the presence of a storage system in a hybrid energy district. Indeed, for small-scale BESS, the capacity payment is not granted and the same will occur if no BESS is present in the energy district. Therefore, for an RES owner that is willing to participate in the capacity market, it is mandatory to purchase a BESS to fulfill the requirements of the Italian authorities. Lastly, as in the previous study, participation in this new service has a very low impact on cycling and degradation. Figure 5 depicts the covered CAPEX for the RES-coupled study case with Algorithm 2. The heatmap produced by the GUI can identify the most cost-effective BESS for the scenario under analysis, namely the 20 MW/60 MWh system.

5.2.3. Stacking Configuration 3 and 4: Energy Arbitrage + Capacity Market + mFRR + FR

Lastly, mFRR and fast reserve provision has a similar influence on the RES-coupled study case to the stand-alone. Indeed, the two services are strictly related to the operation of the BESS rather than the presence of RES power plants.

The best results for the RES-coupled study case are listed in Table 7. The outcomes of the second study case are more impacted by service stacking than the first analysis. Indeed, the capacity prize is larger due to the necessity of coupling RES with a BESS to provide a constant power setpoint. The service stacking improves the covered CAPEX by up to 30% compared to the sole arbitrage provision. Furthermore, as stated in [74], BESS have an important economic advantage in supporting RES systems compared to the stand-alone case. The coupling with power sources reduces the expenditures on the energy sale on the day-ahead market and stabilizes the unpredictability of RES granting the capacity payment.

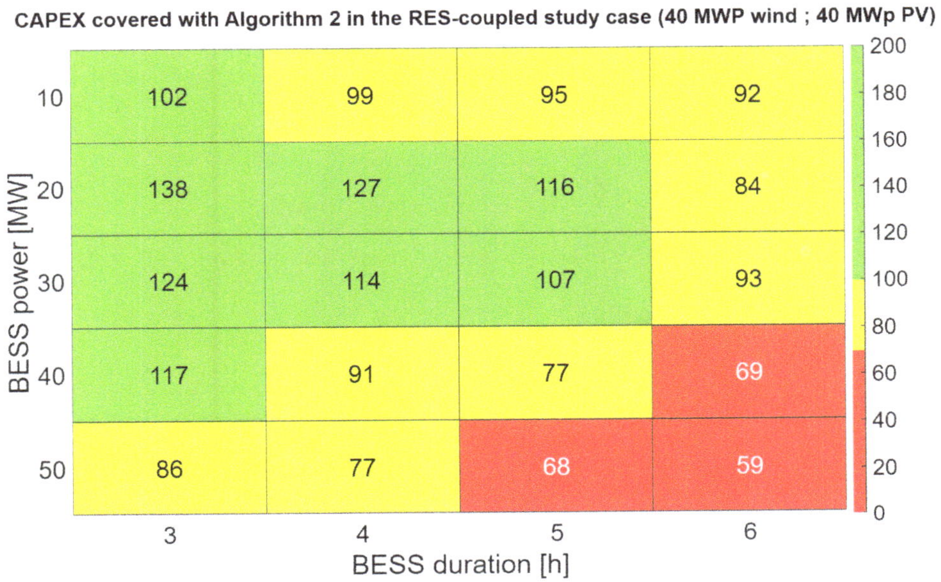

Figure 5. CAPEX covering 20 years with different sizes of BESS for the RES-coupled study case with 40 MWp wind and PV.

Table 7. Main results for RES-coupled study case 2 with 40 MWp for both wind and PV.

Stacking Configuration	Best Configuration	IRR [%]	CAPEX Covered [%]	FEC	Capacity Fade [%]
1	10 MW/30 MWh	5.97	99	6410.3	32.45
2	20 MW/60 MWh	10.46	138	6233.3	30.23
3	20 MW/60 MWh	10.55	139	6180.5	30.17
4	20 MW/60 MWh	10.84	144	6124.2	29.87

6. Conclusions

This article proposes a model-aware analysis to resolve the BESS sizing issue considering different applications that implement service stacking. The economic analysis evaluated the investment throughout a 20-year scenario, exploiting a high-fidelity empirical BESS model equipped with equations that are able to describe both (i) the performance of BESS in different operating and ambient conditions and (ii) the capacity degradation of the asset. Novel algorithms were developed to properly emulate the BESS control strategy participating in the capacity and energy markets. Flexibility and reduced computational effort make the tool valid for the rapid assessment of different system configurations (e.g., stand-alone or RES-integrated), the sizing (e.g., the varying power and duration of the BESS), and the services (i.e., a control strategy block can implement complex algorithms to emulate service provision under different regulatory frameworks).

In this work, two systems were simulated, featuring a stand-alone and RES-coupled BESS, which served as case studies. The provided services are implemented to fit Italian regulation and market conditions. The outcomes highlighted the wide difference between the stand-alone and RES-coupled BESS' return of the investment. A stand-alone system that is cycled once per day could not cover more than 60% of the CAPEX in 20 years. Although the inclusion of capacity payments led to an improvement in terms of economics, the current Italian regulations do not provide sufficient revenue to return on the investment, as per the assumptions made in the study. However, in the future, reductions in the system inertia will require more BESS use for frequency regulation. As a consequence, a new market or an extension of the existing mechanisms (e.g., fast reserve, limited to 1000 h per year as of 2023) could enter the pitch and provide new remuneration schemes for BESS. Conversely, the configuration presenting a BESS coupled with RES shows a positive IRR thanks to the optimal synergy between the storage and the intermittent power production. A comparison between the two study cases highlighted the advantage that BESS has in supporting the RES system.

This can be used to obtain the best result from the energy sale on the day-ahead market and allows for a better integration of the RES + BESS system in the grid. This is testified by the decreased connection power required for the hybrid RES + BESS plant and the exploitation of a larger capacity payment.

Optimal sizing is fundamental to obtaining a positive economic outcome, as highlighted by the results. Moreover, the system configuration play a key role in the economics of the asset. Indeed, the same services and the same power rating with RES can provide a 2.3 times higher return on the investment compared to the stand-alone case.

Additionally, the outcomes highlighted the necessity of service stacking to fully cover the investment costs. In particular, it has been pointed out that the sole energy service (i.e., arbitrage) cannot cover the capital costs of the assets, increasing the interest for BESS participation in capacity remuneration mechanisms or long-term contracts to ensure flexibility. The most efficient service stacking led to an increase in the covered CAPEX that was equal to 11% in the stand-alone study case and an astonishing 45% in the hybrid configuration. This improvement was achieved without increasing the number of cycles performed by the BESS.

The limitations of the study include the fact that the iterative procedure is not an optimization; thus, the optimal solution can be hidden between two tested configurations. Given the possibility of increasing the granularity of the simulations thanks to the low computational effort required, this is not considered a major issue. Additionally, the study was performed using a BESS model from the literature, featuring a Li-NMC battery. It is well-known that, as of 2023, LFP chemistry has overcome Li-NMC, especially in utility-scale applications. The generalization of the procedure to include an LFP battery model is of great interest.

Future works could focus on improving the algorithms proposed in this article. These include the possibility of cycling the BESS more than once per day (to better test the trade-off between BESS life and NPV), the inclusion of an aging-aware algorithm to identify an efficient trade-off between profit and degradation, and the testing of innovative ancillary services that can fulfil the new needs of power systems. Furthermore, the possibility of easily evaluating different BESS technologies will drive the comparison and identification of efficient services for a specific cell chemistry.

Author Contributions: Conceptualization, G.G., M.M., S.M. and G.R.; methodology, F.B. (Filippo Bovera),G.R., M.M. and M.S.; software M.S.; Validation F.B. (Federico Bresciani) and M.S., writing—original draft, M.S.; writing—review and editing G.R., F.B. (Federico Bresciani) and M.M.; visualization M.S.; supervision, M.M., G.G. and S.M.; project administration M.M. and G.G. All authors have read and agreed to the published version of the manuscript.

Funding: This work has been funded by Eni in the context of the Joint Research Center between Eni and Politecnico di Milano (Prot. Nr. 4310481954—Project B3M).

Data Availability Statement: Data will be made available on request.

Conflicts of Interest: The authors declare no conflict of interest.

Nomenclature

Parameters
P_i^{prod}: Power produced by RES at instant i
P^{nom}: BESS nominal power
E^{nom}: BESS nominal energy
POD_{limit}: Point of Delivery maximum power
P_i^{DAM}: Day-ahead market price at instant i
$SOC_0^{initial}$: SOC at the beginning of the day
CDP: Awarded capacity
CMS_i^{input}: Capacity Market Signal
CMS_i^{charge}: Capacity Market charge Signal
P_i^{mFRR}: mFRR price
$P^{qualified}$: Power qualified to provide FRU
P^{FRU}: Power enslaved to FRU provision
FRU^{signal}: FRU participation signal
Decision Variables
P_{cha}: BESS charging power
P_{dis}: BESS discharging power
E^{cha}: Energy associated with the daily charge
E^{dis}: Energy associated with the daily discharge
$P^{chargefree}$: Power without charge from the hybrid energy district
$P^{chargepaid}$: Power bought from the DAM
profit: Profits daily generated by the BESS

References

1. Horowitz, C.A. Paris Agreement. *Materials* **2016**, *55*, 740–755. [CrossRef]
2. Robert, C. Pietzcker, Sebastian Osorio, Renato Rodrigues, Tightening EU ETS targets in line with the European Green Deal: Impacts on the decarbonization of the EU power sector. *Appl. Energy* **2021**, *293*, 116914. [CrossRef]
3. European Commision. The European Green Deal. Bruxelles. June 2021. Available online: https://sustainabledevelopment.un.org/post2015/transformingourworld (accessed on 18 April 2023).
4. Antonelli, M.; Desideri, U.; Franco, A. Effects of large scale penetration of renewables: The Italian case in the years 2008–2015. In *Renewable and Sustainable Energy Reviews*; Elsevier Ltd.: Amsterdam, The Netherlands, 2018; Volume 81, pp. 3090–3100. [CrossRef]
5. Bloomberg; EATON; Statkraft. Flexibility Solutions for High-Renewable Energy Systems. 2018. Available online: https://data.bloomberglp.com/professional/sites/24/2018/11/UK-Flexibility-Solutions-for-High-Renewable-Energy-Systems-2018-BNEF-Eaton-Statkraft.pdf (accessed on 5 June 2023).
6. International Energy Agency. World Energy Outlook 2022. 2022. Available online: www.iea.org/t&c/ (accessed on 20 May 2023).
7. Dunn, B.; Kamath, H.; Tarascon, J.-M. Electrical Energy Storage for the Grid: A Battery of Choices System Power Ratings, Module Size. Available online: www.sciencemag.org (accessed on 20 March 2023).
8. Englberger, S.; Jossen, A.; Hesse, H. Unlocking the Potential of Battery Storage with the Dynamic Stacking of Multiple Applications. *Cell Rep. Phys. Sci.* **2020**, *1*, 100238. [CrossRef]
9. Rancilio, G.; Lucas, A.; Kotsakis, E.; Fulli, G.; Merlo, M.; Delfanti, M.; Masera, M. Modeling a large-scale battery energy storage system for power grid application analysis. *Energies* **2019**, *12*, 3312. [CrossRef]
10. Vykhodtsev, A.V.; Jang, D.; Wang, Q.; Rosehart, W.; Zareipour, H. A review of modelling approaches to characterize lithium-ion battery energy storage systems in techno-economic analyses of power systems. *Renew. Sustain. Energy Rev.* **2022**, *166*, 112584. [CrossRef]
11. Tamilselvi, S.; Gunasundari, S.; Karuppiah, N.; Razak RK, A.; Madhusudan, S.; Nagarajan, V.M.; Sathish, T.; Shamim, M.Z.M.; Saleel, C.A.; Afzal, A. A review on battery modelling techniques. *Sustainability* **2021**, *13*, 42. [CrossRef]
12. Moura, S.J.; Chaturvedi, N.A. Adaptive PDE Observer for Battery SOC/SOH Estimation via an Electrochemical Model. *Am. Soc. Mech. Eng.* **2012**, *45295*, 101–110.
13. Doyle, M.; Newman, J. The use of mathematical modeling in the design of lithium polymer battery system. *Electrochim. Acta* **1995**, *40*, 2191–2196. [CrossRef]
14. Xu, L.; Lin, X.; Xie, Y.; Hu, X. Enabling high-fidelity electrochemical P2D modeling of lithium-ion batteries via fast and non-destructive parameter identification. *Energy Storage Mater.* **2022**, *45*, 952–968. [CrossRef]
15. Allam, A.; Onori, S. Online Capacity Estimation for Lithium-Ion Battery Cells via an Electrochemical Model-Based Adaptive Interconnected Observer. *IEEE Trans. Control Syst. Technol.* **2020**, *29*, 1636–1651. [CrossRef]
16. He, W.; Pecht, M.; Flynn, D.; Dinmohammadi, F. A physics-based electrochemical model for lithium-ion battery state-of-charge estimation solved by an optimised projection-based method and moving-window filtering. *Energies* **2018**, *11*, 2120. [CrossRef]
17. Barzacchi, L.; Lagnoni, M.; Di Rienzo, R.; Bertei, A.; Baronti, F. Enabling early detection of lithium-ion battery degradation by linking electrochemical properties to equivalent circuit model parameters. *J. Energy Storage* **2022**, *50*, 104213. [CrossRef]
18. Rahman, M.A.; Anwar, S.; Izadian, A. Electrochemical model parameter identification of a lithium-ion battery using particle swarm optimization method. *J. Power Sources* **2016**, *307*, 86–97. [CrossRef]
19. Chan, H.L.; Sutanto, D. A new battery model for use with battery energy storage systems and electric vehicles power systems. In Proceedings of the 2000 IEEE Power Engineering Society Winter Meeting. Conference Proceedings (Cat. No.00CH37077), Singapore, 23–27 January 2000; Institute of Electrical and Electronics Engineers Inc.: New York, NY, USA, 2000; pp. 470–475. [CrossRef]
20. He, H.; Xiong, R.; Guo, H.; Li, S. Comparison study on the battery models used for the energy management of batteries in electric vehicles. In *Energy Conversion and Management*; Elsevier Ltd.: Amsterdam, The Netherlands, 2012; pp. 113–121. [CrossRef]

21. Campagna, N.; Castiglia, V.; Miceli, R.; Mastromauro, R.A.; Spataro, C.; Trapanese, M.; Viola, F. Battery models for battery powered applications: A comparative study. *Energies* **2020**, *13*, 85. [CrossRef]
22. Berger, M.; Kocar, I.; Farantatos, E.; Haddadi, A. Modeling of Li-ion battery energy storage systems (BESSs) for grid fault analysis. *Electr. Power Syst. Res.* **2021**, *196*, 107160. [CrossRef]
23. Pichetjamroen, A.; Chindamanee, P.; Teerakawanich, N.; Somakettarin, N. A Study on Performances of Flexible Power Control with Empirical Lithium-ion Battery Modeling in PV Power Systems. In Proceedings of the 19th International Conference on Electrical Engineering/Electronics, Computer, Telecommunications and Information Technology, ECTI-CON 2022, Prachuap Khiri Khan, Thailand, 24–27 May 2022; Institute of Electrical and Electronics Engineers Inc.: New York, NY, USA, 2022. [CrossRef]
24. Perinov; Garniwa, I.; Hudaya, C.; Sudiarto, B. Comparison of Battery Models for Battery Energy Storage System Development. *J. Phys. Conf. Ser.* **2021**, *1858*, 012046. [CrossRef]
25. Hussein, A.A.-H.; Batarseh, I. An overview of generic battery models. In *2011 IEEE Power and Energy Society General Meeting*; IEEE: New York, NY, USA, 2011; pp. 1–6.
26. Nebuloni, R.; Meraldi, L.; Bovo, C.; Ilea, V.; Berizzi, A.; Sinha, S.; Tamirisakandala, R.B.; Raboni, P. A hierarchical two-level MILP optimization model for the management of grid-connected BESS considering accurate physical model. *Appl. Energy* **2023**, *334*, 120697. [CrossRef]
27. Shamarova, N.; Suslov, K.; Ilyushin, P.; Shushpanov, I. Review of Battery Energy Storage Systems Modeling in Microgrids with Renewables Considering Battery Degradation. *Energies* **2022**, *15*, 6967. [CrossRef]
28. Weniger, J.; Tjaden, T.; Quaschning, V. Sizing of residential PV battery systems. In *Energy Procedia*; Elsevier Ltd.: Amsterdam, The Netherlands, 2014; pp. 78–87. [CrossRef]
29. Marchi, B.; Pasetti, M.; Zanoni, S. Life Cycle Cost Analysis for BESS Optimal Sizing. In *Energy Procedia*; Elsevier Ltd.: Amsterdam, The Netherlands, 2017; pp. 127–134. [CrossRef]
30. Etherden, N.; Bollen, M.H.J. Dimensioning of energy storage for increased integration of wind power. *IEEE Trans. Sustain. Energy* **2013**, *4*, 546–553. [CrossRef]
31. Ravishankar, R.; Vrudhula, S.; Rakhmatov, D. Battery Modeling for Energy-Aware System Design. *Computer* **2003**, *36*, 77–87. [CrossRef]
32. Wu, D.; Ma, X. Modeling and Optimization Methods for Controlling and Sizing Grid-Connected Energy Storage: A Review. *Curr. Sustain./Renew. Energy Rep.* **2021**, *8*, 123–130. [CrossRef]
33. Wu, D.; Balducci, P.; Crrawford, A.; Mongrid, K.; Ma, X. Building Battery Energy Storage System Performance Data into economic assesment. In *IEEE Power & Energy Society General Meeting (PESGM)*; IEEE: New York, NY, USA, 2020; pp. 1–5.
34. Nguyen, T.A.; Copp, D.A.; Byrne, R.H.; Chalamala, B.R. Market Evaluation of Energy Storage Systems Incorporating Technology-Specific Nonlinear Models. *IEEE Trans. Power Syst.* **2019**, *34*, 3706–3715. [CrossRef]
35. Hannan, M.A.; Faisal, M.; Ker, P.J.; Begum, R.A.; Dong, Z.Y.; Zhang, C. Review of optimal methods and algorithms for sizing energy storage systems to achieve decarbonization in microgrid applications. In *Renewable and Sustainable Energy Reviews*; Elsevier Ltd.: Amsterdam, The Netherlands, 2020; Volume 131. [CrossRef]
36. Fernández-Blanco, R.; Dvorkin, Y.; Xu, B.; Wang, Y.; Kirschen, D.S. Optimal Energy Storage Siting and Sizing: A WECC Case Study. *IEEE Trans. Sustain. Energy* **2017**, *8*, 733–743. [CrossRef]
37. Shuai, H.; Fang, J.; Ai, X.; Tang, Y.; Wen, J.; He, H. Stochastic optimization of economic dispatch for microgrid based on approximate dynamic programming. *IEEE Trans. Smart Grid* **2019**, *10*, 2440–2452. [CrossRef]
38. Pozo, D. Linear battery models for power systems analysis. *Electric. Power Syst. Res.* **2022**, *212*, 108565. [CrossRef]
39. Akram, U.; Khalid, M.; Shafiq, S. Optimal sizing of a wind/solar/battery hybrid grid-connected microgrid system. *IET Renew. Power Gener.* **2018**, *12*, 72–80. [CrossRef]
40. Javadi, M.; Gong, Y.; Chung, C.Y. Frequency Stability Constrained BESS Sizing Model for Microgrids. *IEEE Trans. Power Syst.* **2023**, 1–13. [CrossRef]
41. Duman, A.C.; Erden, H.S.; Gönül, Ö.; Güler, Ö. Optimal sizing of PV-BESS units for home energy management system-equipped households considering day-ahead load scheduling for demand response and self-consumption. *Energy Build* **2022**, *267*, 112164. [CrossRef]
42. Powell, W. Approximate Dynamic Programming: Solving the Curses of Dimensionality Informs Computing Society Tutorial. 2008. Available online: http://www.castlelab.princeton.edu (accessed on 30 May 2023).
43. Wang, X.; Li, F.; Zhang, Q.; Shi, Q.; Wang, J. Profit-Oriented BESS Siting and Sizing in Deregulated Distribution Systems. *IEEE Trans. Smart Grid* **2023**, *14*, 1528–1540. [CrossRef]
44. Hannan, M.A.; Wali, S.B.; Ker, P.J.; Abd Rahman, M.S.; Mansor, M.; Ramachandaramurthy, V.K.; Muttaqi, K.M.; Mahlia, T.M.I.; Dong, Z.Y. Battery energy-storage system: A review of technologies, optimization objectives, constraints, approaches, and outstanding issues. *J. Energy Storage* **2021**, *42*, 103023. [CrossRef]
45. Lampinen, J. Multiobjective Nonlinear Pareto-Optimization Multiobjective Nonlinear Pareto-Optimization A Pre-Investigation Report LAPPEENRANTA 2000. Available online: https://www.researchgate.net/publication/239560754 (accessed on 12 January 2023).
46. Truong, C.N.; Schimpe, M.; Bürger, U.; Hesse, H.C.; Jossen, A. Multi-use of stationary battery storage systems with blockchain based markets. In *Energy Procedia*; Elsevier Ltd.: Amsterdam, The Netherlands, 2018; pp. 3–16. [CrossRef]

47. Hjalmarsson, J.; Thomas, K.; Boström, C. Service stacking using energy storage systems for grid applications—A review. *J. Energy Storage* **2023**, *60*, 106639. [CrossRef]
48. Yang, Y.; Bremner, S.; Menictas, C.; Kay, M. Battery energy storage system size determination in renewable energy systems: A review. In *Renewable and Sustainable Energy Reviews*; Elsevier Ltd.: Amsterdam, The Netherlands, 2018; Volume 91, pp. 109–125. [CrossRef]
49. Rancilio, G.; Bovera, F.; Merlo, M. Revenue Stacking for BESS: Fast Frequency Regulation and Balancing Market Participation in Italy. *Int. Trans. Electr. Energy Syst.* **2022**, *2022*, 1894003. [CrossRef]
50. Brogan, P.V.; Best, R.; Morrow, J.; Duncan, R.; Kubik, M. Stacking battery energy storage revenues with enhanced service provision. *IET Smart Grid* **2020**, *3*, 520–529. [CrossRef]
51. Alharbi, A.M.; Alsaidan, I.; Gao, W. Optimal Scheduling of Battery Energy Storage System Performing Stacked Services. In Proceedings of the IEEE Green Technologies Conference, Houston, TX, USA, 30 March–1 April 2022; IEEE: New York, NY, USA, 2022; pp. 110–115. [CrossRef]
52. Zhang, Y.; Anvari-Moghaddam, A.; Peyghami, S.; Dragicevic, T.; Li, Y.; Blaabjerg, F. Optimal sizing of behind-the-meter BESS for providing stackable services. In Proceedings of the 2022 IEEE 13th International Symposium on Power Electronics for Distributed Generation Systems (PEDG 2022), Kiel, Germany, 26–29 June 2022; IEEE: New York, NY, USA, 2022. [CrossRef]
53. Watson, D.; Hastie, C.; Gaudette, B.; Rodgers, M. Demonstrating Stacked Services of a Battery in a Wind R&D Park. *IEEE Trans. Power Syst.* **2018**, *33*, 1411–1419. [CrossRef]
54. Hameed, Z.; Træholt, C.; Hashemi, S. Investigating the participation of battery energy storage systems in the Nordic ancillary services markets from a business perspective. *J. Energy Storage* **2023**, *58*, 106464. [CrossRef]
55. Tsagkou, A.S.; Kerasidis, E.D.; Doukas, D.I.; Labridis, D.P.; Marinopoulos, A.G.; Tengnér, T. Stacking Grid Services with Energy Storage Techno-Economic Analysis. In Proceedings of the IEEE Manchester PowerTech, Manchester, UK, 18–22 June 2017; pp. 1–6.
56. Liu, Y.; Du, W.; Xiao, L.; Wang, H.; Bu, S.; Cao, J. Sizing a Hybrid Energy Storage System for Maintaining Power Balance of an Isolated System with High Penetration of Wind Generation. *IEEE Trans. Power Syst.* **2016**, *31*, 3267–3275. [CrossRef]
57. Pelletier, S.; Jabali, O.; Laporte, G.; Veneroni, M. Battery degradation and behaviour for electric vehicles: Review and numerical analyses of several models. *Transp. Res. Part B Methodol.* **2017**, *103*, 158–187. [CrossRef]
58. Moncecchi, M.; Brivio, C.; Mandelli, S.; Merlo, M. Battery energy storage systems in microgrids: Modeling and design criteria. *Energies* **2020**, *13*, 2006. [CrossRef]
59. Ecker, M.; Nieto, N.; Käbitz, S.; Schmalstieg, J.; Blanke, H.; Warnecke, A.; Sauer, D.U. Calendar and cycle life study of Li(NiMnCo)O_2-based 18650 lithium-ion batteries. *J. Power Sources* **2014**, *248*, 839–851. [CrossRef]
60. Ramasamy, V.; Zuboy, J.; O'Shaughnessy, E.; Feldman, D.; Desai, J.; Woodhouse, M.; Basore, P.; Margolis, R.U.S. Solar Photovoltaic System and Energy Storage Cost Benchmarks, With Minimum Sustainable Price Analysis: Q1 2022. 2022. Available online: https://www.osti.gov/biblio/1891204 (accessed on 15 June 2023).
61. Cole, W.; Frazier, A.W.; Augustine, C. Cost Projections for Utility-Scale Battery Storage: 2021 Update. 2021. Available online: https://www.nrel.gov/docs/fy21osti/79236.pdf (accessed on 15 June 2023).
62. Rancilio, G.; Merlo, M.; Lucas, A.; Kotsakis, E.; Delfanti, M. BESS modelling: Investigating the role of auxiliary system consumption in efficiency derating. In Proceedings of the International Symposium on Power Electronics, Electrical Drives, Automation and Motion, Sorrento, Italy, 24–26 June 2020; pp. 1–6.
63. Terna Mercato della Capacità—Disposizioni Tecniche di Funzionamento n.2. 2021. Available online: https://www.terna.it/it/sistema-elettrico/pubblicazioni/news-operatori/dettaglio/CM-Disposizioni-tecniche-di-funzionamento-DTF (accessed on 10 March 2023).
64. Gulotta, F.; Rossi, A.; Bovera, F.; Falabretti, D.; Galliani, A.; Merlo, M.; Rancilio, G. Opening of the Italian Ancillary Service Market to Distributed Energy Resources: Preliminary Results of UVAM project. In Proceedings of the HONET 2020—IEEE 17th International Conference on Smart Communities: Improving Quality of Life Using ICT, IoT and AI, Charlotte, NC, USA, 14–16 December 2020; IEEE: New York, NY, USA, 2020; pp. 199–203. [CrossRef]
65. Bovera, F.; Blaco, A.; Rancilio, G.; Delfanti, M. Assessing the Accuracy of Different Machine Learning Classification Algorithms in Forecasting Results of Italian Ancillary Services Market. In Proceedings of the 2019 16th International Conference on the European Energy Market (EEM), Ljubljana, Slovenia, 18–20 September 2019.
66. Arera. Deliberazione 3 Giugno 2020 200/2020/R/EEL Approvazione del Regolamento, Predisposto da Terna S.P.A., Ai Sensi della Deliberazione Dell'autorità 300/2017/R/EEL, Relativo al Progetto Pilota per L'erogazione del Servizio di Regolazione Ultra-Rapida di Frequenza. 2020. Available online: https://www.arera.it/it/docs/20/200-20.htm (accessed on 25 February 2023).
67. Terna Mercato Della Capacità Rendiconto Degli Esiti—Asta Madre 2024. 2022. Available online: https://www.terna.it/it/sistema-elettrico/pubblicazioni/news-operatori/dettaglio/rendiconto-asta-capacity-market-2024 (accessed on 14 March 2023).
68. Terna Progetto Pilota Fast Reserve—Esiti Asta. 2020. Available online: https://www.terna.it/it/sistema-elettrico/pubblicazioni/news-operatori/dettaglio/esiti-asta-Fast-reserve (accessed on 28 January 2023).
69. RSE Atlante Eolico RSE. Available online: https://atlanteeolico.rse-web.it/ (accessed on 7 June 2023).
70. ARPA Puglia Temperature Puglia. Available online: https://www.arpa.puglia.it/pagina2839_meteo.html (accessed on 7 June 2023).

71. Gestore Mercati Energetici (GME) Esiti Mercato Elettrico. Available online: https://www.mercatoelettrico.org/it/ (accessed on 7 June 2023).
72. Afman, M.; Hers, S.; Scholten, T. Energy and Electricity Price Scenarios 2020-2023-2030. 2017. Available online: chrome-extension://efaidnbmnnnibpcajpcglclefindmkaj/https://cedelft.eu/wp-content/uploads/sites/2/2021/04/CE_Delft_3H58_Energy_and_electricity_price_scenarios_DEF.pdf (accessed on 7 June 2023).
73. Mercier, T.; Olivier, M.; De Jaeger, E. The value of electricity storage arbitrage on day-ahead markets across Europe. *Energy Econ.* **2023**, *123*, 106721. [CrossRef]
74. Lazard LCOE Lazard. 2023. Available online: https://www.lazard.com/research-insights/2023-levelized-cost-of-energyplus/ (accessed on 7 June 2023).

energies

MDPI

Article

Data Augmentation and Feature Selection for the Prediction of the State of Charge of Lithium-Ion Batteries Using Artificial Neural Networks

Sebastian Pohlmann [1,*], Ali Mashayekh [2], Manuel Kuder [3], Antje Neve [1] and Thomas Weyh [2]

1 Institute of Distributed Intelligent Systems, University of the Bundeswehr, Werner-Heisenberg-Weg 39, 85577 Neubiberg, Germany; antje.neve@unibw.de
2 Institute of Electrical Energy Systems, University of the Bundeswehr, Werner-Heisenberg-Weg 39, 85577 Neubiberg, Germany; ali.mashayekh@unibw.de (A.M.); thomas.weyh@unibw.de (T.W.)
3 Bavertis GmbH, Marienwerderstraße 6, 81929 Munich, Germany; manuel.kuder@unibw.de
* Correspondence: sebastian.pohlmann@unibw.de

Abstract: Lithium-ion batteries are a key technology for the electrification of the transport sector and the corresponding move to renewable energy. It is vital to determine the condition of lithium-ion batteries at all times to optimize their operation. Because of the various loading conditions these batteries are subjected to and the complex structure of the electrochemical systems, it is not possible to directly measure their condition, including their state of charge. Instead, battery models are used to emulate their behavior. Data-driven models have become of increasing interest because they demonstrate high levels of accuracy with less development time; however, they are highly dependent on their database. To overcome this problem, in this paper, the use of a data augmentation method to improve the training of artificial neural networks is analyzed. A linear regression model, as well as a multilayer perceptron and a convolutional neural network, are trained with different amounts of artificial data to estimate the state of charge of a battery cell. All models are tested on real data to examine the applicability of the models in a real application. The lowest test error is obtained for the convolutional neural network, with a mean absolute error of 0.27%. The results highlight the potential of data-driven models and the potential to improve the training of these models using artificial data.

Keywords: lithium-ion batteries; state of charge; machine learning; artificial neural networks; data augmentation

Citation: Pohlmann, S.; Mashayekh, A.; Kuder, M.; Neve, A.; Weyh, T. Data Augmentation and Feature Selection for the Prediction of the State of Charge of Lithium-Ion Batteries Using Artificial Neural Networks. *Energies* **2023**, *16*, 6750. https://doi.org/10.3390/en16186750

Academic Editor: Simone Barcellona

Received: 28 July 2023
Revised: 11 August 2023
Accepted: 15 August 2023
Published: 21 September 2023

1. Introduction

As the transportation sector is responsible for a large share of greenhouse gas emissions, it is crucial for the automotive and mobility industry to turn towards renewable energy [1]. Lithium-ion batteries (LIBs) have taken a predominant role as electrochemical energy storage solutions in many applications, ranging from portable consumer electronics to integration in power grids and battery electric vehicles (BEVs) [2]. Next to applications in stationary energy storage systems with particularly high efficiency needs in third-world countries [3], LIBs are best-suited for empowering BEVs due to their high energy densities and their long lifespans [4]. Nevertheless, the operation of LIBs has to be optimized to exceed the performance of internal combustion engine vehicles. The demand for BEVs is growing quickly, and the materials they require are rare. The optimization of BEVs is essential for their worldwide success, especially in developing countries, such as Latin American countries, which have large populations and uncertain future markets [5]. The whole lifecycle of the BEV, including production, operation, recycling, and reuse, has to be considered. Additionally, the price pressure seen regarding BEVs intensifies the need for further enhancements during operation [6]. The battery management system (BMS) of a BEV is responsible for determining the condition of the vehicle's battery. This system monitors and controls the battery cells [7]. During operation, the condition of a battery is influenced by various intertwined parameters and the ambient conditions [8]. Additionally, there are several mechanisms for the degradation of an LIB, all of which directly affect the performance and the state of the battery. These mechanisms are caused by chemical processes, mechanical damage, temperature, and different loading conditions [9].

A key challenge for the application of LIBs is to accurately predict their state of charge (SoC), which is necessary to ensure their safety and facilitate their efficient charge and discharge cycles [10]. Other than physical estimation methods, such as coulomb counting or other electrochemical models [11–13], there are primarily two different approaches: model-based approaches and data-driven approaches. The main representatives of the model-based methods are from the Kalman filter family, including the extended Kalman filter, dual extended Kalman filter, and unscented Kalman filter models [14–16]. The data-driven methods utilize machine learning (ML) or other statistical algorithms to estimate the condition of a battery. Because these algorithms approximate the electrochemical processes inside a battery cell with high levels of accuracy while having decreased levels of complexity, they have gained considerable interest [17]. The main reason for the success of a data-driven method is the data the method is based on, which should be reliable and capture the behavior of the cell [18]. Poor data impede the state estimation for batteries, as the parameters are highly dependent on the loading as well as the ambient conditions and are further internally correlated [19]. The methods used to estimate the SoC of a battery include support vector machines (SVM) [20], regression algorithms [21], and artificial neural networks (ANNs) [22]. Different types of ANNs can be distinguished from one another. The conventional type are multilayer perceptrons (MLP), which are feedforward neural networks [23]. The other applied types are recurrent neural networks (RNNs) [24,25] and convolutional neural networks (CNNs), where the convolution is typically performed along the time axis [26,27]. The aim of the data-driven models is to approximate a function between the measurable parameters of a battery and the nonmeasurable conditions, such as the SoC. As ML models are highly dependent on their input data, it is crucial to have a sufficiently large dataset to replicate the behavior of a battery. For LIBs in particular, which have several working conditions and respond differently to changing ambient conditions, creating an appropriate dataset is a key challenge. Furthermore, battery tests are time- and cost-consuming [28]. A possible solution to overcome these problems is the usage of artificially augmented data. In the last few years, it has been shown that data augmentation techniques can lead to improved results for ML models, thus making it possible to successfully tackle the problem of limited datasets [29–31].

In this work, the SoC of an LIB is estimated using different ML algorithms. To decrease the effort required for time-consuming battery tests to a minimum, a real-world dataset is enriched using artificially augmented data. The goal is to approximate a function for the SoC with the current, voltage, and temperature as input variables. After preprocessing, the data are used to train and test the ML models. The results are compared to a reference model, which is a linear regression model. Two types of ANNs, an MLP and a CNN, are trained and tested to evaluate the impact of the data augmentation technique. The MLP is chosen because of its lower complexity and simpler structure in comparison with other ANNs. Therefore, less computing power is needed to train the model. An advantage of CNNs is their additional data processing step along the time axis, which allows the accuracy of the model to benefit from determining more complex correlations, which are not identifiable using only the rare input data.

2. Materials and Methods

2.1. Data Origin and Data Augmentation

The parameters of a battery are highly intercorrelated. During operation, it is important to accurately monitor the condition of the cells to ensure efficient usage and adequate loading. In the process of developing an ML model, considerable attention must be paid to the selection of features, as the results are highly dependent on the input data. The data processing steps include data collection, data cleaning, and data transformation [32]. Real-world data were obtained by loading an LIB in a battery test system. The experiments were conducted using the Molicel 21700 P42a battery cell, which has a capacity of 4.2 Ah with an end-of-charge voltage of 4.2 V $\pm$ 0.05 and a cut-off voltage of 2.5 V. The battery was cycled in a temperature chamber at a constant temperature of 23 °C. The test system was an OctoStat5000 from Ivium Technologies. The cell was discharged in ten percent intervals from a SoC of 100% to a SoC of 0%. Every 0.5 s, the voltage, current, and temperature of the cell were measured. A discharge cycle of the analyzed battery cell is shown in Figure 1.

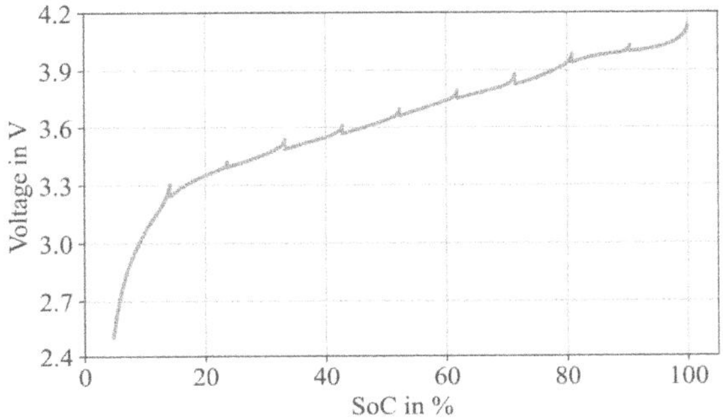

Figure 1. Discharge cycle of the analyzed battery cell with voltage plotted over the SoC. The cell is discharged in 10% intervals and rested after each interval.

Based on the real-world data, a data augmentation technique is applied to enrich the data basis for training the ML models. The input data for the algorithms, also called features, are current, temperature, and voltage. The output of the models, referred to as target value, is the SoC. To artificially create the new data, a whole discharge cycle is used and the current as well as the SoC values are kept constant. Two different regression models are trained for voltage and temperature, using a ridge regression. The loss function L, consisting of the squared error between observed y and predicted values y^*, is supplemented by an L2-penalty for the weight parameters w to decrease the risk of overfitting and to process the highly correlated data. This is summarized in (1).

$$L(w) = \frac{1}{2n} \sum_{i=1}^{n} (y^*(x_{ij}) - y_i)^2 + \lambda \sum_{j=1}^{p} w_j^2 \tag{1}$$

Based on the previous ten time steps, which is equivalent to five seconds cumulated with a sample rate of 0.5 s, the next value for voltage, respectively, temperature, is estimated. Both models are regression algorithms and, therefore, can be demonstrated in a regression formula. The regression equation for the voltage estimation is shown in (2), and for the temperature in (3). The target value is the current value of timestep t for voltage and temperature, respectively.

$$\begin{aligned} V_t = 0.003 &+ 0.078V_{t-10} + 0.079V_{t-9} + 0.080V_{t-8} + 0.085V_{t-7} + 0.090V_{t-6} \\ &+ 0.094V_{t-5} + 0.103V_{t-4} + 0.117V_{t-3} + 0.122V_{t-2} + 0.154V_{t-1} \end{aligned} \tag{2}$$

$$\begin{aligned} T_t = 0.080 &- 0.008T_{t-10} + 0.006T_{t-9} - 0.002T_{t-8} - 0.029T_{t-7} + 0.059T_{t-6} \\ &- 0.022T_{t-5} - 0.023T_{t-4} + 0.040T_{t-3} + 0.055T_{t-2} + 0.924T_{t-1} \end{aligned} \tag{3}$$

To keep the computation time and the processing power low, only every twentieth measurement point is calculated using the regression model. The data points between these reference points are interpolated. At first, the estimation models for voltage and temperature are tested against real measurements to validate their accuracy. Even though these models are highly accurate, a small error benefits the data augmentation as further randomness is included in the process. Following that, the real input data are slightly modified. Therefore, new reference points are estimated, and a new discharge curve is created. Again, the data between the reference points are interpolated. Demonstrated in Figure 2, the accuracy and two artificially created discharge curves are shown as an example for the voltage.

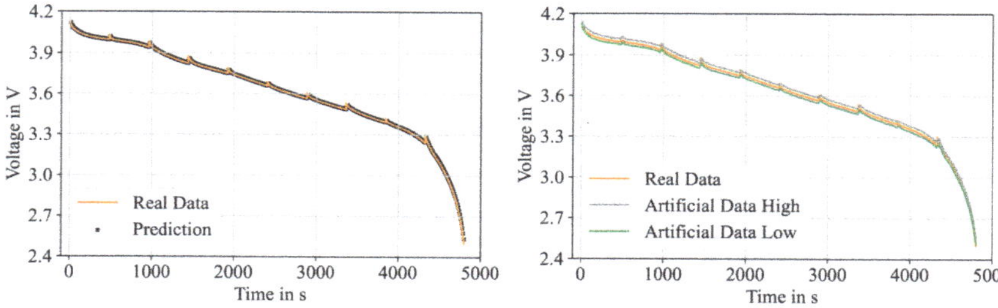

Figure 2. Results of the data augmentation method. On the left, the accuracy of the voltage estimation model for real data is presented. Real measurements are compared to the results of the voltage estimation model. On the right, results of the estimation model with slightly modified input values resulting in two artificially created discharge curves are shown.

Before the data are used in the ML models, they are preprocessed. At first, the data are normalized by using a standard scaler. By means of the mean values $\overline{x}$ and the standard deviation σ, the data are transformed, resulting in a data distribution with zero mean and unit variance [33]. This is performed for each battery parameter.

In a crucial step for the high performance of ML models [34]), the dimension of the data is reduced after the normalization, which can be referred to as the actual feature selection. In context of the curse of dimensionality and to reduce the risk of overfitting, a principal component analysis (PCA) is conducted. In sum, the curse of dimensionality describes the problem of an exponentially growing search space with an increasing amount of features [35]. A PCA is a method to reduce the dimension of a problem by analyzing the variance of the data [36], which is transformed into a new coordinate space with a lower dimension. This is especially useful while working with highly intercorrelated data [37]. The principal components are orthonormal axes, which cover a certain value of variance of the initial data. These components are given by the dominant eigenvectors of the sample covariance matrix. They can be identified by the largest corresponding eigenvalues [38]. The PCA is connected to the singular value decomposition (SVD), which is a matrix factorization that can also be used to reduce the dimension. A SVD is typically computationally more efficient, and because of the relation between the singular and the eigenvalues, the SVD in (4) can be used to calculate the principal components. M is the feature matrix, V is the orthonormal basis of the eigenvectors of $M^\top M$, $\sum$ is a diagonal matrix with the singular values θ_i, which are the square roots of the eigenvalues λ, and U is the orthonormal matrix, which can be calculated using (5) [39].

$$M = U \sum V^\top \tag{4}$$

$$u_i = \frac{1}{\theta_i} M v_i \tag{5}$$

After preprocessing the data, the features are used to train different ML models. The training data consist of 80% of the initial data plus the augmented data, and the test data make up the remaining 20%. These models are introduced in the following.

2.2. Machine Learning Models

2.2.1. Linear Regression

As a reference model, and to compare the results to a less complex model, a linear regression is implemented. Based on weight parameters w_i, which are determined during the training, a linear function between the input features voltage, current, and temperature, as well as the target values, is approximated. The estimated values y^* are calculated using (6) [40].

$$y^* = w_0 + \sum_{i=1}^{n} w_\mathrm{i} x_\mathrm{i} \tag{6}$$

By minimizing the squared error between the real and predicted values, the values for the weight parameters are determined. The corresponding loss function L is shown in (7).

$$L(w) = \frac{1}{2n} \sum_{i=1}^{n} (y^*(x_{ij}) - y_i)^2 \tag{7}$$

The reference model is an ordinary least square approach to determine the SoC based on measurement values of voltage, current, and temperature.

2.2.2. Artificial Neural Networks

Two types of ANNs are used to predict the SoC. The first one is an MLP, which is a simple feedforward neural network. All neurons from a layer are connected to the neurons from the next layer. The information is passed on in one direction to the output layer, where the target value is calculated. The inputs in a neuron are summed up and are then further processed in an activation function, where the output is calculated. The chain of mathematical functions is used to approximate the target values. To determine the appropriate parameters of the model, a four-fold-cross-validation was conducted by dividing the training data into four equally sized groups. One group acts as the validation set and the remaining groups as training data [41]. There are approaches in the literature using an MLP to estimate the SoC of a battery cell [42]. The structure of applied feedforward neural networks in the area of SoC estimation is mainly kept simple, with only a few hidden layers and a low number of neurons [43–45]. The structure of the proposed MLP consists of three hidden layers with ten neurons in both the first and the second hidden layer and five neurons in the third hidden layer. As activation function, the rectified linear unit (ReLU) function is used, which has another advantage of efficient model training. The final learning rate is 0.1 and the Adam optimizer is used to improve the training speed and ensure the accuracy of the estimation results [46].

The second applied model is a CNN. While the main application of CNNs is image processing, it is gaining more and more interest for other areas as well [47,48]. The first approaches to estimate the condition of a battery cell can be found in the literature. Most of the CNN-based models are used in combination with other ANNs [49,50]. The main difference for an MLP is the convolution by means of a kernel function. When analyzing time series data, the convolution is conducted along the time axis [51]. Typically, the kernel filter is followed by a pooling layer, where several points can be pooled in a single data point [52]. Instead of an MLP, where the neurons are fully connected, CNNs reduce and reassemble the feature matrix to learn new and complex patterns of the input data. The structure and the general approach of a CNN is shown in Figure 3 [53].

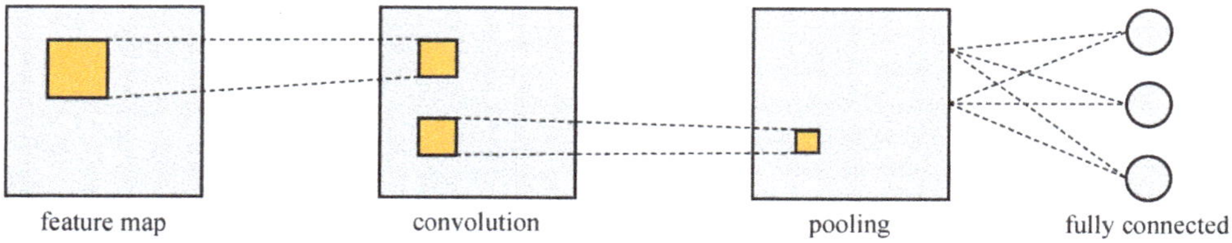

Figure 3. Structure of a CNN with a feature map followed by a convolution, a pooling layer, and a fully connected layer. The filter is moved across the features.

The applied CNN consists of two convolution layers with a pooling layer followed by another two convolutions and a pooling layer. During the convolution, 32 randomly initialized filters are used in the first part and 16 filters are used in the second part. When applying the filters, the window size is five timesteps along the temporal axis and the dimension is not affected during the convolution. The final layer is a fully connected layer, where the value for the SoC is estimated. As a metric during the training of the model, the coefficient of determination R^2 is used, which is shown in (8). The R^2 score is calculated by means of the real values y_i, the predicted values y_i^*, and the mean value $\overline{y}$.

$$R^2 = 1 - \frac{\sum_{i=1}^{n}(y_i - y_i^*)^2}{\sum_{i=1}^{n}(y_i - \overline{y})^2} \tag{8}$$

To validate the ML models, two metrics with the mean absolute error (MAE) and the root mean square error (RMSE) are used. Both use the difference between the real target values y and the estimated values y^* and are shown in (9) and (10).

$$MAE(y_i, y_i^*) = \frac{1}{n}\sum_{i=1}^{n}|y_i - y_i^*| \tag{9}$$

$$RMSE(y_i, y_i^*) = \sqrt{\frac{1}{n}\sum_{i=1}^{n}(y_i - y_i^*)^2} \tag{10}$$

3. Results

Three different ML models are trained, respectively, with and without data augmentation to evaluate the accuracy of the SoC prediction. The raw data are temperature, voltage, and current values. The test data are exclusively real data. To have a benchmark and a comparison for the neural networks, a linear regression model is used as reference. The input data in all models are the same. To analyze the impact of the data augmentation technique, the results are calculated without artificial data, with ten times the initial data, and with 20 times the initial data. All models are retrained five times and the mean values as well as the standard deviation are presented, as the initialization of the data augmentation technique and the neural networks is random and therefore slightly different. The results for the linear regression are shown in Table 1. The MAE and the RMSE are separated for the training and the test of the models. All mean values and the corresponding standard deviation are listed for the three different input datasets.

Table 1. Results of the linear regression with the three different sizes of training datasets. The mean of five times retraining the model and the corresponding standard deviations are shown.

Linear Regression	Training		Test	
	MAE	RMSE	MAE	RMSE
Without augmented data	3.874 (±0.021)%	4.941 (±0.012)%	4.089 (±0.205)%	4.999 (±0.095)%
With augmented data (10×)	3.914 (±0.029)%	4.970 (±0.018)%	4.041 (±0.220)%	4.980 (±0.100)%
With augmented data (20×)	4.004 (±0.027)%	5.066 (±0.036)%	3.977 (±0.050)%	5.044 (±0.052)%

The impact of the data augmentation method on the linear regression is small. In comparison to other ML algorithms, the dependence of a linear regression on a large database is slight, and an impact or a significant improvement was not expected. Accordingly, the results for all three different sizes of training data are similar and no influence of the data augmentation method can be measured. Nevertheless, it is possible to estimate the SoC with a simple linear model with error values below 5%. There is no overfitting and, consequently, the test errors are similar to the training errors. Nevertheless, the main focus of the linear model is to have reference accuracies for the neural networks.

Before analyzing the accuracies of both models, the convergence during the training of the models is examined. During the training, the weights are updated after each epoch. The training phase with the MAE over the epochs for the MLP is shown in Figure 4a and for the CNN in Figure 4b.

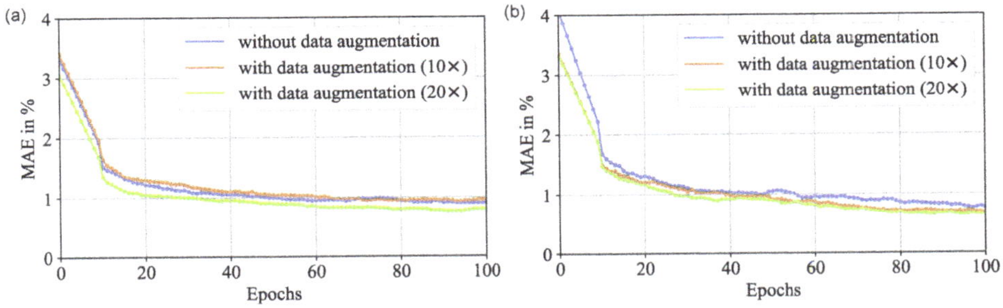

Figure 4. Convergence of the different models. MAE is shown over the epochs for the MLP (**a**) and the CNN (**b**) with and without augmented data.

A comparison of the convergence is drawn between the performance with different input data for both models. The convergence for the model without augmented data is indicated in blue, with ten times the initial data in orange, and with 20 times the initial data in green. As expected, the error is decreasing quickly over the first epochs, and then converges against a certain value. This behavior can be determined for all input data, but there are differences in the number of epochs to reach the final value and the final error itself. For both models, the augmented data are favorable for convergence, as the error decreases faster than it does without the augmented data. The difference is greater for the MLP, but also visible for the CNN. A faster convergence has the advantages of less training effort and a higher robustness, as there is less chance to become stuck in a local minima. Although the difference in the convergence behavior is apparent, it is in a smaller range and is, therefore, not significant for the optimization of the model. Further, regardless of the training dataset, each model converges to a similar value. In comparison, the CNN drops faster below an MAE of 1%, and the differences between the model with and without augmented data are slightly greater. This can also be shown when analyzing the accuracies of each model. The first examined ANN is the MLP. The preprocessed input data are passed through the layers of the MLP and the SoC is estimated. The results are summarized in Table 2. Similar to Table 1, the MAE and RMSE with the corresponding standard deviations are shown for the training and test of both models with the same three different input datasets.

Table 2. Results of the MLP with the three different sizes of training datasets. The mean of five times of retraining the model and the corresponding standard deviations are shown.

MLP	Training		Test	
	MAE	RMSE	MAE	RMSE
Without augmented data	0.828 (±0.292)%	1.072 (±0.329)%	0.553 (±0.051)%	0.805 (±0.072)%
With augmented data (10×)	0.626 (±0.184)%	0.848 (±0.190)%	0.539 (±0.087)%	0.758 (±0.109)%
With augmented data (20×)	0.722 (±0.222)%	0.977 (±0.242)%	0.727 (±0.217)%	0.978 (±0.246)%

Firstly, it can be noted that the model shows significantly higher accuracies than the linear model. Even the test RMSE is below 1% for all three datasets. Despite this notion, the impact of the data augmentation technique is small. The errors of the dataset with ten times the initial data are lower, but they are increasing with a higher amount of data. The standard deviation is also not highly impacted. The general ability to estimate the SoC can be determined. Additionally, the PCA is working efficiently, as no indications for overfitting can be detected. Further optimization with data augmentation is not necessary. Even though it does not deteriorate the performance of the model, the influence on the error and the standard deviation is low. The dimension reduction method improves the ratio between dimension and number of features and, thus, the influence of the data augmentation method on the simpler neural network is low.

In comparison to that, a CNN is used to learn new and complex patterns in the input data. This is conducted by using a convolution filter along the time axis. The results are summarized in Table 3.

Table 3. Results of the CNN with the three different sizes of training datasets. The mean of five times of retraining the model and the corresponding standard deviations are shown.

CNN	Training		Test	
	MAE	RMSE	MAE	RMSE
Without augmented data	0.975 (±0.459)%	1.173 (±0.531)%	0.505 (±0.201)%	0.723 (±0.249)%
With augmented data (10×)	0.371 (±0.269)%	0.494 (±0.315)%	0.315 (±0.140)%	0.478 (±0.124)%
With augmented data (20×)	0.261 (±0.071)%	0.392 (±0.102)%	0.270 (±0.068)%	0.437 (±0.101)%

Several aspects are striking when analyzing the results of the CNN. Without the augmented data, the CNN is slightly worse than the MLP. Additionally, the difference between training and test errors is higher, which is an indicator that there is potential for optimization. Further, the standard deviation is higher. In comparison to the MLP, there is a higher degree of randomness, as the filters for the convolution are randomly initialized. The model learns complex patterns in the data, but is not able to reproduce them with the limited database. Still, the standard deviation is not high and acceptable, but the fluctuation of the CNN is higher. By increasing the amount of training data, the errors are decreasing. While using ten times the initial data, the training MAE could be reduced by over 50% from 0.975% to 0.371%. With 20 times the initial data, the error could again be decreased to 0.261%. Further, the standard deviation is also decreasing, which means that the patterns in the data can be learned regardless of the filters. The filters vary while retraining the model. With a sufficiently large dataset, the impact of the filters and the uncertainty of the model is decreasing. The same behavior can be shown for the test data. The difference between training and test errors is small and, therefore, there is no overfitting. On the contrary, some test errors are slightly lower than the training errors using augmented data. The reason for that is the data augmentation technique, which should reflect a wider range of discharge behavior. Consequently, there is a higher variety in the training dataset, which could lead to higher errors; yet, still, the errors are nearly the same. The direct comparison of the errors of MLP and CNN with the corresponding standard deviations is shown in Figure 5. The error bars displaying the MAE are demonstrated for the different input data and, further, the uncertainty for retraining the model is shown. While the impact on the MLP is small, the optimization potential using augmented data is clearly visible for the CNN.

As stated before, the test data consist of only real data of a discharge cycle. Therefore, a direct comparison between real data and the prediction of the model can be drawn. For the MLP, this is shown in Figure 6a and for the CNN in Figure 6b. For a full discharge cycle, the voltage is shown over the SoC. The real experimental data are indicated in gray. The results of the final model with augmented data are demonstrated in orange.

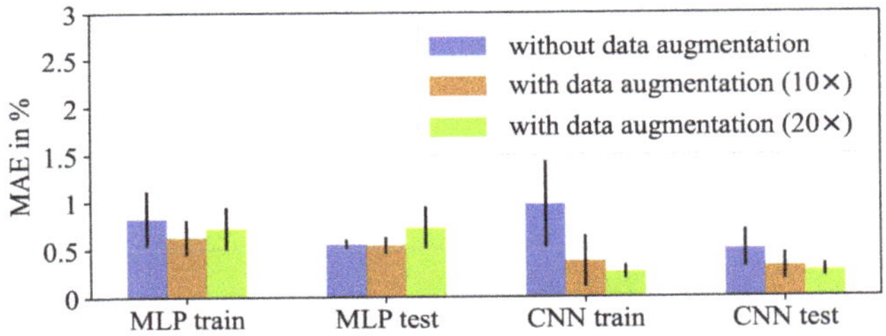

Figure 5. Error bars and standard deviation of the five times retrained models without data augmentation, with ten times the initial data and 20 times the initial data. The MAE is shown for the MLP and the CNN.

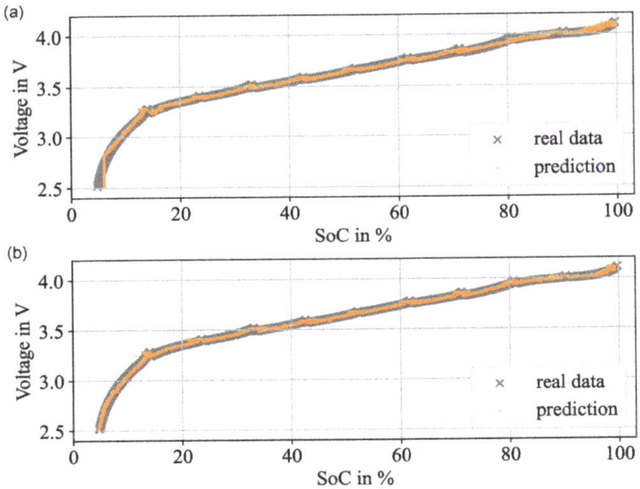

Figure 6. Test results of the SoC estimation model in comparison with the real values for the MLP (**a**) and for the CNN (**b**). The results for the SoC are shown over the last voltage value for the estimation.

The estimations and the real data are mainly overlapping. The drawbacks of the MLP can be seen in the high and low SoC areas and in the transition area from linear to nonlinear, where there are higher fluctuations in the estimation results. In the mainly linear range between a SoC of 20% to 80%, the performance of both models is similar. The CNN shows fewer outliers and, especially, the low SoC area can be accurately approximated. This behavior can explain the differences between a test MAE of the MLP with 0.727% and the CNN with 0.270%. As the edge areas are critical for an accurate state estimation, the CNN shows better results and is more suitable.

Overall, the performance of the CNN is slightly better. The impact of the data augmentation method is huge in relative values, but small in absolute values, as both neural networks show high accuracies by estimating the SoC without further data processing.

4. Discussion

When it comes to LIBs, a major challenge for balanced and safe loading cycles is the accurate determination of the condition of a battery cell. In comparison with conventional modeling techniques, data-driven models require less development time and no electrochemical characterization. Nevertheless, they need a reliable and large enough data basis, which covers the behavior of the cell. Especially for battery cells, for which tests and experiments are time- and cost-intensive, this is a key problem. Data augmentation, which is usually used for image processing, is a method to overcome this challenge by artificially creating new training data. In this case, two time-series forecasting models for voltage and temperature are developed. The accuracy of both models is high, which is shown in Section 2.1, but, moreover, the small error is favorable because of the additional randomness of the results. Therefore, new grid points are estimated and a wider range of input values is covered. Simultaneously, the current and the SoC values are kept constant. This technique relies on creating a whole discharge cycle, but it can also be used to enrich certain SoC areas. Only grid points are estimated because of the computing time. As a compromise between computing costs and accuracy, only every twentieth point is estimated. The points in between are interpolated. This interpolation does not impair the estimation and is hence sufficient for the data augmentation method. At first, a linear model is developed to predict the SoC. The errors are mainly below 5%. Although it could be shown that it is possible to determine the SoC with a linear regression, the accuracies cannot compete with conventional methods. Nevertheless, it is a reference model and a starting point to evaluate the results of the ANNs. In an MLP, the information is processed from input to output layer and a function between the features and the target values is approximated. The data augmentation technique has low impact on the accuracy of the MLP. A reason for that is the preprocessing method. ML algorithms are prone to overfit, when the dimension is equal to or greater than the number of features. This is the case for a feature matrix x_{ij}, where j is equal or greater than i. By reducing the dimension using the PCA, the tendency to overfit could be reduced. This is sufficient for the MLP, and the impact of the approach to artificially create additional data is decreased. Further, the testing data cover a limited range of loading conditions. Hence, analyzing the test errors does not capture the full capabilities of the optimized model. On the contrary, the CNN can determine more complex patterns, in which a huge database is beneficial. When examining the results of the CNN, it becomes apparent that the data augmentation technique increases the accuracy of the CNN. Two other advantages can be observed. First, the model converges earlier with the artificial data. Even though the improvement is small, there is less chance to become stuck in local minima, and fewer iterations are needed to train the model. Second, the standard deviation is decreasing, which results in a more robust model and an improved ability to reproduce the results.

Overall, the CNN shows better results and more potential for optimization. As the different loading conditions and their combinations are infinite in a real application, the ability to capture complex patterns in the input data is a key advantage of the CNN. The data augmentation method leads to improved results and the impact is expected to be higher for testing against several loading conditions. Still, the potential of data augmentation for the optimization of ANNs is evident. Further, the errors below 0.5% show that the CNN is able to accurately estimate the SoC. In comparison to conventional estimation models, such as representatives of the Kalman filter family, the errors could be slightly reduced [54–56]. Next to that, there is no need for an elaborately electrochemical characterization of the cell. This shows the huge potential to determine the condition of a battery cell with data-driven methods.

5. Conclusions

The electrification of the transport sector is inevitable to reach environmental goals. LIBs as electrochemical energy storage systems are a crucial factor for the success of electric vehicles. For efficient and optimized driving cycles, it is important to be able to determine the condition of the battery at all times. As ML algorithms reach high accuracies, are robust, and need less development and computing time, they are a promising alternative to conventional battery models. Nevertheless, they need a reliable and huge dataset to represent reality; however, battery tests are time- and cost-consuming. Therefore, the applicability of data augmentation was examined in this paper to optimize the ML models. The data of real-world experiments with the battery cell Molicel 21700 P42a were used. The training data were enriched by artificially created data by means of linear estimation models for voltage and temperature, and the impact on the ML models was analyzed.

The additional data improve the performance of the models in terms of convergence, robustness, and accuracy. Both neural networks succeed the linear model and are able to estimate the SoC with errors below 1%. The linear model serves as a reference model, but the final results with error values around 4% to 5% cannot compete with the ANNs. Comparing both ANNs, the CNN reaches the lower test error with an MAE of 0.27% and outperforms the MLP with an MAE of 0.539%. Therefore, the CNN is identified as the most suitable model. Further, the optimization methods have a higher impact on the CNN. By means of the data augmentation method, it is possible to nearly halve the test error from 0.505% to the lowest error of 0.27%. Thus, the data augmentation method shows itself as an effective way to optimize the estimation model.

In the future, it is planned to further develop the data augmentation technique. The current method consists of linear estimations for grid points. The values between the grid points are interpolated. It should be examined if the estimations can be improved using other approaches while keeping the computing time and efforts low. Further, the proposed algorithm should be tested against a higher variety of loading conditions to evaluate possible fields of application and to compete with traditional estimation approaches.

Author Contributions: Conceptualization, S.P.; methodology, S.P.; software, S.P.; validation, S.P. and A.N.; formal analysis, S.P. and A.N.; investigation, S.P.; resources, S.P. and A.M.; data curation, A.M.; writing—original draft preparation, S.P.; writing—review and editing, S.P., A.M., M.K., A.N., and T.W.; visualization, S.P.; supervision, M.K., A.N., and T.W.; project administration, M.K., A.N., and T.W.; funding acquisition, M.K., A.N., and T.W. All authors have read and agreed to the published version of the manuscript.

Funding: This research [project MORE] is funded by dtec.bw—Digitalization and Technology Research Center of the Bundeswehr, which we gratefully acknowledge. dtec.bw is funded by the European Union—NextGenerationEU. Further, we acknowledge financial support by the University of the Bundeswehr Munich.

Data Availability Statement: The data presented in this study are available on request from the corresponding author.

Conflicts of Interest: The authors declare no conflict of interest.

Abbreviations

The following abbreviations are used in this manuscript:

ANN	Artificial neural network
BEV	Battery electric vehicle
BMS	Battery management system
CNN	Convolutional neural network
LIB	Lithium-ion battery
MAE	Mean absolute error
ML	Machine learning
MLP	Multilayer perceptron
PCA	Principal component analysis
RMSE	Root mean square error
RNN	Recurrent neural network
SoC	State-of-Charge
SVD	Singular value decomposition
SVM	Support vector machine

References

1. Buberger, J.; Kersten, A.; Kuder, M.; Eckerle, R.; Weyh, T.; Thiringer, T. Total CO_2-equivalent life-cycle emissions from commercially available passenger cars. *Renew. Sustain. Energy Rev.* **2022**, *159*, 112158. [CrossRef]
2. Bernhart, W. Challenges and Opportunities in Lithium-ion Battery Supply. In *Future Lithium-Ion Batteries*; The Royal Society of Chemistry: London, UK, 2019; pp. 316–334. [CrossRef]
3. Schulte, J.; Figgener, J.; Woerner, P.; Broering, H.; Sauer, D.U. Forecast-based charging strategy to prolong the lifetime of lithium-ion batteries in standalone PV battery systems in Sub-Saharan Africa. *Sol. Energy* **2023**, *258*, 130–142. [CrossRef]
4. Stock, S.; Pohlmann, S.; Günter, F.J.; Hille, L.; Hagemeister, J.; Reinhart, G. Early Quality Classification and Prediction of Battery Cycle Life in Production Using Machine Learning. *J. Energy Storage* **2022**, *50*, 104144. .: 10.1016/j.est.2022.104144. [CrossRef]
5. Castro, F.D.; Cutaia, L.; Vaccari, M End-of-life automotive lithium-ion batteries (LIBs) in Brazil: Prediction of flows and revenues by 2030. *Resour. Conserv. Recycl.* **2021**, *169*, 105522. [CrossRef]
6. Boxall, N.J.; King, S.; Cheng, K.Y.; Gumulya, Y.; Bruckard, W.; Kaksonen, A.H. Urban mining of lithium-ion batteries in Australia: Current state and future trends. *Miner. Eng.* **2018**, *128*, 45–55. [CrossRef]
7. Wang, Y.; Tian, J.; Sun, Z.; Wang, L.; Xu, R.; Li, M.; Chen, Z. A comprehensive review of battery modeling and state estimation approaches for advanced battery management systems. *Renew. Sustain. Energy Rev.* **2020**, *131*, 110015. [CrossRef]
8. Lee, J.H.; Lee, I.S. Lithium Battery SOH Monitoring and an SOC Estimation Algorithm Based on the SOH Result. *Energies* **2021**, *14*, 4506. [CrossRef]
9. Shchurov, N.I.; Dedov, S.I.; Malozyomov, B.V.; Shtang, A.A.; Martyushev, N.V.; Klyuev, R.V.; Andriashin, S.N. Degradation of Lithium-Ion Batteries in an Electric Transport Complex. *Energies* **2021**, *14*, 8072. [CrossRef]
10. Bonfitto, A. A Method for the Combined Estimation of Battery State of Charge and State of Health Based on Artificial Neural Networks. *Energies* **2020**, *13*, 2548. [CrossRef]
11. Ng, K.S.; Moo, C.S.; Chen, Y.P.; Hsieh, Y.C. Enhanced coulomb counting method for estimating state-of-charge and state-of-health of lithium-ion batteries. *Appl. Energy* **2009**, *86*, 1506–1511. [CrossRef]
12. Chaoui, H.; Mandalapu, S. Comparative Study of Online Open Circuit Voltage Estimation Techniques for State of Charge Estimation of Lithium-Ion Batteries. *Batteries* **2017**, *3*, 12. [CrossRef]
13. Marcicki, J.; Canova, M.; Conlisk, A.T.; Rizzoni, G. Design and parametrization analysis of a reduced-order electrochemical model of graphite/LiFePO4 cells for SOC/SOH estimation. *J. Power Sources* **2013**, *237*, 310–324. [CrossRef]
14. Hossain, M.; Haque, M.E.; Arif, M.T. Kalman filtering techniques for the online model parameters and state of charge estimation of the Li-ion batteries: A comparative analysis. *J. Energy Storage* **2022**, *51*, 104174. [CrossRef]
15. Shrivastava, P.; Soon, T.K.; Idris, M.Y.I.B.; Mekhilef, S. Overview of model-based online state-of-charge estimation using Kalman filter family for lithium-ion batteries. *Renew. Sustain. Energy Rev.* **2019**, *113*, 109233. [CrossRef]
16. Luo, Y.; Qi, P.; Kan, Y.; Huang, J.; Huang, H.; Luo, J.; Wang, J.; Wei, Y.; Xiao, R.; Zhao, S. State of charge estimation method based on the extended Kalman filter algorithm with consideration of time–varying battery parameters. *Int. J. Energy Res.* **2020**, *44*, 10538–10550. [CrossRef]
17. Sharma, P.; Bora, B.J. A Review of Modern Machine Learning Techniques in the Prediction of Remaining Useful Life of Lithium-Ion Batteries. *Batteries* **2023**, *9*, 13. [CrossRef]
18. Hannan, M.A.; Lipu, M.S.H.; Hussain, A.; Ker, P.J.; Mahlia, T.M.I.; Mansor, M.; Ayob, A.; Saad, M.H.; Dong, Z.Y. Toward Enhanced State of Charge Estimation of Lithium-ion Batteries Using Optimized Machine Learning Techniques. *Sci. Rep.* **2020**, *10*, 4687. [CrossRef]
19. Basia, A.; Simeu-Abazi, Z.; Gascard, E.; Zwolinski, P. Review on State of Health estimation methodologies for lithium-ion batteries in the context of circular economy. *CIRP J. Manuf. Sci. Technol.* **2021**, *32*, 517–528. [CrossRef]

20. Álvarez Antón, J.C.; García Nieto, P.J.; Blanco Viejo, C.; Vilán Vilán, J.A. Support Vector Machines Used to Estimate the Battery State of Charge. *IEEE Trans. Power Electron.* **2013**, *28*, 5919–5926. [CrossRef]
21. Deng, Z.; Hu, X.; Lin, X.; Che, Y.; Xu, L.; Guo, W. Data-driven state of charge estimation for lithium-ion battery packs based on Gaussian process regression. *Energy* **2020**, *205*, 118000. [CrossRef]
22. Cui, Z.; Wang, L.; Li, Q.; Wang, K. A comprehensive review on the state of charge estimation for lithium-ion battery based on neural network. *Int. J. Energy Res.* **2022**, *46*, 5423–5440. [CrossRef]
23. Li, X.; Jiang, H.; Guo, S.; Xu, J.; Li, M.; Liu, X.; Zhang, X.; Bhardwaj, A. SOC Estimation of Lithium-Ion Battery for Electric Vehicle Based on Deep Multilayer Perceptron. *Comput. Intell. Neurosci.* **2022**, *2022*, 3920317. [CrossRef] [PubMed]
24. Li, S.; Ju, C.; Li, J.; Fang, R.; Tao, Z.; Li, B.; Zhang, T. State-of-Charge Estimation of Lithium-Ion Batteries in the Battery Degradation Process Based on Recurrent Neural Network. *Energies* **2021**, *14*, 306. [CrossRef]
25. Jiao, M.; Wang, D.; Qiu, J. A GRU-RNN based momentum optimized algorithm for SOC estimation. *J. Power Sources* **2020**, *459*, 228051. [CrossRef]
26. Qian, C.; Xu, B.; Chang, L.; Sun, B.; Feng, Q.; Yang, D.; Wang, Z. Convolutional neural network based capacity estimation using random segments of the charging curves for lithium-ion batteries. *Energy* **2021**, *227*, 120333. [CrossRef]
27. Bian, C.; Yang, S.; Liu, J.; Zio, E. Robust state-of-charge estimation of Li-ion batteries based on multichannel convolutional and bidirectional recurrent neural networks. *Appl. Soft Comput.* **2022**, *116*, 108401. [CrossRef]
28. Hu, W.; Peng, Y.; Wei, Y.; Yang, Y. Application of Electrochemical Impedance Spectroscopy to Degradation and Aging Research of Lithium-Ion Batteries. *J. Phys. Chem. C* **2023**, *127*, 4465–4495. [CrossRef]
29. Naaz, F.; Herle, A.; Channegowda, J.; Raj, A.; Lakshminarayanan, M. A generative adversarial network-based synthetic data augmentation technique for battery condition evaluation. *Int. J. Energy Res.* **2021**, *45*, 19120–19135. . [CrossRef]
30. Qiu, X.; Wang, S.; Chen, K. A conditional generative adversarial network-based synthetic data augmentation technique for battery state-of-charge estimation. *Appl. Soft Comput.* **2023**, *142*, 110281. [CrossRef]
31. Channegowda, J.; Maiya, V.; Joshi, N.; Raj Urs, V.; Lingaraj, C. An attention-based synthetic battery data augmentation technique to overcome limited dataset challenges. *Energy Storage* **2022**, *4*, e354. [CrossRef]
32. Fayyad, U.; Piatetsky-Shapiro, G.; Smyth, P. From Data Mining to Knowledge Discovery in Databases. *AI Mag.* **1996**, *17*, 37. [CrossRef]
33. Zhang, F.; Lai, T.L.; Rajaratnam, B.; Zhang, N.R. *Cross-Validation and Regression Analysis in High-Dimensional Sparse Linear Models*; Stanford University: Stanford, CA, USA, 2011.
34. Kondo, M.; Bezemer, C.P.; Kamei, Y.; Hassan, A.E.; Mizuno, O. The impact of feature reduction techniques on defect prediction models. *Empir. Softw. Eng.* **2019**, *24*, 1925–1963. [CrossRef]
35. Debie, E.; Shafi, K. Implications of the curse of dimensionality for supervised learning classifier systems: Theoretical and empirical analyses. *Pattern Anal. Appl.* **2019**, *22*, 519–536. [CrossRef]
36. Gruosso, G.; Storti Gajani, G.; Ruiz, F.; Valladolid, J.D.; Patino, D. A Virtual Sensor for Electric Vehicles' State of Charge Estimation. *Electronics* **2020**, *9*, 278. [CrossRef]
37. Yanai, H.; Takeuchi, K.; Takane, Y. *Projection Matrices, Generalized Inverse Matrices, and Singular Value Decomposition*, 1st ed.; Statistics for Social and Behavioral Sciences; Springer: New York, NY, USA, 2011.
38. Tipping, M.E.; Bishop, C.M. Mixtures of Probabilistic Principal Component Analyzers. *Neural Comput.* **1999**, *11*, 443–482. [CrossRef] [PubMed]
39. Zhou, N.; Zhao, X.; Han, B.; Li, P.; Wang, Z.; Fan, J. A novel quick and robust capacity estimation method for Li-ion battery cell combining information energy and singular value decomposition. *J. Energy Storage* **2022**, *50*, 104263. [CrossRef]
40. Joshi, A.V. *Machine Learning and Artificial Intelligence*; Springer International Publishing: Berlin/Heidelberg, Germany, 2019.
41. Yu, Y.; Feng, Y. Modified Cross-Validation for Penalized High-Dimensional Linear Regression Models. *J. Comput. Graph. Stat.* **2013**, *23*, 1009–1027. [CrossRef]
42. Chemali, E.; Kollmeyer, P.J.; Preindl, M.; Emadi, A. State-of-charge estimation of Li-ion batteries using deep neural networks: A machine learning approach. *J. Power Sources* **2018**, *400*, 242–255. [CrossRef]
43. Feng, F.; Teng, S.; Liu, K.; Xie, J.; Xie, Y.; Liu, B.; Li, K. Co-estimation of lithium-ion battery state of charge and state of temperature based on a hybrid electrochemical-thermal-neural-network model. *J. Power Sources* **2020**, *455*, 227935. [CrossRef]
44. He, W.; Williard, N.; Chen, C.; Pecht, M. State of charge estimation for Li-ion batteries using neural network modeling and unscented Kalman filter-based error cancellation. *Int. J. Electr. Power Energy Syst.* **2014**, *62*, 783–791. [CrossRef]
45. Hannan, M.A.; Lipu, M.S.H.; Hussain, A.; Saad, M.H.; Ayob, A. Neural Network Approach for Estimating State of Charge of Lithium-Ion Battery Using Backtracking Search Algorithm. *IEEE Access* **2018**, *6*, 10069–10079. [CrossRef]
46. Kingma, D.P.; Ba, J. Adam: A Method for Stochastic Optimization. *arXiv* **2017**, arXiv:1412.6980.
47. Pedrycz, W.; Chen, S.M. *Interpretable Artificial Intelligence: A Perspective of Granular Computing*, 1st ed.; Studies in Computational Intelligence; Springer International Publishing: Cham, Switzerland, 2021. [CrossRef]
48. Kiranyaz, S.; Avci, O.; Abdeljaber, O.; Ince, T.; Gabbouj, M.; Inman, D.J. 1D convolutional neural networks and applications: A survey. *Mech. Syst. Signal Process.* **2021**, *151*, 107398. [CrossRef]
49. Hu, C.; Ma, L.; Guo, S.; Guo, G.; Han, Z. Deep learning enabled state-of-charge estimation of LiFePO4 batteries: A systematic validation on state-of-the-art charging protocols. *Energy* **2022**, *246*, 123404. [CrossRef]

50. Cui, Z.; Kang, L.; Li, L.; Wang, L.; Wang, K. A hybrid neural network model with improved input for state of charge estimation of lithium-ion battery at low temperatures. *Renew. Energy* **2022**, *198*, 1328–1340. [CrossRef]
51. Hannan, M.A.; How, D.N.T.; Lipu, M.S.H.; Ker, P.J.; Dong, Z.Y.; Mansur, M.; Blaabjerg, F. SOC Estimation of Li-ion Batteries With Learning Rate-Optimized Deep Fully Convolutional Network. *IEEE Trans. Power Electron.* **2021**, *36*, 7349–7353. [CrossRef]
52. Rebala, G.; Ravi, A.; Churiwala, S. *An Introduction to Machine Learning*; Springer International Publishing: San Ramon, CA, USA; San Jose, CA, USA; Hyderabad, Telangana, India , 2019. [CrossRef]
53. Zhang, D.; Zhong, C.; Xu, P.; Tian, Y. Deep Learning in the State of Charge Estimation for Li-Ion Batteries of Electric Vehicles: A Review. *Machines* **2022**, *10*, 912. [CrossRef]
54. Tian, Y.; Xia, B.; Sun, W.; Xu, Z.; Zheng, W. A modified model based state of charge estimation of power lithium-ion batteries using unscented Kalman filter. *J. Power Sources* **2014**, *270*, 619–626. [CrossRef]
55. Li, X.; Huang, Z.; Tian, J.; Tian, Y. State-of-charge estimation tolerant of battery aging based on a physics-based model and an adaptive cubature Kalman filter. *Energy* **2021**, *220*, 119767. [CrossRef]
56. Wang, L.; Lu, D.; Liu, Q.; Liu, L.; Zhao, X. State of charge estimation for LiFePO4 battery via dual extended kalman filter and charging voltage curve. *Electrochim. Acta* **2019**, *296*, 1009–1017. [CrossRef]

energies

MDPI

Article

3D Heterogeneous Model for Electrodes in Lithium-Ion Batteries to Study Interfacial Detachment of Active Material Particles and Carbon-Binder Domain

Mohammadali Mirsalehian [1,*], Bahareh Vossoughi [1], Jörg Kaiser [2] and Stefan Pischinger [1,2]

1 Chair of Thermodynamics of Mobile Energy Conversion Systems (TME), RWTH Aachen University, Forckenbeckstraße 4, 52074 Aachen, Germany; bahareh.vossoughi@rwth-aachen.de (B.V.); pischinger_s@tme.rwth-aachen.de (S.P.)
2 FEV Europe GmbH, Neuenhofstraße 181, 52078 Aachen, Germany
* Correspondence: mirsalehian@tme.rwth-aachen.de

Abstract: Mechanics plays a crucial role in the performance and lifespan of lithium-ion battery (LIB) cells. Thus, it is important to address the interplay between electrochemistry and mechanics in LIBs, especially when aiming to enhance the energy density of electrodes. Accordingly, this work introduces a framework for a fully coupled electro-chemo-mechanical heterogeneous 3D model that allows resolving the inhomogeneities accompanied by electrochemical and mechanical responses of LIB electrodes during operation. The model is employed to numerically study the mechanical degradation of a nickel manganese cobalt (NMC) cathode electrode, assembled in a half-cell, upon cycling. As opposed to previous works, a virtual morphology for a high-energy electrode with low porosity is developed in this study, which comprises distinct domains of active material (AM) particles, the carbon-binder domain (CBD), and the pore domain to resemble real commercial electrodes. It is observed that the mechanical strain mismatch between irregularly and randomly positioned AM particles and the CBD might lead to local contact detachment. This interfacial gap, in combination with the diminishing contact strength over cell cycling, continuously deteriorates the electrode performance upon cycling by impedance rise and capacity drop. In agreement with previous experimental reports, the presented simulation results exhibit that the contact loss mostly takes place in the regions closer to the separator. Eventually, the resulting gradual capacity drop and change in impedance spectrum over cycling, as the consequence of interfacial gap formation, are discussed and indicated.

Keywords: lithium-ion batteries; electrode microstructure; heterogeneous physical model; mechanical degradation; electrochemical impedance spectroscopy

Citation: Mirsalehian, M.; Vossoughi, B.; Kaiser, J.; Pischinger, S. 3D Heterogeneous Model for Electrodes in Lithium-Ion Batteries to Study Interfacial Detachment of Active Material Particles and Carbon-Binder Domain. *Energies* **2023**, *16*, 7391. https://doi.org/10.3390/en16217391

Academic Editors: Daniel-Ioan Stroe and Simone Barcellona

Received: 5 September 2023
Revised: 11 October 2023
Accepted: 24 October 2023
Published: 1 November 2023

1. Introduction

Lithium-ion batteries have established themselves as the predominant energy storage system due to their superior energy and power densities, as well as their extended operational lifespan. Therefore, they have become the prevailing choice for energy storage across a diverse spectrum of applications. These applications encompass electric vehicles (EVs), consumer electronics, energy storage systems, and various eco-friendly industries [1]. Nonetheless, given the ever-growing demand within industries for heightened energy density, enhanced performance, and cost-effective solutions, ongoing optimization investigations remain a vital necessity in the pursuit of advancing LIB technology [2].

Given the substantial time and resources required for the development and characterization of new lithium-ion cell types in both research laboratories and commercial cell manufacturing facilities, computational modeling and simulation emerge as efficient tools, offering fast and in-depth insights at a microscopic level into the intricate processes occurring within lithium-ion electrodes during charging and discharging. In this context, the continuum model, a widely employed physics-based framework [3,4], simplifies the complex electrode microstructure by treating it as a continuum through the application of approximation methods. This simplification substantially reduces computational costs. However, it is essential to acknowledge that the existing insufficiencies inherent in the continuum model limit its capability in a range of applications [5–7].

The microstructural architecture and spatial arrangement of domains constituting Lithium-ion battery electrodes exert profound influence over the complicated interplay between diverse electrochemical and mechanical processes during LIB operation [8]. Consequently, to attain a comprehensive understanding of the effect of electrode structure on LIB performance, it becomes imperative to construct a virtual representation that emulates the actual, heterogeneous electrode structure. Creating a model from tomography imaging is an expensive and time-intensive endeavor, entailing the inevitable destruction of samples, thereby imposing limitations on iterative optimization and model enhancement, especially in the context of prototype electrodes or cells [9–11]. Additionally, the data derived from imaging techniques might be insufficient for comprehending battery electrochemistry and mechanics due to various constraints. In contrast, the computational modeling approach offers a conceptual design in the initial phases of research projects, facilitating optimization with respect to desired functionalities. In other words, virtual material testing and experimentation with diverse electrode properties can be conducted efficiently and cost-effectively [10].

Developing a heterogeneous model provides deep insights into localized non-uniformities within the electrode structure, leading to uneven utilization of active materials. Such inhomogeneity may result in degradation phenomena, like lithium plating on anode material particles [12,13], particularly those near the separator interface during fast charging of electric vehicles. Additionally, this model enables investigations into optimized battery operating parameters and contributes to the advancement of microstructural engineering strategies [5,14–16] aimed at mitigating aging phenomena within the electrode.

Mechanics play a pivotal role in determining both the performance and longevity of lithium-ion batteries. With the growing demand for extended cycle life, fast charging, and increased driving range in EV applications, mechanical degradation is one of the obstacles that directly regulates the mechanisms of capacity deterioration [17,18]. Therefore, the mutual impacts of electrochemistry and mechanics in LIB cells need to be addressed in the pursuit of electrodes with high energy density. In this regard, the study of electrochemical–mechanical interactions within an electrode structure can be effectively conducted through the application of heterogeneous models. This encompasses the examination of the impact of lithiation and delithiation on the evolution of mechanical strain and stress within active material particles [19,20], the influence of the mechanics on electrochemical response [21–24], and the exploration of potential fracture nucleation and propagation phenomena within the electrode structure [25–27]. Accordingly, the model enables the development of measures to suppress mechanical degradation, ranging from the electrode's geometrical features to mechanical properties to operational considerations such as cycling voltage windows.

In addition to the fracture inside active material as one degradation mechanism, the mismatch between the mechanical behaviors of AM particles and inactive materials may cause decohesion at the interfaces between them. The conductive additive, which is typically carbon black, provides the electronic conduction pathways between the active particles and the current collector (CC). The contact between the AM particle and the conductive additive could be lost during cycling, partially hindering the electronic path. As a result, the impedance of the electrode increases, and the capacity retention is reduced, leading to electrode performance degradation. In contrast to the particle fracture, the investigation of mechanical failure at the interface has received comparatively less attention. This disparity arises from the difficulty in experimentally resolving the intricate interfacial interactions between the active materials and the conductive agents. Nevertheless, the scanning electron microscope (SEM) images presented by some researchers confirmed that the continuing deformation of AM particles during cycling can result in decohesion at the interface between the active material and the conductive agents [28]. Figure 1 shows cross-sectional SEM images of an NMC cathode electrode before and after 10 cycles at 5 C-rate to visualize the interfacial debonding. While Figure 1a illustrates the existing cohesion between the particle and surrounding inactive matrix, the incurred interfacial decohesion in the cycled state is highlighted by the red dashed lines in Figure 1b. In the rapidly advancing field of solid-state batteries, the occurrence of delamination between particles and the solid electrolyte (SE) can be more pronounced compared to conventional batteries employing liquid electrolytes. Figure 2 contains SEM images showing the developing bond at the interface between NMC particles and solid electrolyte in three states: the pristine state, after the first cycle, and after 50 cycles. The SEM images distinctly reveal the NMC particles due to their characteristic, nearly spherical morphology. While the NMC particles and the solid electrolyte form a densely packed composite electrode in the pristine state, (Figure 2a,b), the NMC particles are surrounded by gaps after the first charging or after 50 cycles, (Figure 2c–f).

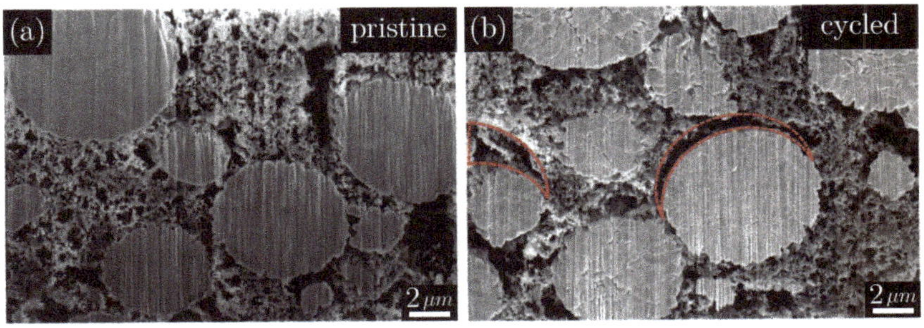

Figure 1. Cross-sectional SEM images of NMC particles and surrounding inactive matrix in (**a**) pristine state and (**b**) cycled state after 10 cycles at 5 C-rate of a lithium-ion cell cathode reported by Xu et al. [28] (shown with publisher's permission).

The mechanistic comprehension of LIB cell degradation faces a critical challenge due to the intrinsic heterogeneity in the structural configuration and the composition of the electrode. While Yang et al. [29] have demonstrated a noteworthy spatial and temporal dependency of damage in a commercial electrode, many prior experimental studies primarily concentrated on an idealized structure, such as thin films and individual particles. Others focused on localized regions within the electrode, characterized by limited dimensions, thereby being unable to provide a statistical representation of the overall chemomechanical behavior exhibited by the electrodes at a large scale. Hence, the exploration of heterogeneous chemomechanics and damage within lithium-ion batteries remains at its early stages of development, with a multitude of questions yet to be elucidated. The above-mentioned experimental challenges also exist in numerical investigations. The common continuum modeling approach and the extended single-particle approach oversimplify and assume free-standing spherical particles of the same size that are not subject to mechanical or electrochemical constraints from the adjacent environment. Recently, Baboo et al. [30] employed a simple single-particle model (SPM) and adjusted the solid diffusivity and specific active surface area to predict the effects of the formation of solid electrolyte interphase (SEI), side reactions, and AM particle fragmentation on the degradation of a LIB full cell performance during cycling. Their investigation was conducted for three different binders with diverse adhesion strengths. However, the simple SPM was unable to replicate the decreasing capacity during cycling, which was observed in the experiment. It can be attributed to the existing simplifications in SPM to express the complicated and spatially-resolved physicochemical interactions taking place inside the LIB cell during operation. Nonetheless, in commercial batteries, both the cathode and anode exhibit a composite nature characterized by substantial heterogeneity on the nano- to microscale, comprising irregularly shaped and sized active particles within a matrix of polymeric binders and conductive additives, along with pores containing electrolyte.

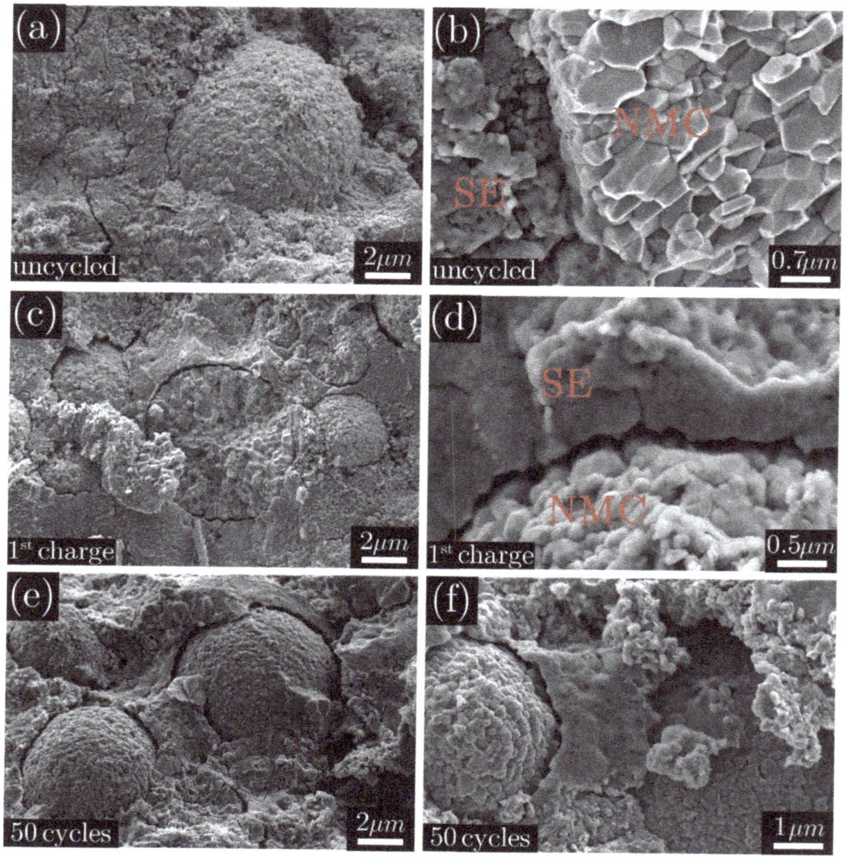

Figure 2. SEM images of the cathode composite of NMC particles embedded in the solid electrolyte (Li_3PS_4) (**a**,**b**) as prepared in pristine state, (**c**,**d**) after single charging at 0.1 C-rate and, (**e**,**f**) after 50 full cycles in the discharged state provided, by Koerver et al. [31] (shown with publisher's permission).

Recently, a few researchers have developed experimental setups and fully coupled electro-chemo-mechanical heterogeneous models to investigate the interface of AM particles and inactive matrix in electrodes. Xu et al. [28] employed experimental tomography to visualize nanoscale interfacial debonding between NMC particles and the conductive matrix. Their observations revealed that the interfacial detachment during cycling primarily occurs in the electrode region near the separator, rather than in the vicinity of the current collector. Moreover, Xu et al. built a microstructural-resolved model utilizing tomographic data to examine the heterogeneous damage within composite cathode particles situated within the binder matrix. However, the volumetric fraction of AM particles in the reconstructed virtual morphology by Xu et al. equates $\varepsilon_{\mathrm{AM}}^{\mathrm{ave}} = 40\%$ and is thus not as dense as the current commercial electrodes with high energy densities. Moreover, the pore and carbon-binder domains in the re-built morphology in their work are simplified and represented by a single integrated composite domain. Later, Liu et al. [32] also investigated the degradation of contacts between the AM particles and the inactive matrix in an NMC half-cell electrode using a reconstructed morphology with similar simplified geometrical features as developed in the work of Xu et al. [28]. Accordingly, Liu et al. adopted a phenomenological approach to reproduce the mechanical fatigue of cohesion along the AM particles and the inactive matrix during cycling. It was observed that the steady decay of the interfacial strength causes a growing interfacial debonding at cycling.

In this work, a 3D microstructural-resolved model is developed to study the impact of degradation of contact strength between the AM particles and CBD in the electrode in LIB cells during cycling. In contrast to previous models, the constructed virtual morphology of the electrode encompasses distinguished domains to represent the individual existing components inside the real electrode. Such a heterogeneous model enables predicting the electrochemical and mechanical behavior of the electrode's constituents more reliably. In the end, the influence of interfacial decohesion upon cycling is studied and discussed in terms of capacity retention and electrochemical impedance spectroscopy (EIS) in the time and frequency domains, respectively.

2. Methods

2.1. Electrode Microstructure

In this work, the heterogeneous model, introduced by the authors in their previous work [2], is employed to build a fully coupled electro-chemo-mechanical model and study the interfacial bonding of NMC622 AM particles and carbon-binder domain in a cathode electrode with conventional liquid electrolyte. The presented model allows to observe the heterogeneous electrochemistry, stress, and interfacial detachment in a commercial cathode electrode. Therefore, the spatial and temporal variations on different levels can be investigated. The offered modeling approach in this work can be further utilized to study the mechanical degradation in solid-state batteries as well.

As opposed to the developed cathode electrode's virtual morphology with low AM domain volumetric fraction by Xu et al. [28] and Liu et al. [32], a denser cathode electrode with $\varepsilon_{AM}^{ave} = 62.83\%$ has been built in this study to represent an electrode with higher energy density. In addition, the generated morphology contains separate distinguished domains of pore and CBD that increase the reliability of simulation results as compared to the assumed simplification made by the previous research to consider these two domains as one integrated domain.

Subsequently, a half-cell representative volume element (RVE) is assembled by incorporating the developed cathode electrode microstructure and two cubic domains. The resulting model explicitly encompasses the crucial half-cell components, including the separator, current collector, and lithium foil, as vividly illustrated in Figure 3. Table 1 includes the considered geometrical properties in the generation of the half-cell RVE.

The simulations are carried out in COMSOL Multiphysics® version 6.0 using an Intel® Xeon® CPU @ 3.70 GHz (2 processors) with 128 GB of RAM. Moreover, COMSOL LiveLink® for MATLAB® is used in the current work to transfer model's mesh data from MATLAB ® version R2020b to COMSOL Multiphysics® software to create the FEM model.

Table 1. Half-cell RVE microstructure generation specifications.

Property	Value
Cathode microstructure dimension	$50 \times 50 \times 25\ \mu m^3$
Average NMC particle radius r_p	2 µm and 5 µm
Particle radius standard deviation	0.2 µm
Active material volumetric fraction (ε_{AM}^{ave})	62.83%
Carbon-binder domain volumetric fraction (ε_{CBD}^{ave})	10.55%
Porosity (ε_e^{ave})	26.6%
Separator thickness	11 µm
Al current collector thickness	6.5 µm

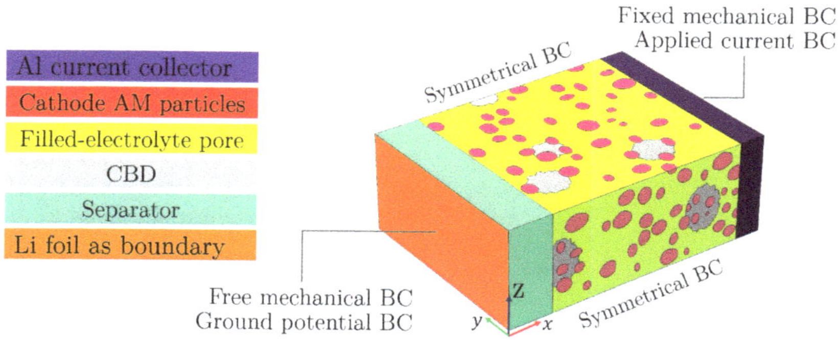

Figure 3. Developed 3D half-cell RVE model and associated boundary conditions (BCs) to study AM-CBD degrading contact.

2.2. Electro-Chemo-Mechanical Model

The governing equations to describe diverse coupled electrochemical and mechanical interactions within various domains, along with their corresponding parameters, are delineated in Table 2, Table 3 and Table 4, respectively. In the equations and associated parameters, the subscripts e and s denote the electrolyte and the solid-phase properties, respectively.

Table 2. List of physics implemented in different domains of the half-cell.

Domain/Boundary	Equation	No.
Electrochemistry		
AM, CBD, CC	$\nabla \cdot (\sigma_s^{\text{eff}\,*} \nabla_{\phi_s}) - j^{\text{total}\,**} = 0$	(1)
AM	$\frac{\partial c_s}{\partial t} = \nabla.(D_s \nabla c_s) - (\frac{D_s \Omega}{RT} \nabla c \cdot \nabla \sigma_h + c \nabla \cdot \nabla \sigma_h)$	(2)
Pore, CBD, separator	$\frac{\partial(\varepsilon_e{}^{***} c_e)}{\partial t} = \nabla \cdot D_e^{\text{eff}} \nabla c_e + \frac{1-t_+^0}{F} j^{\text{total}\,**}$	(3)
	$\nabla.(\kappa^{\text{eff}} \nabla \phi_e) + \nabla \cdot (\kappa_D^{\text{eff}} \nabla \ln(c_e)) + j^{\text{total}\,**} = 0$	(4)
CBD, Separator	$D_e^{\text{eff}} = (\varepsilon_e)^p \cdot D_e^{\text{bulk}}$	(5)
	$\kappa_e^{\text{eff}} = (\varepsilon_e)^p \cdot \kappa_e^{\text{bulk}}$	(6)
CBD	$\sigma_s^{\text{eff}} = (\varepsilon_s)^p \cdot \sigma_s^{\text{bulk}}$	(7)
AM-electrolyte interface	$j^f = A_v \cdot j_0 \left(\exp\left(\frac{\alpha_a F \eta}{RT}\right) - \exp\left(-\frac{\alpha_c F \eta}{RT}\right) \right)$	(8)
	$j_0 = F k c_s^{\alpha_c} (c_{s,\max} - c_s)^{\alpha_a} \left(\frac{c_e}{c_{e,\text{ref}}}\right)^{\alpha_a}$	(9)
	$\eta = \phi_s - \phi_e - U - \frac{\Omega \sigma_h}{F}$	(10)
	$j^c = A_v \cdot \frac{\partial(\phi_s - \phi_e)}{\partial t} \cdot C_{\text{DL}}$	(11)
	$j^{\text{total}} = j^f + j^c$	(12)
Mechanics		
All domains	$\nabla \cdot \sigma = 0$	(13)
	$\sigma = \mathbf{C} : \boldsymbol{\epsilon}_e$	(14)
	$\epsilon = \frac{1}{2}[\nabla \mathbf{u} + (\nabla \mathbf{u})^T]$	(15)
	$\epsilon = \boldsymbol{\epsilon}_\mathbf{e} + \boldsymbol{\epsilon}_\mathbf{Li}^{****}$	(16)
AM	$\boldsymbol{\epsilon}_\mathbf{Li} = \frac{\Omega}{3}(c_s - c_{s,0})\mathbf{I}$	(17)
AM-CBD interface	$K_{\text{int}} = K_{int,0} - k_d(n_{cyc} - 1)$	(18)

* σ_s^{eff} for CBD domain while σ_s^{bulk} in AM and CC domains. ** $j^{\text{total}} = 0$ in CC and separator domains. *** ε_e excluded in equation for pore domain. **** $\epsilon_{\text{Li}} = 0$ for all domains except AM.

2.2.1. Solid Domains

Ohm's law, Equation (1) in Table 2, is solved in the AM, CBD, and CC domains to describe electronic migration. Current collector domains are impermeable to lithium ions, so normal ion flux is set to zero at their boundaries.

At the interfaces between AM and the electrolyte, the Butler–Volmer (BV) equation is used to describe the electrochemical reaction kinetics or the volumetric faradaic current density j^f. In the BV equation, the total overpotential is modified by introducing a stress-biased term to account for the effect of mechanics on the kinetics of the faradic reaction. Accordingly, the fourth term on the right side of Equation (10) in Table 2 represents the stress-biased term when Ω and σ_h denote the partial molar volume of NMC particles and the hydrostatic stress, respectively. In addition, to account for the formation of the double layer (DL) at this interface, a homogeneous volumetric capacitive current density represented as j^c is incorporated alongside the faradaic current density. These components collectively contribute to the determination of the total volumetric current density, denoted as j^{total}, at these interfaces.

To describe solid-state diffusion, the formulation of Fick's second law is modified to account for the stress-induced diffusion flux, Equation (2) in Table 2. The additional terms inside the second bracket on the right side of this equation reflect the stress-driven flux of the lithium inside the AM particles. Then, at the interface between the active material and the electrolyte, the Neumann boundary condition is established to consider the interplay between the faradaic current density and the diffusion processes occurring within the AM particles.

2.2.2. Pore and Separator Domains

It is assumed that the pore domain located between the solid domains is completely saturated with the liquid electrolyte. The separator domain is treated as an electrolyte-permeable media with a porosity of 50%. To account for the separator's porous nature, the Bruggeman relation is utilized to predict the effective transport coefficients of the separator domain.

Also, a rather small porosity of 27.6% is assigned to the CBD to account for the reported porosity factor by Daemi et al. [33]. Accordingly, the electrolyte's effective ionic conductivity and diffusivity in CBD are obtained by the corresponding bulk values scaled by a factor of 0.276, following Boyce et al. [34].

The concentrated solution theory is employed to mathematically describe electrolyte transport. A porosity factor ε_e is introduced to Equations (3) and (4) in Table 2 within domains of the separator and the CBD to consider their porous nature, which is not explicitly included in their re-built modeled morphology.

2.2.3. Lithium Foil Boundary

The lithium foil is modeled as a boundary, and the BV equation is used to compute the volumetric current density associated with lithium metal deposition and dissolution at the interface between the lithium foil and the electrolyte. Here, the Butler–Volmer equation is formulated with zero overpotential in order to not introduce additional impedance. As a result, the calculated impedances are primarily attributed to contributions originating from the porous separator and cathode microstructure, saturated with electrolyte, and a current collector foil attached.

2.2.4. Mechanics

A mechanical analysis is performed to consider the effects of lithiation-induced deformation on both mechanical and electrochemical responses of the various electrode constituent domains. The heterogeneity of the reaction kinetics and the non-uniform mechanical behavior of the active and inactive materials lead to the development of stresses followed by mechanical degradation in the cell. Due to the rather slow kinetics of solid diffusion within AM particles, the mechanical equilibrium equation is solved in the absence of body forces, Equation (13) in Table 2, where σ denotes the Cauchy stress tensor. In the case of the cathodic NMC electrode with limited expansion upon lithiation, Hooke's law, Equation (14), is employed as the constitutive relation to describe the linear elastic material model. Accordingly, $\mathbf{C}$ and ϵ_e refer to the elastic stiffness tensor and the elastic strain, respectively. By assuming small strains and rigid body rotations, the Green-Lagrange strain tensor is reduced by the removal of quadratic terms, resulting in the well-known compatibility relation, Equation (15), which correlates the total strain tensor and the symmetric part of the gradient of the displacement field. The total strain is further decomposed into the elastic and lithiation-induced parts, as expressed in Equation (16). The lithiation-induced strain in the AM particles, $\boldsymbol{\epsilon}_{Li}$, is attained by Equation (17), in analogy to the thermal strain in materials, where **I** is the identity tensor. An initial active material concentration, $c_{s,0}$, is prescribed whilst the electrode and all constituent domains are assumed to be initially in an unstressed state. Equation (17) formulates the contribution of lithiation-induced deformation by the development of hydrostatic stress. As illustrated in Figure 3, while symmetrical boundary condition is applied on the RVE sides in the x-y and x-z planes, fully-clamped and free boundary conditions are allocated for the remaining two sides in the y-z plane at $x = l_{electrode}$ and $x = 0$, respectively.

In order to evaluate the strength along the interfaces between the AM particles and the carbon-binder domain, an interfacial cohesion model is required. Despite the model developed by Müller et al. [35], which considers a serial spring and a dissipative damper to reproduce the mechanical response of the contact between the AM particles and carbon-binder domain, the induced mechanical hysteresis effect is ignored in this work. Accordingly, a spring layer with zero thickness across the electrode represents the interfacial contact between the AM cathode particles and the CBD. The constant value of this spring per unit area in the pristine state is denoted as $K_{int,0}$. When the spring expands along the interface under tensile stress, the contact locally detaches, hindering electron transfer paths and elevated impedance. On the other hand, the bond is maintained for the remaining contacts, which are locally under compression or no stress. The above-mentioned spatial contact loss induces a drop in the local solid potential ϕ_s along the debonded contact, which in turn has a local impact on the charge conservation equation in the solid domain. Moreover, due to the introduced drop in solid potential by detachment, the overall overpotential η, as the driving force for electrochemical reaction, is retarded, and consequently, the local reaction kinetics described via the Butler–Volmer equation is impeded, resulting in degradation of battery cell performance.

Zhu et al. [36] demonstrated that the interfacial strength gradually decreases due to mechanical fatigue during cycling. Similar to the phenomenological approach presented by Liu et al. [32], the fatigue deterioration of contact between AM particles and CBD during cycling is realized in this work by a decreased spring constant along this interface. Equation (18) expresses the considered degradation of the spring stiffness per cycle, where k_d represents the rate at which the spring stiffness decreases per cycle, and n_{cyc} indicates the cycle number.

Table 3. Electrochemical parametrization of the half-cell model.

Parameters	Value	Ref.
NMC622 Particles		
AM solid conductivity (σ_s)	$1.6 \times 10^{-4}\,\mathrm{S \cdot m^{-1}}$	[34]
AM solid diffusivity (D_s)	$f(SoL)\,\mathrm{m^2 \cdot s^{-1}}$	[34]
max AM solid concentration ($c_{s,\max}$)	48,700 $\mathrm{mol \cdot m^{-3}}$	[34]
Initial AM solid concentration ($c_{s,0}$)	500 $\mathrm{mol \cdot m^{-3}}$	
Equilibrium potential (U)	$-324.2 \cdot SoL^8 + 1034.4 \cdot SoL^7 - 129.6 \cdot SoL^6 - 777.8 \cdot SoL^5 - 214.5 \cdot SoL^4 - 9.8 \cdot SoL^3 + 8.2 \cdot SoL^2 - 2.8 \cdot SoL + 4.4\,\mathrm{V}$	[34]
Kinetics		
Reaction rate constant (k)	$2 \times 10^{-11}\,\mathrm{m \cdot s^{-1}}$	[34]
Transfer coefficients (α_a, α_c)	0.5	
Surface double layer capacitance (C_{DL})	$0.2\,\mathrm{F \cdot m^{-2}}$	[28]
Bruggeman exponent (p)	1.0	[34]
Current collector		
Conductivity (σ_s)	$3.7 \times 10^{7}\,\mathrm{S \cdot m^{-1}}$	[34]

Table 3. *Cont.*

Parameters	Value	Ref.
Carbon-binder domain		
Conductivity (σ_s)	$375\,\mathrm{S \cdot m^{-1}}$	[34]
Electrolyte		
Initial electrolyte concentration ($c_{e,0}$)	$1000\,\mathrm{mol \cdot m^{-3}}$	[34]
Conductivity (κ_e)	$f(c_e)\,\mathrm{S \cdot m^{-1}}$	[37]
Diffusivity (D_e)	$f(c_e)\,\mathrm{m^2 \cdot s^{-1}}$	[37]
Activity ($\partial \ln f / \partial \ln c_e$)	0.43	[37]
Transference (t_+^0)	0.37	[37]

Table 4. Mechanical parameterization of the half-cell model.

Parameter	Value	Ref.
NMC622 particles		
Partial molar volume (Ω)	$1.8 \times 10^{-6}\,\mathrm{m^3 \cdot mol^{-1}}$	[34]
Young's modulus (E)	140 GPa	[34]
Current collector		
Young's modulus (E)	70 GPa	[34]
Carbon-binder domain		
Young's modulus (E)	0.3 GPa	[34]
Electrolyte		
Instantaneous shear modulus (G)	0.3 MPa	[38]
AM-CBD interface		
Spring constant per unit area in pristine state ($K_{int,0}$)	$2 \times 10^{15}\,\mathrm{N \cdot m^{-3}}$	[28]
Decreasing rate of spring constant per cycle (k_d)	$0.375 \times 10^{15}\,\mathrm{N \cdot m^{-3}}$	[28]

3. Results and Discussion

To visualize the expansion of NMC particles upon lithiation and the subsequent temporally and spatially resolved detachment along the contacts between the AM particles and the CBD, the half-cell's first discharge is simulated in the pristine state. The half-cell is initially in the fully charged state and is then galvanostatically discharged at 1 C-rate to 3 V. The applied current at different C-rates can be calculated as expressed below.

$$I_{\mathrm{app}} = \frac{c_{s,\max} \cdot \mathrm{V}_{\text{particles}}^{\text{total}} \cdot \mathrm{F} \cdot \mathrm{C}_{\text{rate}}}{3600\ \mathrm{s}}$$

where $\mathrm{V}_{\text{particles}}^{\text{total}}$ represents the total volume of the NMC particles in the reconstructed cathode electrode. While Figure 4a,b illustrate the state of lithiation (SoL) in the NMC particles across the cathode electrode at the beginning and end of the discharge process, their corresponding deformations are depicted in Figure 4c,d. At the beginning of discharge, SoL looks significantly uniform in the particles across the electrode, and as the discharge continues, non-uniformity becomes more noticeable. As can be seen in Figure 4b, some particles are still not highly lithiated at the end of the discharge process, indicating partially isolated particles, as also observed in previous experimental works on real electrodes [39–41]. Such heterogeneous electrochemical response in the electrode is due to the spatial arrangement and orientation of the different constituent domains in the electrode, resulting in different reaction kinetics.

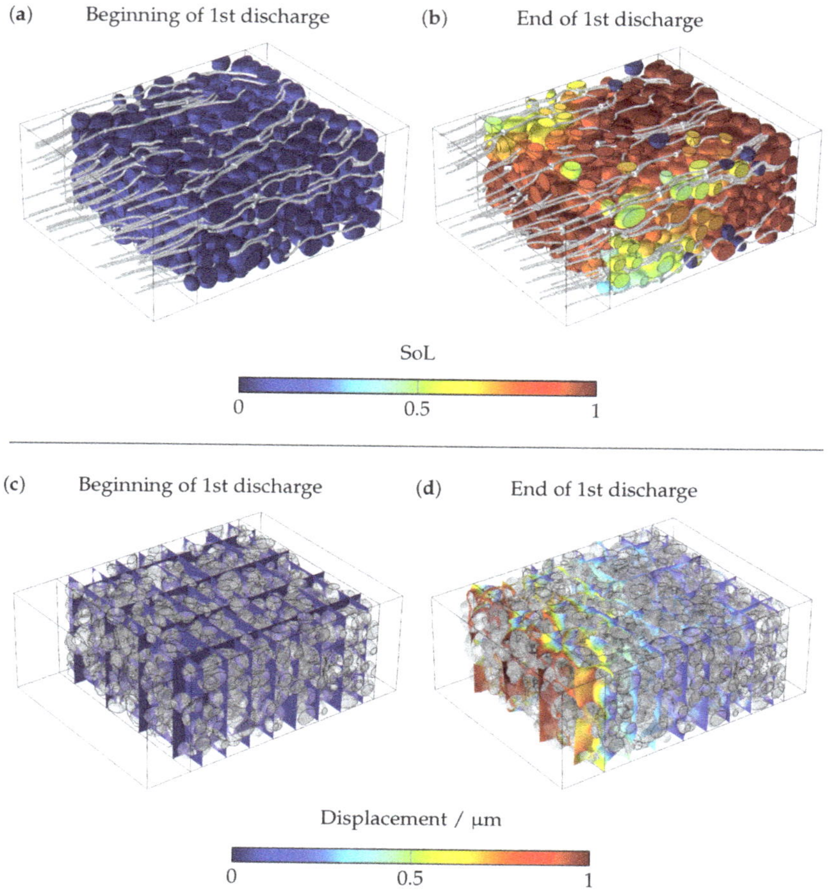

Figure 4. (**a**,**b**) State of lithiation (SoL) of NMC particles and lithium flux stream and, (**c**,**d**) mechanical displacement of the electrode at the beginning (time = 50 s) and end (time = 3220 s) of 1st discharge at 1 C-rate.

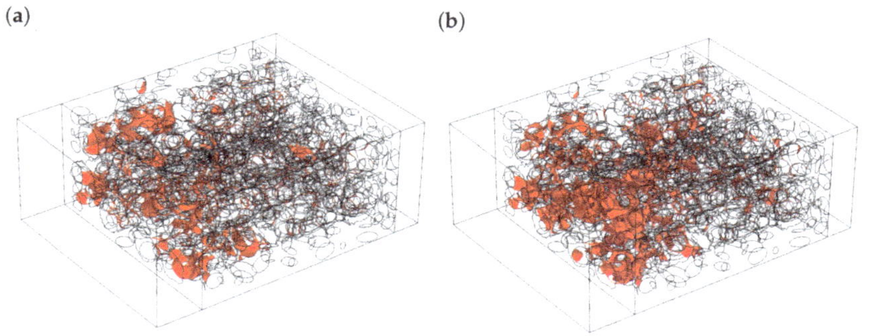

Figure 5. Evolution of the contact area undergoing interfacial detachment outlined by red at SoL = 50 % (**a**) at 1st cycle and (**b**) 5th cycle during discharge at 1 C-rate.

Figure 4d shows the deformed electrode at the end of discharge compared to the almost undeformed electrode at the beginning of the lithiation process, Figure 4c. The applied free boundary condition in the model, as shown in Figure 3, results in the expansion and contraction of the electrode in the thickness direction. Such displacement can form tension stress perpendicular to the contact boundaries between AM particles and CBD, depending on the spatial positioning and arrangement of the constituent domains. Accordingly, Figure 5a exhibits the interfacial area, which undergoes detachment at SoL = 50% upon the first discharge process. In agreement with what has been observed by means of the X-ray contrast tomography technique conducted by Xu et al. [28] and X-ray computed tomography (XCT) by Parks et al. [42], both the fractured AM particles and the detached contacts take place mostly close to the separator upon cycling. Moreover, the portion of contacts experiencing debonding directly relies on the intensity of the spatial mismatch between the mechanical response of the variously positioned and orientated AM particles and the CBD at each lithiated state of NMC particles.

To study the effect of the degrading contacts between AM particles and CBD on electrochemical performance during cycling, the fully charged half-cell is galvanostatically cycled at 1 C-rate for five sequential times in a defined cell voltage window of 3–4.3 V. Figure 5b exhibits the contact area experiencing decohesion at SoL = 50% during discharge at the fifth cycle. As noted, compared to the pristine state, a larger area undergoes detachment due to the degraded contact strength between AM particles and CBD during cycling. This outcome is in agreement with what was observed in SEM images conducted by Xu et al. [28], shown already in Figure 1, where the gap development between active and inactive materials occurs in the electrode during cycling. Moreover, Figure 6 illustrates the evolution of the cell characteristics during cycling. While Figure 6a exhibits the cell voltage and OCV, Figure 6b,c display the evolution of the averaged values for the state of lithiation and volumetric strain over the whole NMC particles during the cycling. In Figure 6d, the averaged interfacial gap associated with the detached contacts at the interfaces between AM particles and CBD is shown. As can be observed, the degrading contact during cycling, which is defined by Equation (18), results in a growing detachment as cycling proceeds. In other words, the diminishing spring constant through cycling leads to not only a larger detached area (Figure 5), but also to the formation of a larger decohesion along these contacts. Furthermore, the interfacial gap in Figure 6d shows an irreversible growth of detachment during cycling. This can be explained by the degradation of interfacial strength during cycling. The calculated interfacial gap maxima in Figure 6d varies from almost 10 nm in the first cycle to 20 nm in the fifth cycle. Liu et al. [32] performed a numerical study and attained a range of 50–65 nm for an NMC electrode cycled at 1 C-rate for five times. This discrepancy primarily originates from two different assumptions in the model built by Liu et al.: a fully clamped boundary condition was assigned to all RVE sides, and the domains of pore and CBD were simplified and represented by a single integrated composite domain with considerably high mechanical stiffness. While the former results in higher absolute values of interfacial gap during cycling, the latter leads to a narrower interfacial gap range compared to the computed values in this work.

The effect of interfacial detachment on the polarization of the cell, and hence on the capacity retention, is better illustrated in Figure 7, which shows the delivered discharge capacities, normalized by the theoretical capacity, during the cycling. The gradual decrease in discharge capacity during cycling is due to increasing impedance caused by increasing detachment. However, the more significant decrease in capacity from the first to the second discharges is mainly due to the higher initial cell voltage and more complete lithium insertion in the first discharge.

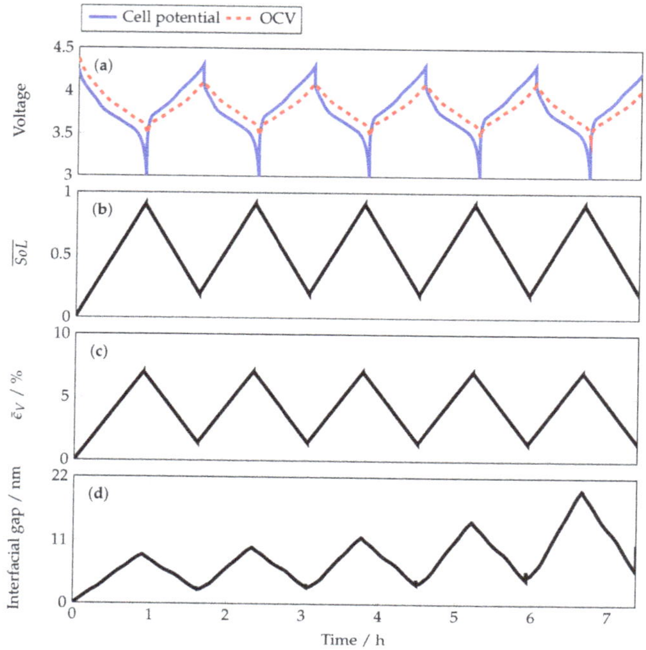

Figure 6. Evolution of (**a**) cell voltage and open circuit voltage (OCV), (**b**) averaged lithiation state ($\overline{SoL}$), (**c**) averaged volumetric strain ($\bar{\varepsilon}_V$) and, (**d**) growing averaged interfacial gap along AM particles and CBD, as cycling at 1 C-rate.

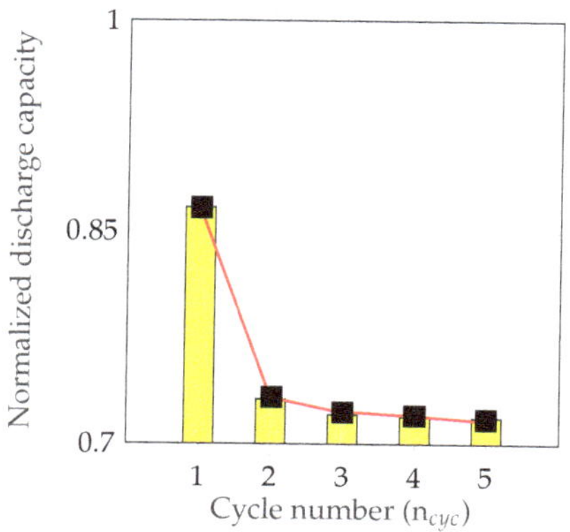

Figure 7. Evolution of normalized deliverable discharge capacity over cycling.

Electrochemical Impedance Spectroscopy (EIS)

As noted earlier, the development of such spatially resolved gaps along the contacts hinders the electron transfer paths and impedes the reaction kinetics. The resulting elevated overpotential caused by the interfacial debonding after only five cycles and its subsequent capacity retention, depicted in Figure 7, indicates that the delivered discharge capacity drops smoothly over the successive cycles. On the other hand, since electrochemical interactions are biased by various degradation mechanisms, electrochemical impedance spectroscopy is often used to characterize the deterioration of the battery's performance. Therefore, to quantify the effect of interfacial decohesion of AM particles and CBD on the performance of the half-cell, the impedance spectra are also numerically computed in different cycles. The EIS simulation is conducted around SoL = 50% as the equilibrium state. Then, a harmonically oscillating voltage signal with varying frequency and small magnitude is introduced to perturb the half-cell. While the input signal magnitude is set to 10 mV, its frequency is varied from 1 mHz to 100 kHz.

Figure 8 illustrates the computed impedance spectra for different studied cases. The calculated impedance spectrum is composed of two semicircles in the high- and mid-frequency ranges, followed by a sloping tail in the low frequencies. To quantify the effects of interfacial detachment on the dynamic response, the EIS simulation was initially carried out under the two conditions "with interfacial gap" and "without interfacial gap" (Figure 8a). As can be observed, when interfacial detachment is accounted for, the spectrum differs from the case with well-bonded contact. This variation primarily includes a shift of the spectrum to the right and also a larger dimension of the high-frequency semicircle, both due to the existing debonded area and hindered electronic transport. Moreover, the mid-frequency semicircle enlarges slightly, which is explained by the impeded reaction kinetics caused by the introduced local drop in the solid-phase potential at the particle surfaces detaching from the CBD. On the other hand, Figure 8b exhibits a similar but significantly more gradual change in the impedance spectra during cycling, which is due to the mechanical deterioration of the interfacial cohesion. Similar to the observed slight capacity decrease in Figure 7, the impedance spectrum slightly alters while cycling. It is certain that the strength of the interface would continue to decrease by further cycling after the fifth cycle, resulting in greater detachment at the interface and thus higher impedance.

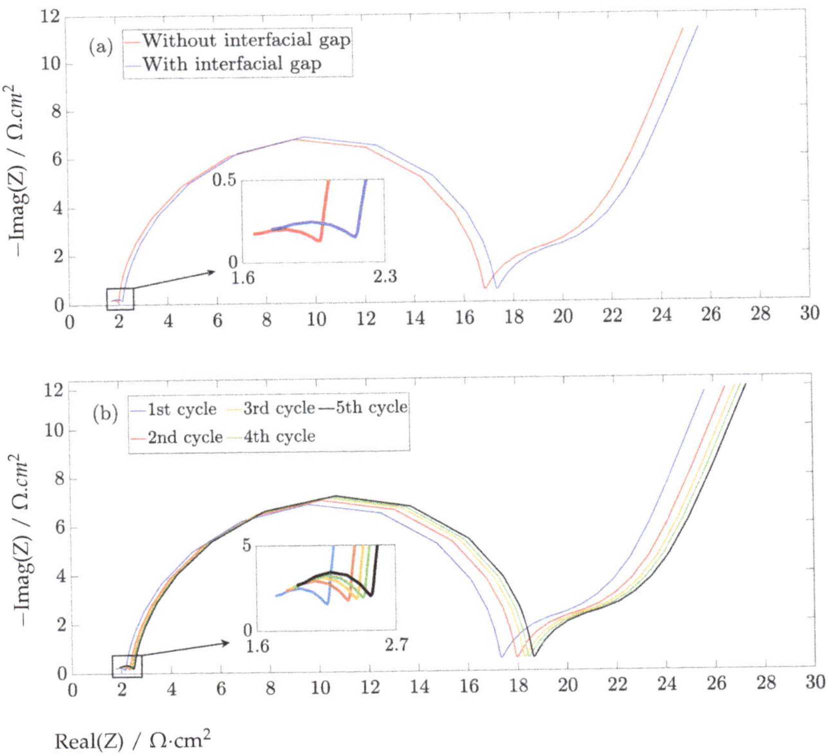

Figure 8. Nyquist plots of the impedance spectra of the half-cell calculated (**a**) with and without interfacial gaps in the pristine state and (**b**) upon cycling with growing area undergoing interfacial detachment.

4. Conclusions

Mechanical degradation is an obstacle on the way to high-energy-density electrodes. Different types of mechanisms lead to mechanical degradation of the LIB cell by contributing to the capacity fade and impedance increase. The numerical methods provide a low-cost and fast alternative to investigate mechanical degradation as compared to experimental tomography techniques. Such methods play an even more important role in the case of cyclic mechanical fatigue, which requires more resources for experimental testing.

In this context, a 3D half-cell's representative volume element (RVE) was built using the virtual morphology of the NMC622 cathode electrode to allow for exploration of the degrading contact of the AM particles and CBD in the electrodes. The half-cell was cycled and its characteristics during cycling were visualized. The degrading interfacial strength over five cycles was quantified in terms of the evolution of the interfacial gap in the time domain and the increased impedance in the frequency domain by means of numerical electrochemical impedance spectroscopy. In agreement with previous experimental tomography images, it was observed that most of the detachments occur in the regions close to the separator. Moreover, it was found that by cycling and degradation of contact between the AM particles and CBD, the impedance spectrum shifts to the right and the high-frequency semicircle enlarges. Furthermore, the mid-frequency semicircle's dimension expands as well. The former changes in the spectrum are due to the more retarded electronic transport paths, while the latter reflects the introduced drop in the solid-phase potential, hindering the reaction kinetics.

The framework presented in this work can be further employed to investigate different interfacial characteristics in LIBs. For instance, it enables determining the minimum required CBD cohesion strength, the optimized CBD volumetric content, and the optimized CBD morphology. Additionally, the framework can be used to study the critical role of delamination of the solid electrolyte (SE) in solid-state batteries. Such investigations provide a conceptual design in the early stages of a research project that allows optimization concerning desired functionality. In other words, virtual material testing and experimenting with various properties of the electrode constituents is possible in short time and with little cost.

Author Contributions: Conceptualization, M.M. and B.V.; methodology, M.M. and B.V.; software, M.M. and B.V.; validation, M.M. and J.K.; formal analysis, M.M. and B.V.; investigation, M.M.; resources, S.P.; data curation, M.M.; writing—original draft preparation, M.M. and B.V.; writing—review and editing, J.K.; visualization, M.M. and B.V.; supervision, J.K.; project administration, S.P. All authors have read and agreed to the published version of the manuscript.

Funding: This research received no external funding.

Data Availability Statement: Not applicable.

Conflicts of Interest: The authors declare no conflict of interest.

Abbreviations

The following abbreviations are used in this manuscript:

Latin Letters		Unit
A_v	Specific active surface area	$[\mathrm{m}^{-1}]$
$\mathbf{C}$	Elastic stiffness tensor	[Pa]
C_{DL}	Specific double layer capacitance	$[\mathrm{F}\cdot\mathrm{m}^{-2}]$
c_e	Electrolyte concentration	$[\mathrm{mol}\cdot\mathrm{m}^{-3}]$
c_s	Solid concentration	$[\mathrm{mol}\cdot\mathrm{m}^{-3}]$
$D_e^{\mathrm{bulk}}, D_e^{\mathrm{eff}}$	Bulk and effective electrolyte diffusivity	$[\mathrm{m}^2\cdot\mathrm{s}^{-1}]$
D_s	Solid diffusivity	$[\mathrm{m}^2\cdot\mathrm{s}^{-1}]$
F	Faraday constant	$[\mathrm{C}\cdot\mathrm{mol}^{-1}]$
$\mathbf{I}$	Identity tensor	
I_{app}	Applied current	[A]

j_0	Exchange current density	$[\mathrm{A}\cdot\mathrm{m}^{-2}]$
j^f	Volumetric faradaic current density	$[\mathrm{A}\cdot\mathrm{m}^{-3}]$
j^c	Volumetric capacitive current density	$[\mathrm{A}\cdot\mathrm{m}^{-3}]$
k	Reaction rate constant	$[\mathrm{m}\cdot\mathrm{s}^{-1}]$
$K_{int,0}$	Spring constant per unit area	$[\mathrm{N}\cdot\mathrm{m}^{-3}]$
K_d	Diminishing rate of spring constant	$[\mathrm{N}\cdot\mathrm{m}^{-3}]$
$L_{electrode}$	Electrode length	[m]
p	Bruggeman exponent	
R	Gas constant	$[\mathrm{J}\cdot\mathrm{mol}^{-1}\cdot\mathrm{K}^{-1}]$
r_p	Average particle radius	[m]
T	Temperature	[K]
U	Open circuit voltage	[V]
$\mathbf{u}$	Deformation field	[m]
Greek Letters		
α_a, α_c	Symmetry coefficient	
ε	Volumetric fraction	
ϵ_e	Elastic strain tensor tensor	
ϵ_{Li}	Lithium-induced strain tensor	
Ω	Partial molar volume	$[\mathrm{m}^3\cdot\mathrm{mol}^{-1}]$
ϕ_e	Electrolyte electrochemical potential	[V]
ϕ_s	Solid electrical potential	[V]
σ	Cauchy stress tensor	[Pa]
σ_h	Hydrostatic stress	[Pa]
$\kappa_e^{bulk}, \kappa_e^{eff}$	Bulk and effective electrolyte conductivity	$[\mathrm{S}\cdot\mathrm{m}^{-1}]$
κ_D^{eff}	Effective electrolyte diffusional conductivity	$[\mathrm{A}\cdot\mathrm{m}^2\cdot\mathrm{mol}^{-1}]$
$\sigma_s^{bulk}, \sigma_s^{eff}$	Bulk and effective solid conductivity	$[\mathrm{S}\cdot\mathrm{m}^{-1}]$
η	Overpotential	[V]
Abbreviations		
AM	Active material	
BC	Boundary condition	
BV	Butler–Volmer	
CBD	Carbon-binder domain	
CC	Current collector	
DL	Double layer	
EIS	Electrochemical impedance spectroscopy	
EV	Electric vehicle	
LIB	Lithium-ion battery	
RVE	Representative volume element	
SEM	Scanning electron microscope	
SoL	State of lithiation	
XCT	X-ray computed tomography	
Subscripts		
s, e	Solid, electrolyte	

References

1. Ansah, S.; Hyun, H.; Shin, N.; Lee, J.S.; Lim, J.; Cho, H.H. A modeling approach to study the performance of Ni-rich layered oxide cathode for lithium-ion battery. *Comput. Mater. Sci.* **2021**, *196*, 110559. [CrossRef]
2. Mirsalehian, M.; Vossoughi, B.; Kaiser, J.; Pischinger, S. 3D Heterogeneous Model for Electrodes in Lithium-Ion Batteries and Its Application to a Modified Continuum Model. *Batteries* **2023**, *9*, 298. [CrossRef]
3. Doyle, M.; Fuller, T.F.; Newman, J. Modeling of Galvanostatic Charge and Discharge of the Lithium/Polymer/Insertion Cell. *J. Electrochem. Soc.* **1993**, *140*, 1526–1533. [CrossRef]
4. Fuller, T.F.; Doyle, M.; Newman, J. Simulation and Optimization of the Dual Lithium Ion Insertion Cell. *J. Electrochem. Soc.* **1994**, *141*, 1–10. [CrossRef]
5. Fang, R.; Ge, H.; Wang, Z.; Li, Z.; Zhang, J. A two-dimensional heterogeneous model of lithium-ion battery and application on designing electrode with non-uniform porosity. *J. Electrochem. Soc.* **2020**, *167*, 130513. [CrossRef]
6. Tjaden, B.; Cooper, S.J.; Brett, D.J.; Kramer, D.; Shearing, P.R. On the origin and application of the Bruggeman correlation for analysing transport phenomena in electrochemical systems. *Curr. Opin. Chem. Eng.* **2016**, *12*, 44–51. [CrossRef]

7. Garcıa, R.E.; Chiang, Y.M.; Carter, W.C.; Limthongkul, P.; Bishop, C.M. Microstructural modeling and design of rechargeable lithium-ion batteries. *J. Electrochem. Soc.* **2004**, *152*, A255. [CrossRef]
8. Mistry, A.N.; Smith, K.; Mukherjee, P.P. Secondary-phase stochastics in lithium-ion battery electrodes. *ACS Appl. Mater. Interfaces* **2018**, *10*, 6317–6326. [CrossRef]
9. Tu, V. Modeling and Finite Element Simulation of the Bifunctional Performance of a Microporous Structural Battery Electrolyte. Master's Thesis, Chalmers University of Technology, Gothenburg, Sweden, 2019.
10. Feinauer, J.; Westhoff, D.; Kuchler, K.; Schmidt, V. 3D Microstructure Modeling and Simulation of Materials in Lithium-ion Battery Cells. In Proceedings of the Simulation Science: First International Workshop, SimScience 2017, Göttingen, Germany, 27–28 April 2017; pp. 128–144. [CrossRef]
11. Nguyen, T.T.; Demortière, A.; Fleutot, B.; Delobel, B.; Delacourt, C.; Cooper, S.J. The electrode tortuosity factor: Why the conventional tortuosity factor is not well suited for quantifying transport in porous Li-ion battery electrodes and what to use instead. *NPJ Comput. Mater.* **2020**, *6*, 1–12. [CrossRef]
12. Danner, T.; Singh, M.; Hein, S.; Kaiser, J.; Hahn, H.; Latz, A. Thick electrodes for Li-ion batteries: A model based analysis. *J. Power Source* **2016**, *334*, 191–201. [CrossRef]
13. He, X.; Bresser, D.; Passerini, S.; Baakes, F.; Krewer, U.; Lopez, J.; Mallia, C.T.; Shao-Horn, Y.; Cekic-Laskovic, I.; Wiemers-Meyer, S.; et al. The passivity of lithium electrodes in liquid electrolytes for secondary batteries. *Nat. Rev. Mater.* **2021**, *6*, 1036–1052. [CrossRef]
14. Lu, X.; Bertei, A.; Finegan, D.P.; Tan, C.; Daemi, S.R.; Weaving, J.S.; O'Regan, K.B.; Heenan, T.M.M.; Hinds, G.; Kendrick, E.; et al. 3D microstructure design of lithium-ion battery electrodes assisted by X-ray nano-computed tomography and modelling. *Nat. Commun.* **2020**, *11*, 2079. [CrossRef] [PubMed]
15. An, F.; Zhou, W.; Li, P. A comparison of model prediction from P2D and particle packing with experiment. *Electrochim. Acta* **2021**, *370*, 137775. [CrossRef]
16. Chouchane, M.; Rucci, A.; Lombardo, T.; Ngandjong, A.C.; Franco, A.A. Lithium ion battery electrodes predicted from manufacturing simulations: Assessing the impact of the carbon-binder spatial location on the electrochemical performance. *J. Power Source* **2019**, *444*, 227285. [CrossRef]
17. Zhao, Y.; Stein, P.; Bai, Y.; Al-Siraj, M.; Yang, Y.; Xu, B.X. A review on modeling of electro-chemo-mechanics in lithium-ion batteries. *J. Power Source* **2019**, *413*, 259–283. [CrossRef]
18. Iqbal, N.; Choi, J.; Lee, C.; Khan, A.; Tanveer, M.; Lee, S. A Review on Modeling of Chemo-mechanical Behavior of Particle–Binder Systems in Lithium-Ion Batteries. *Multiscale Sci. Eng.* **2022**, *4*, 79–93. [CrossRef]
19. Renganathan, S.; Sikha, G.; Santhanagopalan, S.; White, R.E. Theoretical analysis of stresses in a lithium ion cell. *J. Electrochem. Soc.* **2009**, *157*, A155. [CrossRef]
20. Cheng, Y.T.; Verbrugge, M.W. Evolution of stress within a spherical insertion electrode particle under potentiostatic and galvanostatic operation. *J. Power Source* **2009**, *190*, 453–460. [CrossRef]
21. Gao, X.; Lu, W.; Xu, J. Modeling framework for multiphysics-multiscale behavior of Si–C composite anode. *J. Power Source* **2020**, *449*, 227501. [CrossRef]
22. Liu, B.; Jia, Y.; Li, J.; Jiang, H.; Yin, S.; Xu, J. Multiphysics coupled computational model for commercialized Si/graphite composite anode. *J. Power Source* **2020**, *450*, 227667. [CrossRef]
23. Wang, M.; Xiao, X.; Huang, X. A multiphysics microstructure-resolved model for silicon anode lithium-ion batteries. *J. Power Source* **2017**, *348*, 66–79. [CrossRef]
24. Wu, W.; Xiao, X.; Wang, M.; Huang, X. A microstructural resolved model for the stress analysis of lithium-ion batteries. *J. Electrochem. Soc.* **2014**, *161*, A803. . jes. [CrossRef]
25. Ahmadi, M. A hybrid phase field model for fracture induced by lithium diffusion in electrode particles of Li-ion batteries. *Comput. Mater. Sci.* **2020**, *184*, 109879. [CrossRef]
26. Lee, S.; Yang, J.; Lu, W. Debonding at the interface between active particles and PVDF binder in Li-ion batteries. *Extrem. Mech. Lett.* **2016**, *6*, 37–44. [CrossRef]
27. Grantab, R.; Shenoy, V.B. Pressure-gradient dependent diffusion and crack propagation in lithiated silicon nanowires. *J. Electrochem. Soc.* **2012**, *159*, A584. [CrossRef]
28. Xu, R.; Yang, Y.; Yin, F.; Liu, P.; Cloetens, P.; Liu, Y.; Lin, F.; Zhao, K. Heterogeneous damage in Li-ion batteries: Experimental analysis and theoretical modeling. *J. Mech. Phys. Solids* **2019**, *129*, 160–183. [CrossRef]
29. Yang, Y.; Xu, R.; Zhang, K.; Lee, S.J.; Mu, L.; Liu, P.; Waters, C.K.; Spence, S.; Xu, Z.; Wei, C.; et al. Quantification of heterogeneous degradation in Li-ion batteries. *Adv. Energy Mater.* **2019**, *9*, 1900674. [CrossRef]
30. Baboo, J.P.; Yatoo, M.A.; Dent, M.; Hojaji Najafabadi, E.; Lekakou, C.; Slade, R.; Hinder, S.J.; Watts, J.F. Exploring different binders for a LiFePO4 battery, battery testing, modeling and simulations. *Energies* **2022**, *15*, 2332. [CrossRef]
31. Koerver, R.; Aygün, I.; Leichtweiß, T.; Dietrich, C.; Zhang, W.; Binder, J.O.; Hartmann, P.; Zeier, W.G.; Janek, J. Capacity fade in solid-state batteries: Interphase formation and chemomechanical processes in nickel-rich layered oxide cathodes and lithium thiophosphate solid electrolytes. *Chem. Mater.* **2017**, *29*, 5574–5582. [CrossRef]
32. Liu, P.; Xu, R.; Liu, Y.; Lin, F.; Zhao, K. Computational modeling of heterogeneity of stress, charge, and cyclic damage in composite electrodes of Li-ion batteries. *J. Electrochem. Soc.* **2020**, *167*, 40527. [CrossRef]

33. Daemi, S.R.; Tan, C.; Volkenandt, T.; Cooper, S.J.; Palacios-Padros, A.; Cookson, J.; Brett, D.J.; Shearing, P.R. Visualizing the carbon binder phase of battery electrodes in three dimensions. *ACS Appl. Energy Mater.* **2018**, *1*, 3702–3710. [CrossRef]
34. Boyce, A.M.; Martínez-Pañeda, E.; Wade, A.; Zhang, Y.S.; Bailey, J.J.; Heenan, T.M.; Brett, D.J.; Shearing, P.R. Cracking predictions of lithium-ion battery electrodes by X-ray computed tomography and modelling. *J. Power Source* **2022**, *526*, 231119. [CrossRef]
35. Müller, S.; Pietsch, P.; Brandt, B.E.; Baade, P.; De Andrade, V.; De Carlo, F.; Wood, V. Quantification and modeling of mechanical degradation in lithium-ion batteries based on nanoscale imaging. *Nat. Commun.* **2018**, *9*, 2340. [CrossRef] [PubMed]
36. Zhu, J.; Zeng, K.; Lu, L. Cycling effects on surface morphology, nanomechanical and interfacial reliability of $LiMn_2O_4$ cathode in thin film lithium ion batteries. *Electrochim. Acta* **2012**, *68*, 52–59. [CrossRef]
37. Valøen, L.O.; Reimers, J.N. Transport properties of LiPF6-based Li-ion battery electrolytes. *J. Electrochem. Soc.* **2005**, *152*, A882. [CrossRef]
38. Xiao, X.; Wu, W.; Huang, X. A multi-scale approach for the stress analysis of polymeric separators in a lithium-ion battery. *J. Power Source* **2010**, *195*, 7649–7660. [CrossRef]
39. Appiah, W.A.; Park, J.; Byun, S.; Cho, I.; Mozer, A.; Ryou, M.H.; Lee, Y.M. A coupled chemo-mechanical model to study the effects of adhesive strength on the electrochemical performance of silicon electrodes for advanced lithium ion batteries. *J. Power Source* **2018**, *407*, 153–161. [CrossRef]
40. Li, P.; Zhao, Y.; Shen, Y.; Bo, S.H. Fracture behavior in battery materials. *J. Phys. Energy* **2020**, *2*, 022002. [CrossRef]
41. Iqbal, N.; Lee, S. Mechanical failure analysis of graphite anode particles with PVDF binders in Li-ion batteries. *J. Electrochem. Soc.* **2018**, *165*, A1961. [CrossRef]
42. Parks, H.C.; Boyce, A.M.; Wade, A.; Heenan, T.M.; Tan, C.; Martínez-Pañeda, E.; Shearing, P.R.; Brett, D.J.; Jervis, R. Direct 4D Observations of Electrochemically Induced Intergranular Cracking in NMC811 Particles. *ChemRxiv* **2023**. [CrossRef]

energies

Article

Load Capacity of Nickel–Metal Hydride Battery and Proton-Exchange-Membrane Fuel Cells in the Fuel-Cell-Hybrid-Electric-Vehicle Powertrain

Ireneusz Pielecha *, Filip Szwajca and Kinga Skobiej

Institute of Combustion Engines and Powertrains, Faculty of Civil and Transport Engineering, Poznan University of Technology, 60-965 Poznan, Poland; filip.szwajca@put.poznan.pl (F.S.); kinga.skobiej@put.poznan.pl (K.S.)
* Correspondence: ireneusz.pielecha@put.poznan.pl

Abstract: This article investigates the impact of loading on the hybrid powertrain of the FCAT-30 model, equipped with a proton-exchange-membrane fuel cell (PEMFC) and a nickel–metal hydride (NiMH) battery. This study involves analyzing structural component performance based on voltage and current measurements of the fuel cell, battery, and powertrain. Tests conducted under different load conditions reveal significant differences in battery current and fuel-cell voltage, highlighting the crucial role of the battery in the powertrain. External loading induces cyclic operation of the fuel cell, generating peak power. The energy balance analysis demonstrates that, under no-load conditions, the vehicle consumes 37.3% of its energy from the fuel cell, with a total energy consumption of 3597 J. Under load, the energy from the battery is significantly utilized, resulting in a constant fuel-cell share of approximately 19%, regardless of the vehicle's load. This study concludes that the battery predominantly drives the powertrain, with the fuel cell acting as a secondary energy source. These findings provide valuable insights into the power distribution and energy balance in the hybrid powertrain. Using a load driving profile reduced the fuel-cell-stack energy contribution by 6.85% relative to driving without an external load.

Keywords: battery; fuel-cell vehicle; load capacity; hybrid powertrain; electric propulsion system; energy balance assessment; external load; chassis dynamometer

Citation: Pielecha, I.; Szwajca, F.; Skobiej, K. Load Capacity of Nickel–Metal Hydride Battery and Proton-Exchange-Membrane Fuel Cells in the Fuel-Cell-Hybrid-Electric-Vehicle Powertrain. *Energies* **2023**, *16*, 7657. https://doi.org/10.3390/en16227657

Academic Editor: Simone Barcellona

Received: 19 October 2023
Revised: 15 November 2023
Accepted: 16 November 2023
Published: 19 November 2023

1. Introduction

With constant technological advances and growing environmental awareness, the transportation sector is constantly looking for innovative solutions that meet the growing needs of mobility and minimize the negative effects on the environment. The share of vehicles using electric propulsion is steadily increasing in the global market [1–3]. Fuel-cell hybrid electric vehicles (FCHEVs) are a promising alternative, combining the advantages of electric vehicles with a proton-exchange-membrane (PEM) fuel cell [2].

Reducing dependence on fossil fuels, such as oil and natural gas, has gained worldwide acceptance. As a result, the development and intensive use of renewable energy has become an inherent trend. However, most renewable energy sources are characterized by intermittent access, which requires the development of efficient energy storage and generation systems [4,5]. Today, the use of renewable energy to produce hydrogen is gaining popularity and recognition, further supporting the development of fuel-cell technology.

Hydrogen production is expected to triple by 2050, driven by its falling cost [6]. In 2021, a total of 51,437 FCHEVs were registered worldwide [7]. There were 729 hydrogen refueling stations in operation at the time [6]. The results of the analysis by Samsun et al. [8] clearly indicate a very favorable trend in the development of fuel-cell vehicles and hydrogen refueling stations in 2021. The authors in their article [9] point out that for any vehicle with a range greater than 160 km (100 miles), fuel cells are superior to batteries in terms of weight, energy efficiency, and life-cycle costs.

A comparative analysis of internal-combustion-engine vehicles (ICEVs) and FCHEVs [10] shows that emissions, maintenance, operating costs, and efficiency are much more favorable for FCHEVs.

Fuel cells, unlike many traditional energy sources, do not emit pollutants during the power generation process. The electrochemical process in fuel cells is based on a chemical reaction between hydrogen and oxygen, generating clean electricity, with the only byproduct being water. This makes fuel cells one of the greenest solutions for electricity generation [11–14]. The battery in the vehicle performs the function of managing the dynamic response of the vehicle under varying load conditions [10].

The fuel cells used in the first prototype vehicles (in 2002) achieved a volumetric power factor of 1.0 kW/dm^3 with a mass power factor of 0.75 kW/kg [15]. In the FCHV model (in 2008), these ratios were 1.45 kW/dm^3 and 0.9 kW/kg, respectively. The first-generation Toyota Mirai, equipped with nickel–metal hydride (NiMH) batteries, had values of 3.1 kW/dm^3 and 2 kW/kg, while the new generation of the Mirai (Li-ion battery) vehicle achieves 5.4 kW/dm^3 (4.4 kW/kg excluding end plates) and 5.4 kW/kg, respectively [16,17].

Honda used 103 kW fuel cells in the Clarity model, for which the volumetric and mass power factors were 3.1 kW/dm^3 and 2.0 kW/kg, respectively [18,19]. The system uses a Li-ion battery with a capacity of 25.5 kWh [20]. The Hyundai Nexo uses a 95 kW cell and a 40 kWh battery [21,22]. The BMW iX5 Hydrogen is equipped with a 125 kW cell and a battery with a very small capacity of 2 kWh [23,24].

The development of Li-ion batteries can be seen in terms of changes in specific energy: from 90 Wh/kg in 1990 to more than 250 Wh/kg today [25–27]. Their cost has now been reduced from $1000 to about $250 per kilowatt-hour (kW/h) [25,28].

A very important aspect is the study of PEM fuel cells during different loads from the perspective of optimizing their performance under different operating conditions, which directly affects the efficiency of the entire FCHEV system [29].

Optimal operating conditions for a fuel cell are crucial for its efficient operation. In this context, maintaining an optimal temperature plays an important role. Fuel cells, especially those based on PEM and solid-oxide-fuel-cell (SOFC) technology, show their best performance in a tightly controlled temperature range. For PEM fuel cells, maintaining a temperature of 60–80 °C is crucial, while SOFCs require higher temperatures, typically above 500 °C [30–33].

In addition, it is important to maintain appropriate humidity levels. Fuel cells, especially PEMs, are sensitive to ambient humidity, and ensuring proper humidity levels helps maintain proper proton conductivity, which affects the efficiency of electrochemical processes inside the cell [34,35]. The moisture content of the input gases supplied to the anode and cathode has been shown to have a significant impact on fuel-cell performance. Yan et al. [36] showed that lowering the humidity at the cathode negatively affects the steady-state and dynamic performance of the fuel cell. In the context of humidity control, in addition to the effect of humidity on the proton conductivity of membranes, it is worth noting that large-scale commercialization of proton-exchange fuel cells requires achieving higher power and current densities. Nonetheless, at high operating current densities, liquid water accumulation can lead to flooding problems and impede gas diffusion, which accelerates cell performance degradation [37,38]. Therefore, it is important to improve water management capabilities to achieve better cell efficiency.

Controlling the cleanliness of the fuel supplied to the cell also has an important role, as impurities, such as sulfur and particulate matter, can negatively affect the performance of the electrodes, resulting in reduced cell performance [39]. Ensuring the quality and purification of the fuel is key to maintaining optimal operating conditions [40]. Nonetheless, cells show great flexibility due to their ability to use a variety of fuels, such as hydrogen, methanol, natural gas, and biogas [41,42]. Despite the significant manufacturing cost of fuel cells, the benefits of this advanced solution are significant enough to make investment in them economically viable [43,44].

This research aims to determine how the external load on a chassis dynamometer affects the control strategy of a hybrid powertrain equipped with a fuel cell (low-temperature) and a nickel–metal hydride battery. By analyzing similar values of the energy stored in the battery and fuel cell, changes in the shares of energy coming from the two drive sources are identified. Due to the availability of the test object, this article examines a vehicle using nickel-based batteries. The presented research evaluates the impact of external loading on the drivetrain of the FCAT-30 model, a hydrogen hybrid vehicle equipped with a PEM fuel cell and NiMH battery. This study aims to understand the performance of structural components under different load conditions and extends the analysis to include the energy balance. The vehicle was subjected to tests with and without external loading on a chassis dynamometer, allowing for the measurement of voltage and current of the battery, fuel cell, and drivetrain.

2. Research Methodology

2.1. Study Object

This research was conducted using a model vehicle equipped with a hybrid drive system: a PEM-type fuel cell along with a NiMH battery (Figure 1). The remotely controlled 4 × 4 vehicle uses a low-temperature PEM-fuel-cell stack (Table 1) with two hydrogen storage tanks of 10 dm^3 each, with a maximum hydrogen pressure of 30 bar. Mounted measurement systems allow real-time recording of typical vehicle movement parameters along with data acquisition of battery and fuel-cell voltage and current.

Figure 1. FCAT30 hydrogen model of hybrid powertrain with NiMH battery (photo owner).

Table 1. Vehicle model technical parameters.

Parameter	Unit	Value
	Fuel cell	
Fuel-cell type	–	PEM
Number of cells	–	14
Power	W	30
Hydrogen pressure	MPa	0.045–0.055
Cell-stack mass	g	280
H_2 flow at maximum Ne	dm^3/min	0.42
System efficiency	%	40 (at max power)
	Battery	
Type	–	NiMH
Max output voltage	V	7.2
Electric capacity	mAh	4200

Table 1. *Cont.*

Parameter	Unit	Value
	Hydrogen storage	
Tank capacity	Ndm^3	2 × 10
Purity	%	≥99.995
Form of storage	–	AB5—metal hydrides
Tank pressure	MPa	3.0
Tank dimensions	mm × mm	ϕ22 × 88

A static analysis of the energy flow (Figure 2) shows comparable values for the energy stored in the battery and hydrogen tanks, with a difference of about 20%, in favor of the NiMH battery. The battery used directly interacts with the energy management system, unlike the hydrogen energy, which is converted to electricity using a PEM fuel cell.

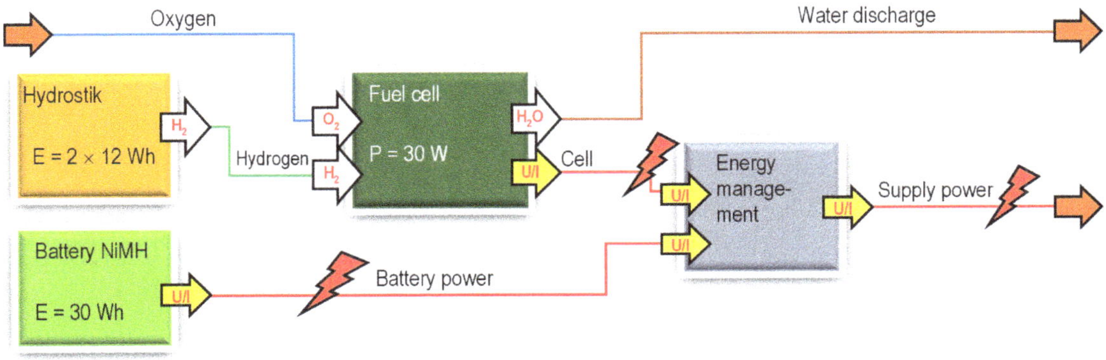

Figure 2. Diagram of energy flow in a hybrid drive system.

The detailed scheme of the system is shown in Figure 3. The vehicle was propelled by the Mabuchi RS-540SH-7520 electric motor, operating within a voltage range of 4.8–7.2 V. The motor attains its peak efficiency of 67% under the following operational parameters: P = 63.2 W, n = 19740 rpm, I = 13 A, Mo = 30.6 mNm. The metering system incorporates a fuel cell, battery, and energy flow controller from CREA Technologie. Owing to proprietary "know-how", the complete system control details remain undisclosed. Arduino circuits were employed to manage the energy flow. The hybrid system is dependent on the chemical-energy battery, and its functionality is possible without the fuel-cell system. As depicted in Figure 3, there is the capability to recharge the electrochemical battery. Measurement data can be directly showcased in LabView or, alternatively, stored on an SD card for subsequent offline analysis on a computer.

This implies the possibility of a much higher load on the battery compared with the fuel cell. It follows that the share of the battery in variable load situations should be greater than that of the fuel cell. The energy flow diagram indicates a parallel drive system, which directly affects the voltage and current values in the results.

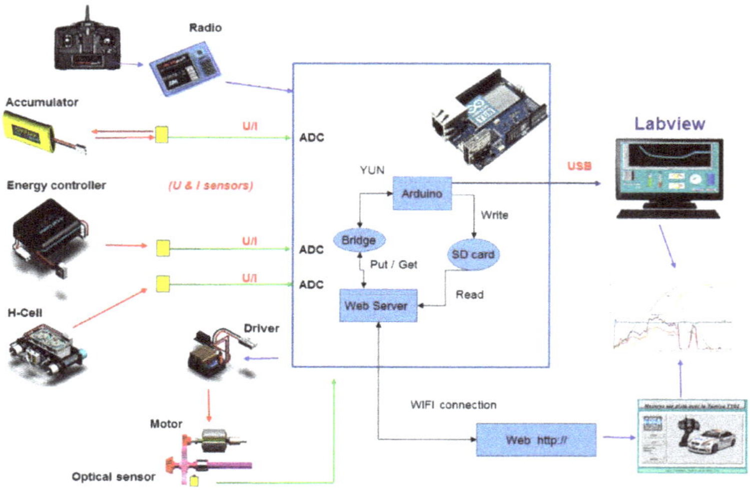

Figure 3. A view of the components of the hybrid propulsion system, including the data acquisition and visualization system.

2.2. Chassis Dynamometer

Tests on the operation of the battery and fuel cell under varying load conditions were carried out on a chassis dynamometer (Figure 4). The dynamometer model used in the experiment has two axles, each equipped with two rollers. Load generation for the drivetrain is carried out through a friction brake, which acts directly on one of the rollers of the vehicle's rear axle. To ensure an even load on all wheels, the axles are connected by a toothed belt. The load generated is transmitted as the force with which the friction lining is pressed against the roller in direct contact with the vehicle's tire. In the context of the analysis, the load on the system is expressed in the basic SI unit of force (N). The movement of the friction lining is controlled by a servo motor, and the force is measured by an additional sensor.

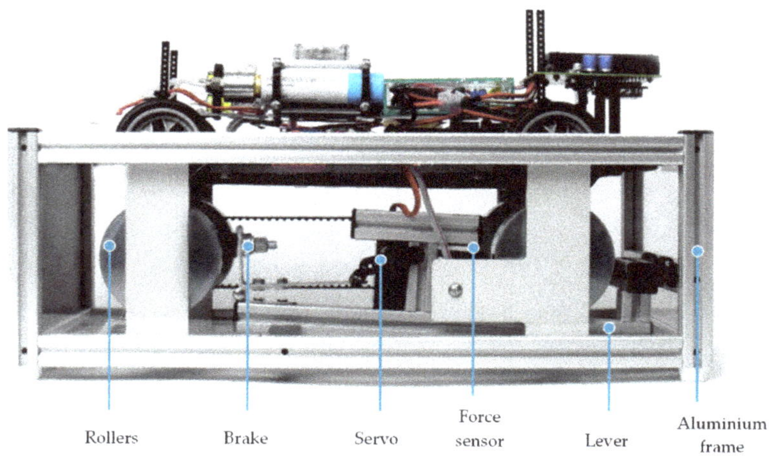

Figure 4. View of the chassis dynamometer with its most important components.

Electrical parameters and some mechanical parameters are recorded using the measurement system. The measurement frequency was about 50 Hz during the acquisition of all 10 measurement data (Figure 5). The program allows recording electrical parameters while visualizing the values of excess voltage, battery current, and driving speed (set using the Setpoint function—Figure 5). The software was developed in the LabView environment using the Arduino platform. Data recording takes place on an SD card mounted in the vehicle, but the application functions wirelessly, using Bluetooth data transmission.

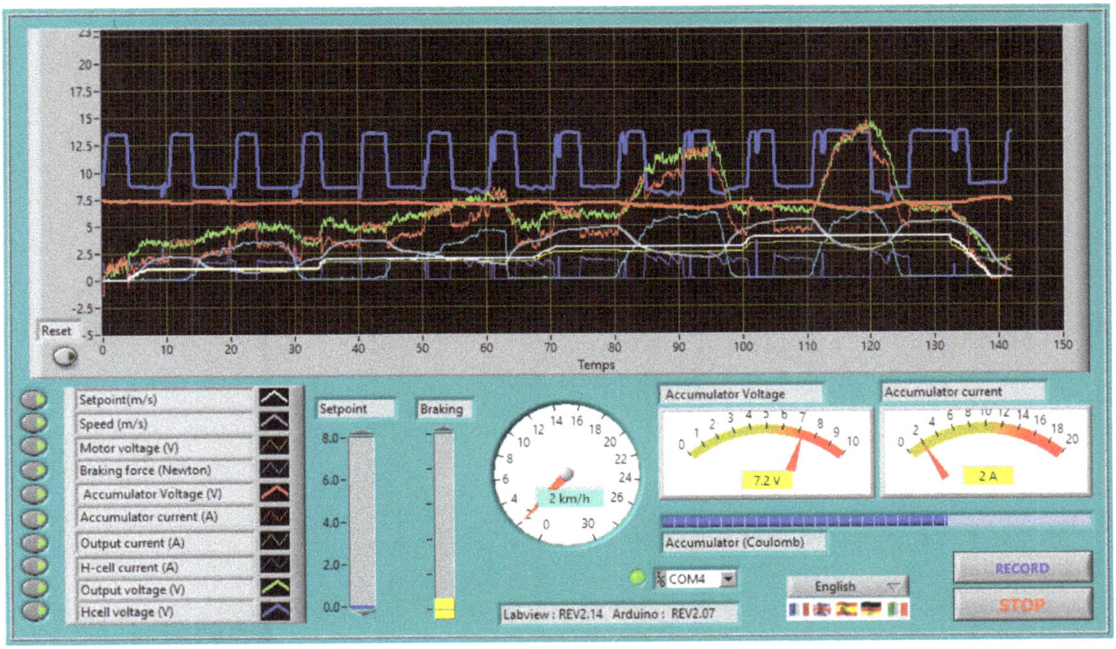

Figure 5. View of the chassis dynamometer software screen.

3. Scope of the Study

Research on the FCAT-30 vehicle's powertrain was carried out in two variants: (a) without external loading and (b) with external loading. The load was introduced by using a chassis dynamometer. In the first variant, the rollers were not additionally braked; the load resulted solely from the moment of inertia of the system driven by the dynamometer.

Tests on the hybrid model vehicle were carried out according to the profiles shown in Figure 6. These profiles are shown as a function of percentage of elapsed time due to the different durations of the test. The no-load speed profile includes several jumps in speed on the chassis dynamometer, initially with an increasing and then decreasing trend (see Figure 6a). The profile with external load included only an incremental increase in driving speed (see Figure 6b) with a simultaneous cyclic increase in load. In both cases, the driving speed was limited to 6 m/s, and the maximum external load occurring was 6.2 N. The test durations without load and with load were 100 and 140 s, respectively. The goal was not to directly compare the two tests but only to evaluate the behavior of the drivetrain (energy flow) under these conditions. For this reason, differences in distance or test duration were ignored.

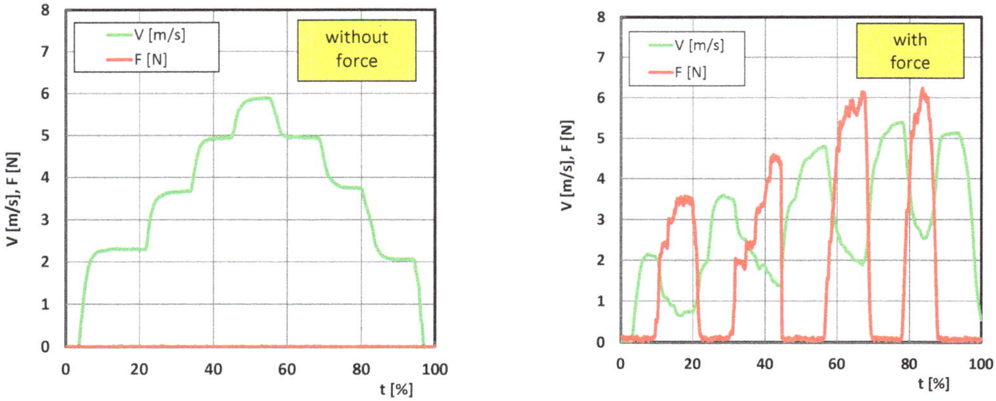

(a) (b)

Figure 6. Vehicle speed profiles on the chassis dynamometer: (**a**) without load and (**b**) with load.

4. Hybrid Propulsion Research

The speed profiles presented earlier (Figure 6) clearly show the differences between external load and no load. Eight parameters were measured during the tests, and their course as a function of test duration is shown in Figure 7. A direct comparison of the results for the two cases analyzed (with load and without load) reveals significant differences in the values of battery current (I_BATT) and cell voltage (U_FC).

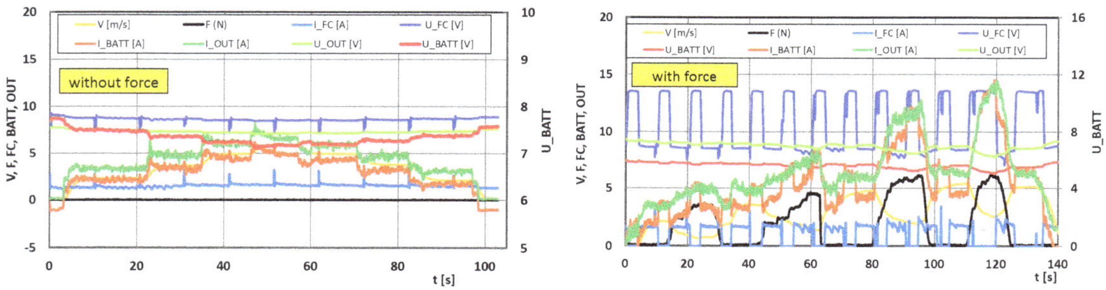

(a) (b)

Figure 7. Test results of measured quantities during hybrid drive tests: (**a**) without load and (**b**) with load.

Tests of the drivetrain with no load show small changes in the battery voltage (U_BATT) in the range of 7.0 to 7.6 V, with simultaneous changes in the battery current (I_BATT) in the range of up to 5 A (see Figure 7a). Under load, the voltage change is similar, but the current change ranges from zero to almost 15 A at the highest brake load (see Figure 7b). These changes are proportional to the strength of the brake load.

At no brake load, a fuel cell operating at about 8–9 V generates about 1–1.5 A. When loaded, the voltage values of U_FC are much higher, exceeding 13 V, and the current is about 2 A at the minimum voltage value of U_FC.

In both cases, the output voltage is close to the voltage U_BATT, and the current I_OUT is related to the sum of both intensities: the battery and the cell. In the case of additional load with force F, cyclic operation of the fuel cell is observed, manifested in the absence of current (i.e., no-load operation). In this case, the resultant current intensity is close to that of the battery.

According to the analysis shown in Figure 7, noticeably higher current values are obtained from the battery than from the fuel cell. This suggests that the fuel cell is not the main source of energy for vehicle propulsion. Such operating conditions differ from standard powertrains, such as in the Toyota Mirai [15] and Hyundai Nexo models. According to the research described in [45,46], the fuel-cell energy contribution is significantly higher, which is a direct result of the high-powered cell compared with the low energy of the battery.

5. Power Share of Propulsion System Components

An analysis of the power transferred to the drivetrain, broken down into the power of the battery, the cell, and the output power of the system in no-load and external-load modes, is shown in Figure 8. The power was calculated as the product of the voltage and current on each component. Under no-load conditions (Figure 8a), the powertrain obtains a maximum of 40 watts from the battery and about 12 watts from the fuel cell. Loading the system (Figure 8b) increases the power of the battery to 90 W and the cell to a maximum of 15 W (continuous power). Finally, without load, the system obtains about 50 W of power, while during load, it obtains about 90 W.

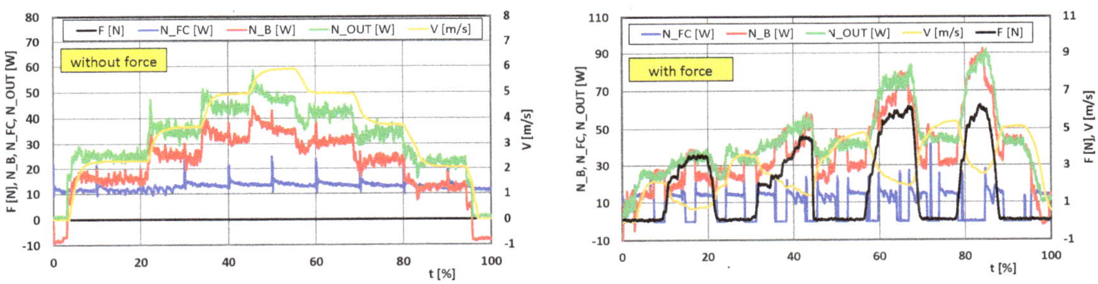

(a) (b)

Figure 8. Magnitudes of cell and battery power during speed profile analysis: (**a**) without load and (**b**) with load.

Under no-load conditions, the power is the resultant power of the battery and the cell, with the battery contributing significantly. During load, the battery's share also dominates. The fuel-cell share in both cases remains at about 15 W.

The fuel-cell value of 15 watts represents 50% of its claimed power. Under no-load conditions, the fuel-cell share is 12:26 W, which is less than 50%. At high vehicle speeds, the share drops to 12 W out of 50 W, or 24%. So, the fuel-cell share is not particularly high. With an external load, these values max out at a ratio of 15 W to 90 W, which is 16% (at maximum load, t = 85%). At low load (t = 5%), the values are 15 W to 28 W, respectively, which is 53%. The described method of controlling the hybrid powertrain makes it possible to protect the fuel cell from power peaks that can contribute to the degradation.

The conclusion is that the drivetrain mainly uses battery power at significant vehicle speed, with the fuel cell acting as a secondary drive source. Nevertheless, the system controller always uses both energy sources: the battery and the fuel cell.

6. Analysis of Fuel-Cell Operation

In the current chapter, operating conditions were analyzed for the fuel cell only, ignoring the other components of the drive train (Figure 9). No load (Figure 9a) results in small changes in the fuel-cell voltage. During the no-load operation of the cell, there is a noticeable drop in current between the peaks that occur, probably related to the drop in ionic conductivity. The decrease in conductivity occurs with the accumulation of water and nitrogen in the anode channels, which are removed when the anode is flushed with hydrogen. The flushing phenomenon is accompanied by the aforementioned voltage and current peaks, which occur cyclically, and loss of hydrogen [47]. The same phenomenon occurs when the vehicle's drivetrain is externally loaded (Figure 9b). This process has also been described by other researchers [47,48].

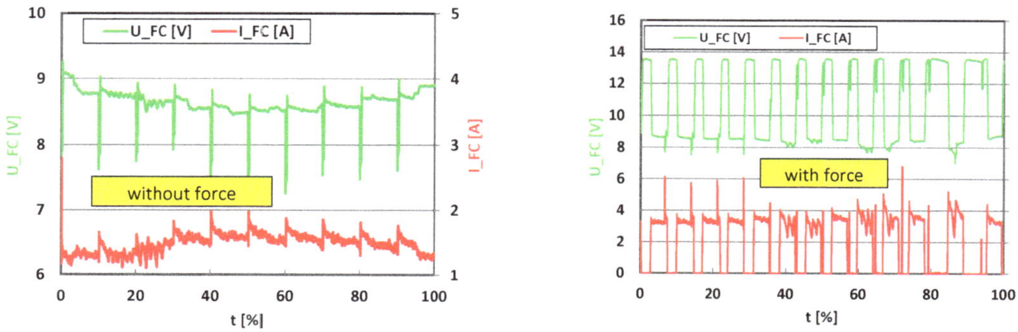

(a) (b)

Figure 9. Changes in voltage and current of the cell under operating conditions: (**a**) without load and (**b**) with load.

An externally loaded fuel cell exhibits sequential operation, which is manifested by the existence of a specific operating frequency. This can be compared to a PWM signal, where most of the time is active time. This means limiting the voltage while maintaining a high value of the fuel-cell current (Figure 8b). Visible current peaks appeared only at a low F-force load during the initial phase of the test. The sudden increase in voltage at no load on the cell (I_FC) is consistent with the polarization characteristics of the PEM fuel cell. This is related to the losses occurring in the cell during operation at different load states [49].

The current–voltage characteristics of the fuel-cell stack differ in the case of no load and with external load (Figure 10). Operation of the cell without a load (Figure 10a) causes small changes in voltage and current. The points with voltage U_FC below 8.4 V refer to the purging of anode channels and activation of the fuel cell, while the rest correspond to typical operating conditions. The operating points of an externally loaded cell are shown in Figure 10b. In this case, the points above 8.4 V are characteristic of the fuel-cell activation and termination states. The applied load increases the load on the fuel cell, increasing the current. The apparent voltage step changes are due to the dynamics of the load and the adjustment of the fuel-cell operation.

Static loading of the cell would result in typical cell operating conditions and result in a current–voltage characteristic in which each load point corresponds to a single voltage value. However, the dynamic nature of the fuel-cell operation means that such static characteristics are "filled in" by dynamic changes in the operating conditions. The lack of current (Figure 10b) indicates typical no-load operation under dynamic conditions since the change in recorded voltage ranges from 9 V to about 13.5 V, which exceeds the rated voltage of the cell used.

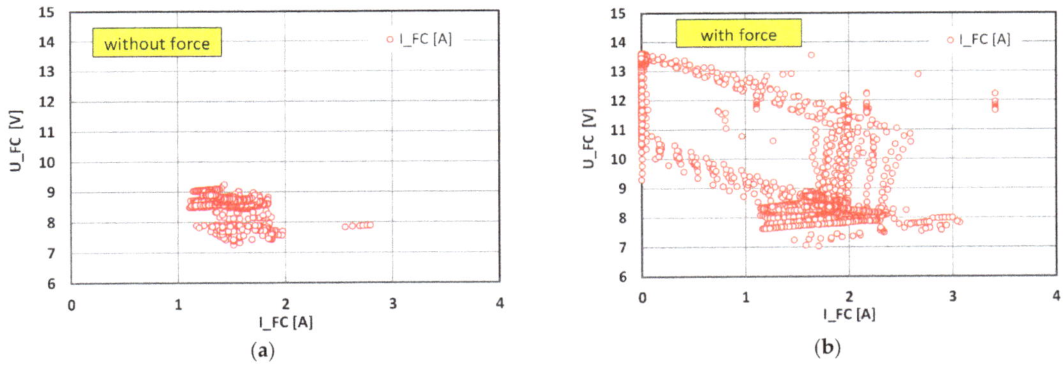

Figure 10. Voltage–current characteristics of the fuel cell: (**a**) without load and (**b**) with load.

7. Battery Operation Indicators

A partial analysis of battery operation is included in Chapter 4. The voltage–current characteristics of battery operation are shown in Figure 11. Operation without an external load on the drivetrain is characterized by slightly higher voltage values and significantly lower current values (Figure 11a) than with a load (Figure 11b). The no-load characteristic is only a part of the typical current–voltage characteristic of a battery. As a result of applying a variable external load, the drivetrain makes significant use of the energy provided by the battery, resulting in a significant increase in the operating field to a current I_BATT of just under 15 A. There are battery charging states in both tests, but they represent a negligible part of the characteristics. It should be noted that the model system tested is not equipped with braking energy recovery.

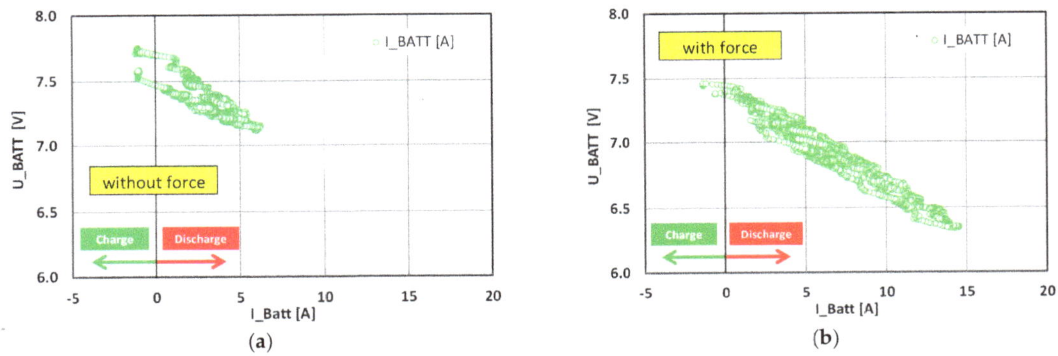

Figure 11. Voltage–current characteristics of the battery: (**a**) without load and (**b**) with load.

8. Indicators of Hybrid Powertrain Operation

The operating parameters of the hybrid powertrain are shown in Figure 12. Analysis of the no-load powertrain (Figure 12a) shows that the battery contributes almost 75% (maximum). However, its share is always greater than 50%. The battery share is much higher at high speed than at low speed. During the early and final stages of the drive at no speed, the battery was recharged, as indicated by the negative value of the battery share.

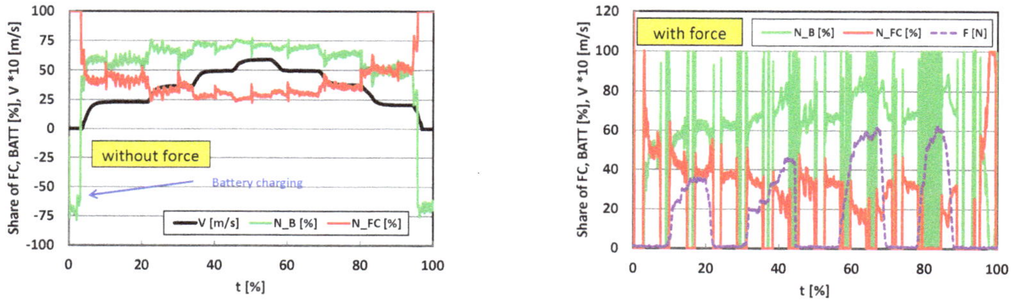

(a) (b)

Figure 12. Power shares of hybrid drive system components under operating conditions: (**a**) without load and (**b**) with load.

Battery charging conditions were not recorded under vehicle load conditions (Figure 12b). When the fuel cell is deactivated and the anode channels are purging, the battery share is about 100%. As the external load increases, the share of the cell decreases, and the share of battery power increases. As the load and speed increase (in both the initial and final phases of the test), the fuel-cell share drops below 50%. The higher the load, the higher the share of the battery. At high external load, the fuel-cell share decreases to around 30%.

An analysis of the total energy consumed is shown in Figure 13. Under no-load conditions (Figure 13a), 2254 J of energy was consumed from the battery, while only 1343 J was consumed from the fuel cell. This shows that the average proportion of energy from the fuel cell is 37%. Despite the lack of braking of the dynamometer rollers, i.e., no external load, the vehicle consumed a total of 3597 J.

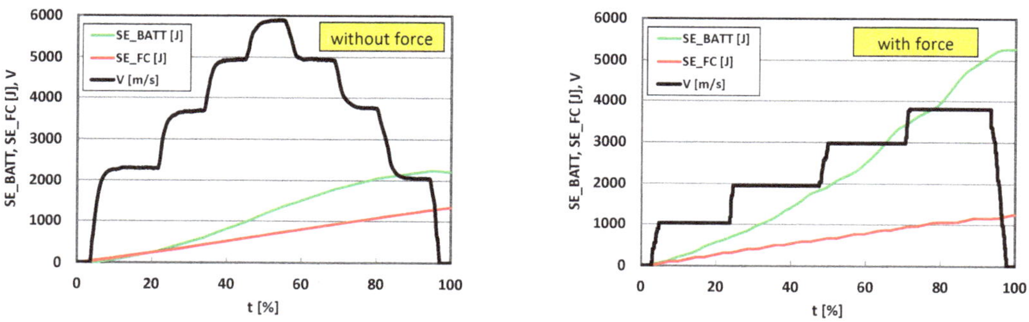

(a) (b)

Figure 13. Changes in energy of the battery and fuel cell under operating conditions: (**a**) without load and (**b**) with load.

Under the load conditions of the vehicle's drivetrain (Figure 13b), the consumed energies from the battery and the cell are 5277 J and 1251 J, respectively. This means that the fuel-cell share is practically constant and does not depend on the vehicle's load. During the load analysis, the fuel-cell share decreased significantly and is only 19%.

The study concludes that the battery predominantly drives the powertrain, with the fuel cell acting as a secondary energy source.

9. Conclusions

1. The experimental study presented here concerns the evaluation of the effect of the load on the drivetrain of the FCAT-30 model hybrid vehicle, equipped with a PEM fuel cell, on the performance of selected structural components. The voltage and current of the battery (BATT), fuel cell (FC), and drivetrain (OUT) were selected as the directly measured parameters analyzed. Based on the measured parameters, the performance of the components was evaluated, and the analysis was extended to include the energy balance.
2. It was pointed out that there is a variation in power distribution with respect to the applied load. When operating the system without an external load, the cell generates an approximately constant power of about 12 W during the test, which is between 20% and 50% of the power transferred to the drive, depending on the speed of the vehicle. The use of an external dynamic load results in cyclic operation of the cell with a peak power of 15 W, where the fuel-cell-stack contribution ranges from 0% (off state) to 38%.
3. Regardless of the test conditions, there is a process of flushing the anode channels, manifested by momentary jumps in cell voltage and current. For dynamic load conditions of current decay (deactivation of cell operation), clear jumps in cell-stack voltage from 8.5 V to 13.7 V were recorded. The no-load fuel-cell-stack operation area was indicated as a voltage below 8.4 V for no-load operation and above 8.4 V for the dynamic-external-load test. The results of the analysis of the energy flow within the NiMH battery indicate a small share of charging from the fuel cell with current in the 0–2 A range. The external-load test significantly increases the power demand, which puts a significant strain on the battery, which is the main energy source.
4. Analysis of the energy balance shows that there are no situations where the drivetrain uses only the fuel cell, except at the beginning and end of the test, where the wheel speed is 0 m/s. For the no-load test, the drivetrain consumed 3597 J, of which 37.3% was energy from the fuel cell. For the external-load test, the vehicle consumed 6528 J, of which the energy of the cell accounted for 19.2%. It was noted that there was no significant effect in the way the drivetrain was loaded on the amount of energy produced by the fuel cell. In the overall balance, the difference between the test with and without load was 6.85%.

Author Contributions: Conceptualization, I.P., F.S. and K.S.; methodology, I.P., F.S. and K.S.; software, I.P. and F.S.; validation, I.P., F.S. and K.S.; formal analysis, F.S. and K.S.; investigation, I.P., F.S. and K.S.; resources, I.P., F.S. and K.S.; data curation, I.P. and F.S.; writing—original draft preparation, I.P., F.S. and K.S.; writing—review and editing, I.P., F.S. and K.S.; visualization, I.P., F.S. and K.S.; supervision, I.P., F.S. and K.S.; project administration, I.P., F.S. and K.S.; funding acquisition, I.P., F.S. and K.S. All authors have read and agreed to the published version of the manuscript.

Funding: This research received no external funding.

Institutional Review Board Statement: Not applicable.

Informed Consent Statement: Not applicable.

Data Availability Statement: The data presented in this study are available on request from the corresponding author.

Conflicts of Interest: The authors declare no conflict of interest.

References

1. Szumska, E.; Jurecki, R. Technological developments in vehicles with electric drive. *Combust. Engines* **2023**, *194*, 38–47. [CrossRef]
2. Luo, Y.; Wu, Y.; Li, B.; Qu, J.; Feng, S.; Chu, P.K. Optimization and cutting-edge design of fuel-cell hybrid electric vehicles. *Int. J. Energy Res.* **2021**, *45*, 18392–18423. [CrossRef]
3. Liu, Z.; Wu, Y.; Feng, J. Competition between battery switching and charging in electric vehicle: Considering anticipated regret. *Environ. Dev. Sustain.* **2023**, 1–22. [CrossRef]

4. Sun, C.; Negro, E.; Vezzù, K.; Pagot, G.; Cavinato, G.; Nale, A.; Herve Bang, Y.; Di Noto, V. Hybrid inorganic-organic proton-conducting membranes based on SPEEK doped with WO_3 nanoparticles for application in vanadium redox flow batteries. *Electrochim. Acta* **2019**, *309*, 311–325. [CrossRef]
5. Hussein, M.M.; Mohamed, T.H.; Mahmoud, M.M.; Aljohania, M.; Mosaad, M.I.; Hassan, A.M. Regulation of multi-area power system load frequency in presence of V2G scheme. *PLoS ONE* **2023**, *18*, e0291463. [CrossRef] [PubMed]
6. Matla, J.; Kaźmierczak, A.; Haller, P.; Trocki, M. Hydrogen as a fuel for spark ignition combustion engines–state of knowledge and concept. *Combust. Engines* **2023**. [CrossRef]
7. Global Hydrogen Review 2021, IEA. Available online: https://www.iea.org/reports/global-hydrogen-review-2021 (accessed on 17 September 2023).
8. Samsun, R.; Rex, M.; Antoni, L.; Stolten, D. Deployment of fuel cell vehicles and hydrogen refueling station infrastructure: A global overview and perspectives. *Energies* **2022**, *15*, 4975. [CrossRef]
9. Thomas, C. Fuel cell and battery electric vehicles compared. *Int. J. Hydrogen Energy* **2009**, *34*, 6005–6020. [CrossRef]
10. Li, Z.P.; Liu, B.H.; Arai, K.; Suda, S. A Fuel cell development for using borohydrides as the fuel. *J. Electrochem. Soc.* **2003**, *150*, A868. [CrossRef]
11. Lewis, A.C. Optimising air quality co-benefits in a hydrogen economy: A case for hydrogen-specific standards for NOx emissions. *Environ. Sci. Atmos.* **2021**, *1*, 201–207. [CrossRef]
12. Dincer, I. Environmental and sustainability aspects of hydrogen and fuel cell systems. *Int. J. Energy Res.* **2006**, *31*, 29–55. [CrossRef]
13. Stambouli, A.B. Fuel cells: The expectations for an environmental-friendly and sustainable source of energy. *Renew. Sustain. Energy Rev.* **2011**, *15*, 4507–4520. [CrossRef]
14. Van Mierlo, J.; Maggetto, G.; Lataire, P. Which energy source for road transport in the future? A comparison of battery, hybrid and fuel cell vehicles. *Energy Convers. Manag.* **2006**, *47*, 2748–2760. [CrossRef]
15. Konno, N.; Mizuno, S.; Nakaji, H.; Ishikawa, Y. Development of compact and high-performance fuel cell stack. *SAE Int. J. Alt. Power* **2015**, *4*, 123–129. [CrossRef]
16. Yoshizumi, T.; Kubo, H.; Okumura, M. Development of high-performance FC stack for the new MIRAI. In *SAE Technical*; SAE International: Warrendale, PA, USA, 2021. [CrossRef]
17. Nonobe, Y. Development of the fuel cell vehicle Mirai. *IEEJ Trans. Electr. Electron. Eng.* **2017**, *12*, 5–9. [CrossRef]
18. Kimura, K.; Kawasaki, T.; Ohmura, T.; Atsumi, Y.; Shimizu, K. Development of New Fuel Cell Vehicle Clarity Fuel Cell. *Honda R&D Technical Review*. April 2016. Available online: http://www.uninsim.com/data/ueditor/php/upload/file/20191118/1574059716840217.pdf (accessed on 17 September 2023).
19. Matsunaga, M.; Fukushima, T.; Ojima, K. Powertrain system of Honda FCX Clarity fuel cell vehicle. *World Electr. Veh. J.* **2009**, *3*, 820–829. [CrossRef]
20. Robledo, C.B.; Oldenbroek, V.; Abbruzzese, F.; van Wijk, A.J.M. Integrating a hydrogen fuel cell electric vehicle with vehicle-to-grid technology, photovoltaic power and a residential building. *Appl. Energy* **2018**, *215*, 615–629. [CrossRef]
21. 2023 Hyundai NEXO Fuel Cell | Specs. Available online: https://www.hyundaiusa.com/us/en/vehicles/nexo (accessed on 30 August 2023).
22. Gómez, J.A.; Santos, D.M.F. The status of on-board hydrogen storage in fuel cell electric vehicles. *Designs* **2023**, *7*, 97. [CrossRef]
23. Rubio, O. BMW i Hydrogen NEXT está en Fase de Pruebas en Carreteras de Europa. Available online: https://greenracingnews.com/es/bmw-i-hydrogen-next-esta-en-fase-de-pruebas-en-carreteras-de-europa/ (accessed on 30 August 2023).
24. 2023 BMW IX5 Hydrogen First Drive Review: More than Just a Science Project. Available online: https://www.cnet.com/roadshow/news/2023-bmw-ix5-hydrogen-first-drive-review/ (accessed on 30 April 2023).
25. Hocking, M.; Kan, J.; Young, P.; Terry, C.; Begleiter, D. Welcome to the Lithium-Ion Age. Deutsche Bank Markets Research. 2016. Available online: http://www.belmontresources.com/LithiumReport.pdf (accessed on 30 August 2023).
26. Placke, T.; Kloepsch, R.; Dühnen, S.; Winter, M. Lithium ion, lithium metal, and alternative rechargeable battery technologies: The odyssey for high energy density. *J. Solid State Electrochem.* **2017**, *21*, 1939–1964. [CrossRef]
27. Zhang, J.; Wang, C.; Zhang, A. Experimental study on temperature and performance of an open-cathode PEMFC stack under thermal radiation environment. *Appl. Energy* **2022**, *311*, 118646. [CrossRef]
28. Curry, C. Lithium-Ion Battery Costs: Squeezed Margins and New Business Models. Bloomberg New Energy Finance. Available online: https://about.bnef.com/blog/lithium-ion-battery-costs-squeezed-margins-newbusiness-models/ (accessed on 14 July 2023).
29. Wang, T.; Li, Q.; Wang, X.; Qiu, Y.; Liu, M.; Meng, X.; Li, J.; Chen, W. An optimized energy management strategy for fuel cell hybrid power system based on maximum efficiency range identification. *J. Power Sources* **2020**, *445*, 227333. [CrossRef]
30. Arsalis, A. A comprehensive review of fuel cell-based micro-combined-heat-and-power systems. *Renew. Sustain. Energy Rev.* **2019**, *105*, 391–414. [CrossRef]
31. Giddey, S.; Badwal, S.; Kulkarni, A.; Munnings, C. A comprehensive review of direct carbon fuel cell technology. *Prog. Energy Combust.* **2012**, *38*, 360–399. [CrossRef]
32. Olabi, A.; Wilberforce, T.; Sayed, E.T.; Elsaid, K.; Abdelkareem, M.A. Prospects of fuel cell combined heat and power systems. *Energies* **2020**, *13*, 4104. [CrossRef]
33. Song, Z.; Pan, Y.; Chen, H.; Zhang, T. Effects of temperature on the performance of fuel cell hybrid electric vehicles: A review. *Appl. Energy* **2021**, *302*, 117572. [CrossRef]

34. Ogungbemi, E.; Wilberforce, T.; Ijaodola, O.; Thompson, J.; Olabi, A.G. Review of operating condition, design parameters and material properties for proton exchange membrane fuel cells. *Int. J. Energy Res.* **2020**, *45*, 1227–1245. [CrossRef]
35. Ji, M.; Wei, Z. A review of water management in polymer electrolyte membrane fuel cells. *Energies* **2009**, *2*, 1057–1106. [CrossRef]
36. Yan, Q.; Toghiani, H.; Causey, H. Steady state and dynamic performance of proton exchange membrane fuel cells (PEMFCs) under various operating conditions and load changes. *J. Power Sources* **2006**, *161*, 492–502. [CrossRef]
37. Yan, S.; Yang, M.; Sun, C.; Xu, S. Liquid water characteristics in the compressed gradient porosity gas diffusion layer of proton exchange membrane fuel cells using the Lattice Boltzmann Method. *Energies* **2023**, *16*, 6010. [CrossRef]
38. Yu, H.; Chen, D.; Ni, X.; Qing, P.; Yan, C.; Wei, W.; Ma, J.; Ji, X.; Chen, Y.; Chen, L. Reversible adsorption with oriented arrangement of a zwitterionic additive stabilizes electrodes for ultralong-life Zn-ion batteries. *Energy Environ. Sci.* **2023**, *16*, 2684–2695. [CrossRef]
39. Cheng, X.; Shi, Z.; Glass, N.; Zhang, L.; Zhang, J.; Song, D.; Liu, Z.S.; Wang, H.; Shen, J. A review of PEM hydrogen fuel cell contamination: Impacts, mechanisms, and mitigation. *J. Power Sources* **2007**, *165*, 739–756. [CrossRef]
40. Rosli, R.; Sulong, A.; Daud, W.; Zulkifley, M.; Husaini, T.; Rosli, M.; Majlan, E.; Haque, M. A review of high-temperature proton exchange membrane fuel cell (HT-PEMFC) system. *Int. J. Hydrogen Energy* **2017**, *42*, 9293–9314. [CrossRef]
41. Cropper, M.A.; Geiger, S.; Jollie, D.M. Fuel cells: A survey of current developments. *J. Power Sources* **2004**, *131*, 57–61. [CrossRef]
42. Binti Awang Mat, Z.; Madya Kar, Y.B.; Hasmady Bin Abu Hassan, S.; Azrina Binti Talik, N. Proton exchange membrane (PEM) and solid oxide (SOFC) fuel cell based vehicles—A review. In Proceedings of the 2017 2nd IEEE International Conference on Intelligent Transportation Engineering (ICITE), Singapore, 1–3 September 2017. [CrossRef]
43. Bar-On, I.; Kirchain, R.; Roth, R. Technical cost analysis for PEM fuel cells. *J. Power Sources* **2002**, *109*, 71–75. [CrossRef]
44. Thompson, S.T.; James, B.D.; Huya-Kouadio, J.M.; Houchins, C.; DeSantis, D.A.; Ahluwalia, R.; Wilson, A.R.; Kleen, G.; Papageorgopoulos, D. Direct hydrogen fuel cell electric vehicle cost analysis: System and high-volume manufacturing description, validation, and outlook. *J. Power Sources* **2018**, *399*, 304–313. [CrossRef]
45. Pielecha, I.; Szałek, A.; Tchorek, G. Two generations of hydrogen powertrain—An analysis of the operational indicators in real driving conditions (RDC). *Energies* **2022**, *15*, 4734. [CrossRef]
46. Pielecha, I.; Szwajca, F. Cooperation of a PEM fuel cell and a NiMH battery at various states of its charge in a FCHEV drive. *Eksploat. Niezawodn.* **2021**, *23*, 468–475. [CrossRef]
47. Liu, Z.; Chen, J.; Liu, H.; Yan, C.; Hou, Y.; He, Q.; Zhang, J.; Hissel, D. Anode purge management for hydrogen utilization and stack durability improvement of PEM fuel cell systems. *Appl. Energy* **2020**, *275*, 115110. [CrossRef]
48. Eom, T.-H.; Kang, J.-W.; Kim, J.; Shin, M.-H.; Lee, J.-H.; Won, C.-Y. Improved voltage drop compensation method for hybrid fuel cell battery system. *Electronics* **2018**, *7*, 331. [CrossRef]
49. Karami, N.; Outbib, R.; Moubayed, N. Fuel flow control of a PEM fuel cell with MPPT. In Proceedings of the 2012 IEEE International Symposium on Intelligent Control, Dubrovnik, Croatia, 3–5 October 2012; pp. 289–294. [CrossRef]

energies

MDPI

Article

State of Health Estimation Procedure for Lithium-Ion Batteries Using Partial Discharge Data and Support Vector Regression

Emil Petkovski, Iacopo Marri, Loredana Cristaldi * and Marco Faifer

Department of Electronics, Information and Bioengineering, Politecnico di Milano, 20133 Milan, Italy; emil.petkovski@polimi.it (E.P.); iacopomarri@gmail.com (I.M.); marco.faifer@polimi.it (M.F.)
* Correspondence: loredana.cristaldi@polimi.it

Abstract: Battery aging is a complex phenomenon, and precise state of health (SoH) monitoring is essential for effective battery management. This paper presents a data-driven method for SoH estimation based on support vector regression (SVR), utilizing features built from both full and partial discharge capacity curves, as well as battery temperature data. It provides an in-depth discussion of the novel features constructed from different voltage intervals. Moreover, three combinations of features were analyzed, demonstrating how their efficacy changes across different voltage ranges. Successful results were obtained using the full discharge capacity curves, built from the full interval of 2 to 3.4 V and achieving a mean R^2 value of 0.962 for the test set, thus showcasing the adequacy of the selected SVR strategy. Finally, the features constructed from the full voltage range were compared with ones built from 10 small voltage ranges. Similar success was observed, evidenced by a mean R^2 value ranging between 0.939 and 0.973 across different voltage ranges. This indicates the practical applicability of the developed models in real-world scenarios. The tuning and evaluation of the proposed models were carried out using a substantial dataset created by Toyota, consisting of 124 lithium iron phosphate batteries.

Keywords: lithium-ion battery; battery degradation; prognostics; machine learning; SoH

Citation: Petkovski, E.; Marri, I.; Cristaldi, L.; Faifer, M. State of Health Estimation Procedure for Lithium-Ion Batteries Using Partial Discharge Data and Support Vector Regression. *Energies* **2024**, *17*, 206. https://doi.org/10.3390/en17010206

Academic Editor: Ramon Costa-Castelló

Received: 31 October 2023
Revised: 29 November 2023
Accepted: 28 December 2023
Published: 30 December 2023

1. Introduction

Lithium-ion batteries have become the preferred choice due to their high energy density, long cycle life, and low self-discharge rate [1]. This makes them especially favorable for electric vehicles, where minimizing the weight of the battery pack while maintaining the desired performance and range is crucial. Furthermore, the increasing presence of photovoltaic and wind systems in the grid has led to the deployment of massive stationary battery storage systems to address the intermittency of these renewable sources and to provide essential support services to the grid [2]. Battery cells are very small units that are a part of battery packs for the majority of storage applications. The pack necessitates the presence of a Battery Management System (BMS) to ensure safe and efficient operation. It monitors and controls the charging and discharging process and implements battery cell balancing. To accomplish this task, the BMS must accurately estimate key battery metrics such as State of Charge (SoC), State of Health (SoH), and remaining useful life (RUL).

Battery SoH is a measure of the battery's ability to hold its charge. It is commonly defined as the ratio of the maximum available capacity of a used battery compared with its original or brand-new state. On the other hand, battery RUL refers to the estimated amount of time or number of charge–discharge cycles that a battery can operate for before experiencing failure or unacceptable performance. Battery degradation is a highly variable process, dependent on ambient conditions, cell chemistry, use patterns, and the BMS. Consequently, a wide variety of model-based [3–6] and data-driven battery aging methods used for SoH and End of Life prediction of batteries exist in literature.

Model-based methods rely on physical principles and mathematical equations to describe the behavior of batteries. These methods are characterized by very high accuracy;

however, their development can be very resource-intensive and time-consuming, requiring both significant computational power and expert knowledge.

With the abundance of data and advancements in the machine learning field, data-driven methods have become particularly popular. By analyzing large datasets, data-driven methods can uncover patterns and relationships that may not be easily identifiable using traditional analytical techniques. Numerous features, also called health indicators (HIs), have been used to realize different ML strategies. They are very commonly extracted from voltage, current, and temperature curves during the charging and discharging processes; for example, the authors of [7] extracted 11 features in the voltage range of 3.8 to 4.1 V. In [8], eight features were extracted from the discharge voltage, current, temperature, and the elapsed time of the discharge process, while in [9], features were built from the same quantities during the constant current (CC) and constant voltage (CV) charging process. In [10], unique features were constructed from the dependence of battery voltage and discharge capacity (Q (V), and the integral of battery temperature, through statistical analysis. Furthermore, incremental capacity analysis (ICA) is another commonly used method for feature extraction, representing the ratio between an increment of capacity and a fixed-voltage increment ($\mathrm{d}Q/\mathrm{d}V$) for a full or partial cycle. The resulting differential curves have a peak that becomes less pronounced as the battery ages. Many works have generated various valuable features from these curves [11–13]. The only drawback of these curves is that they can be easily affected by noise.

Numerous machine learning methods have been proposed in literature regarding State of Health (SoH) estimation. These can range from classical machine learning techniques such as modified linear regression, Gaussian processes regression (GPR), support vector regression (SVR), and random forest to different neural network (NN)-based approaches.

GPR can provide mean capacity estimates as well as probabilistic bounds using various features [14]. In [15], the authors use GPR to estimate the SoH and RUL of lithium-ion batteries while also considering uncertainty in two different attempts: first, by using a known parametric battery degradation model to exploit prior knowledge, and secondly, by developing a model based on experimental capacity measurements of battery cells. Ref. [16] presents an approach where probability prediction together with a GPR model is used to estimate the SoH, which is then used with three other indirect health indicators to forecast the RUL of the batteries.

Refs. [17,18] apply regression to model battery aging patterns; additionally, the RUL prediction capabilities of two fitting functions are analyzed, which are the third-degree polynomial and a custom hybrid function. An SVR strategy is presented in [19] based on curves of battery voltage as a function of charging capacity (V-Q). In [20,21], an SVR strategy based on partial voltage charging curves is proposed, while in [22], a comparison between linear regression, SVR, and random forest methods is provided.

Finally, many works have focused on different methods based on neural networks because of their flexibility, adaptability, and ability to learn complex patterns. Refs. [23,24] employ back-propagation neural networks (BPNN) using various health indicators, resulting in accurate SoH estimation. Ref. [25] proposes a solution based on the convolutional neural network (CNN), incorporating the concepts of transfer learning and network pruning. In [26], an Echo State Network is used together with a single exponential function to predict the SoH evolution curve of the tested batteries in different cycles. From the generated curves, the RUL is also inferred. Long Short-Term Memory (LSTM) networks are a type of neural network architecture that are particularly effective at processing long sequences of data [27–31]. In [28], the proposed method is based on LSTM NN and signal processing methods for SoH and RUL prediction of lithium-ion batteries, while [29] created an SoH estimation method based on LSTM and transfer learning.

This work presents a data-driven method for SoH estimation based on SVR and three features obtained from full and partial discharge capacity curves as well as battery temperature data. SVR was selected instead of the recently popular NN-based strategies because it offers a good trade-off between low computational burden, applicability, and accuracy of results, while also providing higher model interpretability. The performance of all models was evaluated using a substantial dataset, consisting of 124 battery cells. A discussion is provided on the unique extraction procedure of the features, their combination, as well as the parameter tuning process. Results demonstrate that the SVR models can successfully use features built from partial discharge capacity curves for accurate SoH estimation, highlighting the method's applicability in real-life scenarios.

2. Dataset

Toyota, in cooperation with Stanford University and MIT, created a dataset of 124 commercial lithium–iron phosphate battery cells [32], with 1.1 Ah nominal capacity and 3.3 V nominal voltage, cycled under fast charging conditions until their end of life (EoL) is reached. The cycling was carried out in a fixed chamber temperature of 30 °C, applying two steps of constant current charging, according to a policy defined by the format "$C1(Q1) - C2$", where $C1$ and $C2$ are the charging current values, and $Q1$ is the SoC level at which the current is switched. A total of 72 different policies with different values for charging current and switching step are used across the dataset, while the discharge is performed at a constant current of 4.4 A. While the chamber temperature is controlled, the cell temperature can vary by up to 10 °C within a cycle and between cells due to the vastly different charging policies and internal impedance values.

The number of cycles to failure for the batteries spans from approximately 150 to 2300, and the average end of life number is 806. Figure 1 shows the batteries ranked by cycle life, while Figure 2 shows the various SoH evolution profiles. The heterogeneous charging profiles generate a lot of variety within the dataset, allowing for deep analysis and insight into the behavior of the batteries and degradation patterns. For every battery cycle, the dataset includes a range of measured quantities including voltage, charge and discharge capacity, temperature, and internal resistance. In [14], the dataset authors created an in-depth analysis of the data and proposed many data-driven features, using which they developed a linear regression model to predict battery RUL.

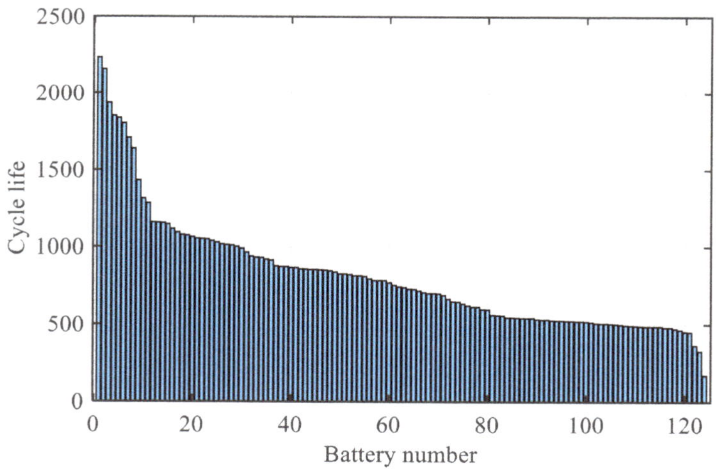

Figure 1. Dataset ordered by battery lifespan expressed in number of cycles.

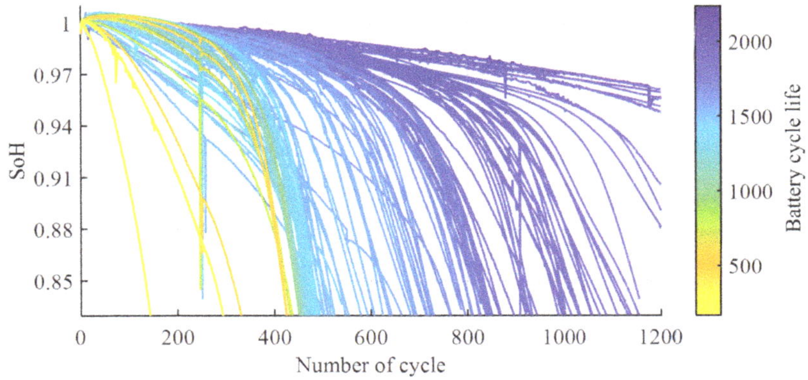

Figure 2. SoH over number of cycles for a set of representative batteries of the dataset.

3. Proposed Approach

This paper proposes a solution to estimate the SoH, which is based on a feature extraction procedure, where features that are considered strong battery health indicators are extracted from the discharge phase of every cycle. In the second step, those features are used as predictors to fit an SVR model, which is well suited to battery degradation modeling since it is strongly resistant to overfitting and can handle non-linear data. The SVR model is tuned with a cross-validation procedure and then used to perform SoH estimation.

3.1. Support Vector Regression

Support vector machine (SVM) is a very powerful algorithm for classification and regression tasks, which has been widely and successfully used in various fields such as image recognition, text classification, and prognostic and diagnostic processes. SVM works by finding the hyperplane that separates the data into classes and maximizes the minimum distance between data points and the hyperplane itself (maximizes the margin). SVM is effective with highly non-linear data thanks to the kernel trick: it can be used to transform the input (non-linear) data into higher-dimensional feature spaces, where linear classification can be performed.

SVR is a version of SVM specifically tailored for regression tasks. Similarly to SVM, it works by finding the hyperplane (line) that best fits the data and minimizes the error by using a tunable value ε as the minimum considered error. Any point lying inside the ε-tube is not considered in the cost function (1). A high value of ε leads to a more tolerant model, prone to underfitting, while a low value of ε leads to a more accurate but likely overfitted model. ξ_n, $\xi_n{}^*$ are the slack variables used to account for the positive and negative error of the points outside the ε-tube (2). The sum of ξ_n and $\xi_n{}^*$ is minimized in the loss function, weighted by a hyperparameter C, called a box constraint, which helps balance the complexity of the model. Finally, w and w' are the weights arrays, normal and transposed, respectively; Y_n is the target value; $X_n{}'$ is the transposed descriptor array; and b is the bias.

$$\min \frac{1}{2} w'w + C \sum_{n=1}^{N} (\xi_n + \xi_n^*) \tag{1}$$

$$\begin{cases} Y_n - (X_n'w + b) \leq \varepsilon + \xi_n \ \forall n \\ (X_n'w + b) - Y_n \leq \varepsilon + \xi_n^* \ \forall n \end{cases} \tag{2}$$

3.2. Feature Selection Procedure

In [10], the authors propose and study a set of features computed over the Toyota-MIT dataset, which they use as predictors to fit a linear regression model and provide an early

RUL prediction at cycle 100 of the batteries. To do so, they extract one value of each feature per battery, using the measurements made between cycles 10 and 100. The extraction approach is revisited in this work and adapted to compute three of the features at every cycle. The values of the features are then used to feed the SVR model to obtain a continuous estimation of the SoH at every cycle, using the feature data of that cycle. The first two features are related to the discharge capacity curves of the batteries as a function of voltage (Q (V)), i.e., their total moved charge (Q) during each discharge phase.

Figure 3 shows these discharge capacity curves at different cycles of battery life. The area under each curve can be seen as the total energy output by the battery during that cycle, and it gradually drops as the battery ages. The discharge capacity curve at cycle 10 is taken as a reference to compute the values of the features. The reference discharge curve is then subtracted by the discharge curve of every cycle after it to obtain a measure of the drop in performance between cycle 10 and the actual cycle (k).

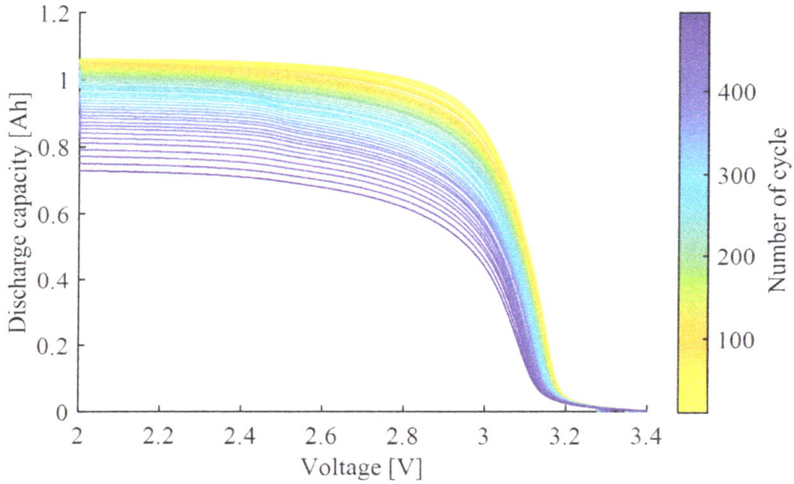

Figure 3. Discharging capacity as a function of voltage for all cycles of a sample battery.

The full voltage range is equal for all the cycles (and batteries) and constitutes a solid base for comparisons. However, before it could be utilized, interpolation of the different curves over a common set of voltage values had to be performed so that their subtraction could be made point by point. In Figure 4, the difference curves are shown for cycle 200 of all batteries, which are the result of the procedure explained above. Blue lines that are almost flat show no significant drop in performance from cycle 10. Yellow lines, on the other hand, indicate a major degradation. The first feature ($Ftr1$) is computed as the common logarithm of the variance of the difference curves, while the second feature ($Ftr2$) is computed as the common logarithm of the minimum of the difference curves. Both the features are computed for every battery and every cycle except cycles from 1 to 10, for which the value is considered equal to the 11th. The result, for both features, is a feature space on which the data are almost linearly distributed over the battery life span, and it is plotted for cycle 200 of every battery in Figure 5a for $Ftr1$ and Figure 5b for $Ftr2$. This property is maintained over all the aging cycles. While the results for both features seem identical, drastic differences will come to the surface when partial discharge curves are discussed in Section 4.2.

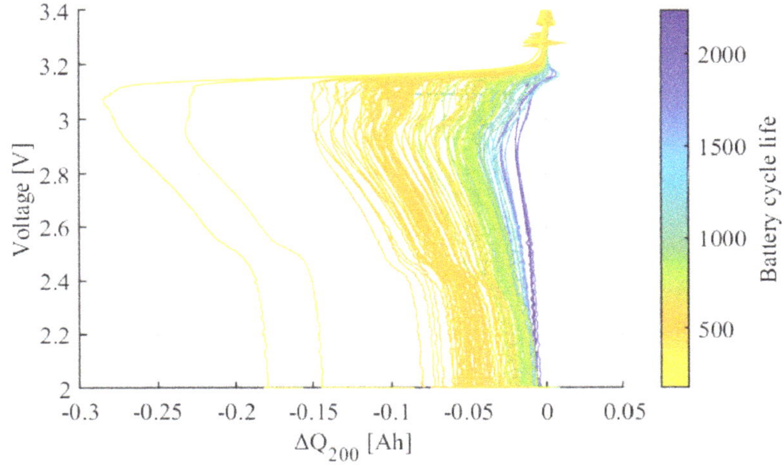

Figure 4. Difference in discharge capacity curves between reference cycle 10 and cycle 200, calculated for all the batteries.

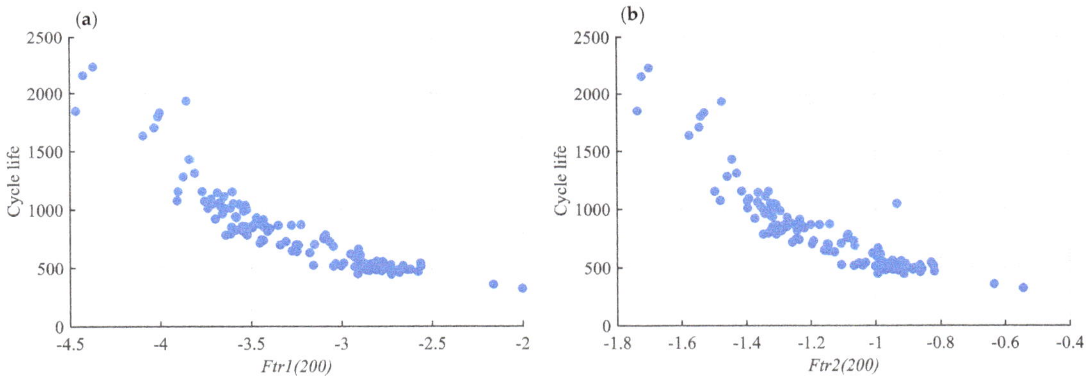

Figure 5. (**a**) Common logarithm of the variance in the difference between capacity curves at cycle 200 and cycle 10 for all batteries ($Ftr1(200)$). (**b**) Common logarithm of the minimum of the difference between capacity curves at cycle 200 and cycle 10 for all batteries ($Ftr2(200)$).

Finally, the third feature ($Ftr3$) is computed as the sum of temperature: for every cycle k of each battery, the average temperature of each cycle ($\overline{T}_{cycle}$) is summed from the first up to the k-th cycle.

In Equation (3), the three features are defined and computed for each cycle of every battery. More specifically, Q_{10} (V) and Q_k (V) represent the discharge capacities at reference cycle 10 and cycle k, respectively, as a function of voltage. ΔQ_k (V) is the difference of those two curves, while $\overline{\Delta Q_k}$ (V) represents the average value of that difference for cycle k.

Parameter p refers to the number of points in the curve. Finally, T_i represents a singular temperature measurement.

$$\begin{cases} Ftr1(k) = \log(|\frac{1}{p-1}\sum_{i=1}^{p}(\Delta Q_k(\mathrm{V}) - \overline{\Delta Q_k}(\mathrm{V}))^2|) \\ Ftr2(k) = \log(|\min(\Delta Q_k(V))|) \\ Ftr3(k) = \sum_{cycle=1}^{k} \overline{T}_{cycle} \\ \Delta Q_k(\mathrm{V}) = Q_k(\mathrm{V}) - Q_{10}(\mathrm{V}) \\ \overline{\Delta Q_k}(\mathrm{V}) = \frac{1}{p}\sum_{i=1}^{p}\Delta Q_{k,i}(\mathrm{V}) \end{cases} \quad (3)$$

4. Results and Discussion

4.1. Experimental Procedure and Full Discharge Window Approach

The voltage operating range of the batteries in the used dataset is approximately from 2 to 3.4 V. Initially, the features were extracted from the whole interval according to the procedure defined in Section 3.2. From the 124 available batteries, 15 were randomly selected to form the test set, while all of the cycles of the remaining 109 batteries formed the training and validation set. The SVR models were trained using solely the data of the 109 batteries. Initial parameter tuning was performed, followed by two sets of five-fold cross-validation (CV) campaigns to determine the best feature sets and then optimize the hyperparameters of the final model. In all stages, the predicted and measured values of SoH were compared after every cycle of every battery of the training and validation set.

To achieve a fully standardized and comparable results framework, the following procedure was developed and strictly followed:

The accuracy of the various proposed models was evaluated using the coefficient of determination (R^2). It is often defined as the amount of variability in the data explained or accounted for by the model, and it is represented by Equation (4).

$$R^2 = 1 - \frac{\sum_{i=1}^{n}(y_i - f(x_i))^2}{\sum_{i=1}^{n}(y_i - \overline{y})^2} \quad (4)$$

where

n—number of observations;
y_i—target value of SoH;
$f(x_i)$—predicted value of SoH;
$\overline{y}$ —mean of the target values.

Initial tuning of the SVR model hyperparameters (box constraint, epsilon, and kernel scale) was performed using MATLAB 2021b's built-in tool for hyperparameter optimization with the following properties: Bayesian optimizer, 60 optimization iterations, and loss metric RMSE. Due to the limited computational power, the number of optimization iterations had to be limited as well. This step is required to find an acceptable starting point from which to proceed with subsequent refinements.

Afterward, one of three sets of features is selected, where each feature set is a different combination of the three constructed features. Five-fold cross-validation (CV) is applied to the three potential feature sets, and the one with the highest mean CV R^2 is chosen. The considered feature sets along with their mean CV R^2 scores for the full discharge window approach are shown in Table 1. The highest R^2 value was achieved for feature set B, consisting of the first and third features.

Once the set of features is fixed, the model is further tuned by performing a five-fold CV for a range of values for each single hyperparameter, where the range depends on the results of the initial tuning of the parameters. This procedure is hierarchical, meaning that it starts from the most impactful hyperparameter (epsilon), then moves to the second most impactful (box constraint), and so on. The CV R^2 of the final optimized model is 0.976.

The hyperparameters of the final model considering the full window approach are a box constraint (BC) equal to 0.0055, epsilon equal to 0.0021, and a gaussian kernel and kernel scale (KS) of 1.

Table 1. Feature sets and achieved R^2 values.

Feature Set	Features	Mean CV R^2
A	$Ftr1, Ftr2, Ftr3$	0.968
B	$Ftr1, Ftr3$	0.970
C	$Ftr2, Ftr3$	0.961

Finally, the test set is fed to the tuned model. In every cycle, the model provides a predicted value of the SoH based on the input data of the same cycle. The R^2 value is calculated for every test battery individually, considering the measured and predicted value of the SoH after each cycle. The best fit is achieved for battery 11 (T11) of the test set, shown in Figure 6 with R^2 = 0.999; the worst fit for battery 4 (T4), shown in Figure 7 with R^2 = 0.794; and the mode is battery 5 (T5), shown in Figure 8 with an R^2 = 0.981. The average R^2 value for all batteries of the test set is also very high and equal to 0.962, indicating that, for the most part, the model fits the SoH evolution of the test set very well.

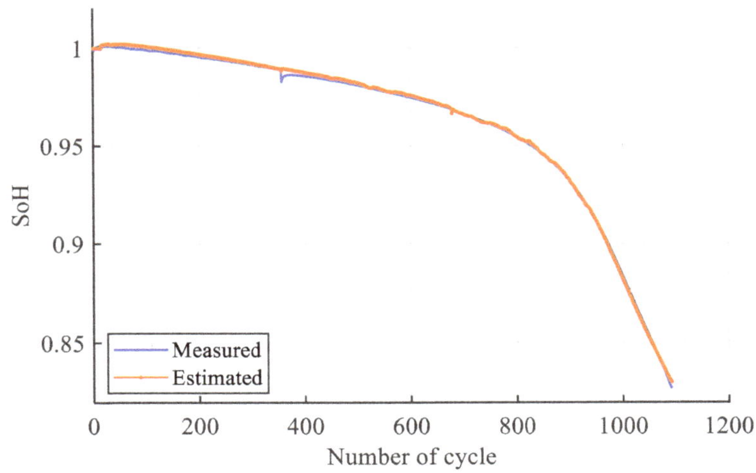

Figure 6. Best SoH estimation results for test battery 11.

4.2. Partial Discharge Window Approach

The full window approach is subject to limitations in real use cases. The typical use of battery-powered devices prevents the battery cycle from adequately covering the entire voltage range since batteries are rarely fully charged and discharged. Therefore, further analysis was carried out, adopting the same experimental procedure described in Section 4.1. to study how performance is affected by shrinking and moving the adopted voltage window over the full voltage span.

Ten voltage intervals were selected and a five-fold CV procedure was performed for every voltage interval to choose the best feature set. The results in Table 2 demonstrate how the efficacy of the three feature sets changes for the different voltage intervals based on their R^2 values. Moreover, the full voltage interval is also present in the table for comparison. For the high voltage intervals of 3.15 to 3.4 V and especially 3.25 to 3.4 V, which are highlighted in red, the R^2 value is low for all feature sets. The explanation for this low accuracy can be found in Figure 3, which shows that between such high voltage limits the moved charge is low and does not profoundly change as the number of cycles increase, thereby resulting in suboptimal features. Table 2 also shows that feature set B performs poorly for voltage intervals with an upper limit of less than 3.1 V. One of the features of this set is related to the variance in the difference between the curves of cycle k and the reference cycle 10. While this set was the best performer in the case of a full voltage window, it can be seen in Figure 4 that these difference curves have a constant value for voltages lower than 3.1 and, thus, the variance is no longer a good indicator. On the other hand, this is not the case for the minimum of the difference curve; thus, feature sets C and A continue to perform well. For each voltage interval, the feature set that results in the highest R^2 value is selected. The chosen feature set along with the final tuned value of the hyperparameter for each model are shown in Table 3.

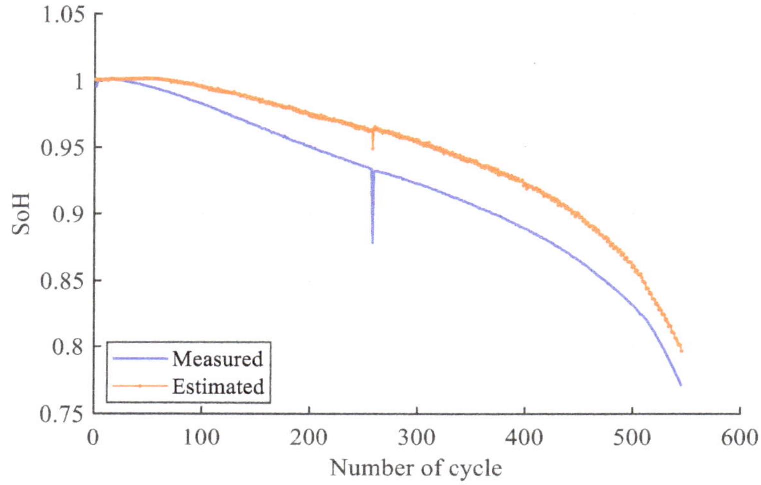

Figure 7. Worst SoH estimation results for test battery 4.

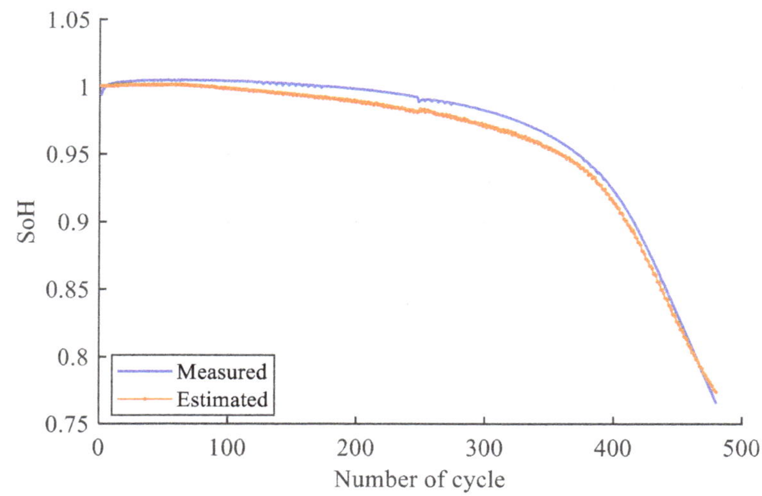

Figure 8. Mode SoH estimation results for test battery 5.

Table 2. Mean cross-validation R^2 values for feature set selection.

Voltage Range (V)	Feature Set A	Feature Set B	Feature Set C
Full window (2–3.4)	0.9686	0.9705	0.9620
3–3.4	0.8852	0.9572	0.9538
3.15–3.4	0.8992	0.8306	0.8777
3.25–3.4	0.8517	0.7894	0.8195
3–3.2	0.9624	0.9645	0.9536
3–3.1	0.9429	0.3904	0.9531
3–3.05	0.9608	0.3883	0.9646
2.8–3	0.9691	0.4715	0.9705
2.9–3	0.9657	0.3932	0.9705
2.4–2.6	0.9774	0.3944	0.9699
2.2–2.4	0.9772	0.3269	0.9741

Table 3. Final model hyperparameters for every voltage interval.

Voltage Range (V)	BC	KS	Epsilon	Kernel	Feature Set
Full window (2–3.4)	0.0055	1.0	0.0021	Gaussian	B
3–3.4	0.0055	1.0	0.0005	Gaussian	B
3.15–3.4	1.100	2.0	0.0001	Gaussian	A
3.25–3.4	0.022	1.0	0.0021	Gaussian	A
3–3.2	0.0055	1.0	0.0001	Gaussian	B
3–3.1	0.0055	1.0	0.0010	Gaussian	C
3–3.05	0.0055	1.0	0.0021	Gaussian	C
2.8–3	0.0055	1.0	0.00005	Gaussian	C
2.9–3	0.0055	1.0	0.0010	Gaussian	C
2.4–2.6	0.1100	2.0	0.0010	Gaussian	A
2.2–2.4	0.1100	2.0	0.0005	Gaussian	A

Table 4 shows a five-fold CV procedure applied on the final models. The results show very repeatable R^2 values for all folds and all voltage intervals except for 3.15–3.4 V and 3.25–3.4 V. For all other intervals, the minimum R^2 value of any fold is higher than 0.91, and the mean R^2 value across all folds ranges between 0.957 and 0.982. These results demonstrate that for the training and validation data, the models are of appropriate complexity and have low bias and variance. This is further corroborated when the models are applied to the remaining test battery data, as shown in Table 5.

Table 4. Five-fold CV R^2 values of the final model for every voltage interval.

Voltage Range (V)	Fold 1	Fold 2	Fold 3	Fold 4	Fold 5	Mean
Full window (2–3.4)	0.9578	0.9682	0.9891	0.9836	0.9816	0.9761
3–3.4	0.9337	0.9554	0.9799	0.9651	0.9701	0.9609
3.15–3.4	0.8672	0.9109	0.8994	0.9135	0.9301	0.9042
3.25–3.4	0.8063	0.8206	0.9146	0.8717	0.8495	0.8525
3–3.2	0.9479	0.9589	0.9846	0.9745	0.9707	0.9673
3–3.1	0.9453	0.9113	0.9784	0.9749	0.9735	0.9567
3–3.05	0.9509	0.9614	0.9816	0.9753	0.9742	0.9687
2.8–3	0.9585	0.9652	0.9857	0.9804	0.9802	0.9740
2.9–3	0.9581	0.9653	0.9855	0.9800	0.9797	0.9737
2.4–2.6	0.9784	0.9722	0.9842	0.9893	0.9866	0.9821
2.2–2.4	0.9760	0.9718	0.9893	0.9850	0.9865	0.9818

Table 5. R^2 value of all 15 test batteries for all selected voltage intervals.

	Full Window	3–3.4	3.25–3.4	3.15–3.4	3–3.2	3–3.1	3–3.05	2.8–3	2.9–3	2.4–2.6	2.2–2.4
T1	0.9834	0.9434	0.618	0.5622	0.9514	0.9548	0.9644	0.9744	0.9752	0.9874	0.9839
T2	0.9832	0.9371	0.9826	0.8765	0.9527	0.9512	0.9530	0.9665	0.9660	0.9913	0.9970
T3	0.9945	0.9901	0.9915	0.9734	0.9908	0.9914	0.9906	0.9914	0.9915	0.9954	0.9919
T4	0.7938	0.7817	0.7788	0.7452	0.7598	0.7689	0.7855	0.8038	0.8043	0.8103	0.8133
T5	0.9813	0.9869	0.9851	0.8784	0.9835	0.9887	0.9869	0.9853	0.9856	0.9782	0.9704
T6	0.9771	0.9773	0.9678	0.7470	0.8824	0.9737	0.9717	0.9709	0.9718	0.9905	0.9948
T7	0.9480	0.8937	0.7668	0.8838	0.9007	0.9216	0.9305	0.9465	0.9471	0.9924	0.9910
T8	0.9976	0.9771	0.6677	0.9613	0.9978	0.9783	0.9771	0.9966	0.9952	0.9987	0.9996
T9	0.9945	0.9825	0.9500	0.9521	0.9900	0.9896	0.9896	0.9924	0.9924	0.9970	0.9977
T10	0.9963	0.9796	0.8633	0.8468	0.9775	0.9925	0.9937	0.9967	0.9967	0.9881	0.9995
T11	0.9992	0.9985	0.8278	0.9964	0.9992	0.9990	0.9988	0.9993	0.9992	0.9981	0.9993
T12	0.9709	0.9533	0.8923	0.961	0.9361	0.9427	0.9455	0.9554	0.9537	0.9962	0.9891
T13	0.8234	0.8126	0.5329	0.6732	0.7952	0.8133	0.8196	0.8133	0.8089	0.8735	0.8706
T14	0.9925	0.9763	0.9929	0.9503	0.9782	0.9796	0.9805	0.9886	0.9885	0.9984	0.9987
T15	0.9944	0.9809	0.8917	0.9505	0.9856	0.9835	0.9856	0.9872	0.9880	0.9958	0.9954
Mean	0.9620	0.9447	0.8473	0.8639	0.9387	0.9486	0.9515	0.9579	0.9576	0.9727	0.9728

The columns of Table 5 are the different voltage intervals used to build the features of the final models, while the rows are each test battery. As was the case for the full voltage window, the R^2 value shows the accuracy with which the predicted value of SoH—as a function of the number of cycles—compares with the measured one, for every battery individually. Unsurprisingly, for the voltage ranges of 3.15–3.4 V and 3.25–3.4 V, the accuracy of the models is low for most batteries. On the other hand, for the rest of the intervals, the R^2 values are high and similar to the full window, and to the R^2 values obtained during the five-fold CV campaign using the training and validation data. More specifically, the mean R^2 for all batteries is in the range of 0.939 and 0.973. Therefore, the first conclusion is that the models are accurate and not just overfitting to the training data. The second, but just as important, conclusion is that partial voltage ranges can be used to build the features of the model, thus giving real-life applicability to the constructed models.

Finally, it can be noted that for batteries 13 (T13) and 4 (T4), highlighted in red in Table 5, the model accuracy is not very high, regardless of the voltage range used. This is because they follow a very different SoH evolution trend. To highlight this difference, for every battery, the SoH as a function-normalized number of cycles was defined by dividing the value of number of cycles by the value of total number of cycles. Therefore, the normalized number of cycles value for all cells is equal to 0 at the beginning and 1 when the battery reaches the end of life. Afterwards, the batteries of the training and validation set are grouped by their SoH value at every percent of the normalized number of cycles. The blue curve in Figure 9 is the 50th percentile for the SoH as a function of the normalized number of cycles, while the dotted blue lines are the 25th and 75th percentile. The same procedure is repeated for the test set, omitting T4 and T13. The yellow line is the 50th percentile of the SoH evolution for the test set, while the yellow dotted lines are the 25th and 75th percentile. Based on the small width of the percentile curves, as well as the mostly overlapping blue and yellow curves, it can be concluded that the batteries of the train and test set follow a similar SoH evolution trend, regardless of their vastly different cycle life. On the contrary, batteries 4 and 13, represented by the two red curves in Figure 9, follow a vastly different pattern of degradation, having a more sudden and almost linear decline in SoH rather than a slow descent. This is most likely due to some physical internal battery issue or extreme usage conditions, making these two cells outliers of the dataset and resulting in lower accuracy of SoH estimation by the models.

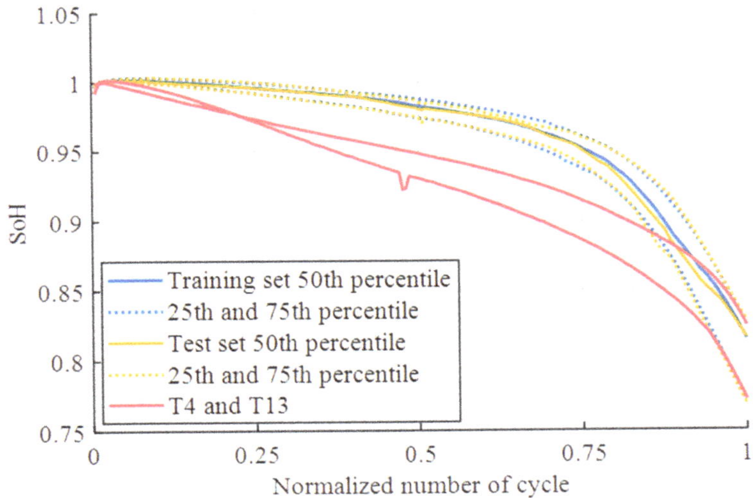

Figure 9. Distribution of the SoH evolution curves for the training data (blue), test data (yellow), and mis-predicted batteries 4 and 13 (red).

5. Conclusions

Accurate battery SoH estimation is crucial to achieve optimal performance, ensure safety, and minimize cost and environmental impact. This paper presents an SoH estimation method based on SVR, which offers a favorable compromise between applicability, computational efficiency, and accuracy of results. The dataset used in this work consists of 124 batteries. A total 109 batteries formed the training and validation set and the remaining 15 batteries formed the test set, which was used only in the final stage, to evaluate the performance of the developed models. Three features were selected, including two derived from discharge capacity curves and the third one based on battery temperature.

Moreover, three combinations of features were analyzed to determine how their effectiveness changes across different voltage ranges. When full discharge capacity curves are considered, all three feature combinations are successful, with the best results obtained when considering the logarithm of the variance of the difference curves and the temperature integral features, reaching a mean CV R^2 value of 0.976 for the training set and 0.962 for the test set. Additionally, the same features were built from 10 small voltage intervals, and the results show the considerations that must be made when following this approach—namely, models using features constructed from the upper voltage limits of 3.15–3.4 V or 3.25–3.4 V demonstrate poor accuracy because at such high voltage limits the moved charge is low and does not profoundly change as the number of cycles increases. Furthermore, feature set B performs poorly for voltage intervals with an upper limit of less than 3.1 V because the difference curves have a constant value for voltages lower than 3.1 V; thus, the feature related to variance is no longer a good indicator. This is not the case for the minimum of the difference curve feature; thus, feature sets A and C continue to perform well. The models that incorporate these limitations are very successful, reaching a CV R^2 value in the range of 0.957 and 0.982, and a mean R^2 value for the test set in the range of 0.939 and 0.973, depending on the selected voltage interval. These values are on par with the values achieved considering the full voltage window. Therefore, it can be concluded that partial voltage ranges can be used to build the features of the model, thus giving real-life applicability to the constructed models.

Author Contributions: Conceptualization, E.P., I.M. and L.C.; methodology, E.P. and I.M.; software, I.M. and E.P.; validation, I.M. and M.F.; formal analysis, E.P.; investigation, I.M.; resources, E.P.; data curation, L.C. and E.P.; writing—original draft preparation, E.P. and I.M; writing—review and editing, E.P. and I.M.; visualization, E.P. and I.M.; supervision, L.C., M.F. and E.P.; project administration, L.C. and M.F.; funding acquisition, L.C. and M.F. All authors have read and agreed to the published version of the manuscript.

Funding: The research received no external funding.

Data Availability Statement: Data are contained within the article.

Conflicts of Interest: The authors declare no conflicts of interest.

References

1. Pelletier, S.; Jabali, O.; Laporte, G.; Veneroni, M. Battery Degradation and Behaviour for Electric Vehicles: Review and Numerical Analyses of Several Models. *Transp. Res. Part B Methodol.* **2017**, *103*, 158–187. [CrossRef]
2. Cristaldi, L.; Faifer, M.; Laurano, C.; Ottoboni, R.; Petkovski, E.; Toscani, S. Power Generation Control Algorithm for the Participation of Photovoltaic Panels in Network Stability. *IEEE Trans. Instrum. Meas.* **2023**, *72*, 9000809. [CrossRef]
3. Li, X.; Fan, G.; Rizzoni, G.; Canova, M.; Zhu, C.; Wei, G. A Simplified Multi-Particle Model for Lithium Ion Batteries Via a Predictor-Corrector Strategy and Quasi-Linearization. *Energy* **2016**, *116*, 154–169. [CrossRef]
4. Petit, M.; Prada, E.; Sauvant-Moynot, V. Development of an Empirical Aging Model for Li-Ion Batteries and Application to Assess the Impact of Vehicle-to-Grid Strategies on Battery Lifetime. *Appl. Energy* **2016**, *172*, 398–407. [CrossRef]
5. Barcellona, S.; Colnago, S.; Dotelli, G.; Latorrata, S.; Piegari, L. Aging Effect on the Variation of Li-Ion Battery Resistance as Function of Temperature and State of Charge. *J. Energy Storage* **2022**, *50*, 104658. [CrossRef]
6. Xu, B.; Oudalov, A.; Ulbig, A.; Andersson, G.; Kirschen, D.S. Modeling of Lithium-Ion Battery Degradation for Cell Life Assessment. *IEEE Trans. Smart Grid* **2018**, *9*, 1131–1140. [CrossRef]
7. Li, X.; Yuan, C.; Li, X.; Wang, Z. State of Health Estimation for Li-Ion Battery Using Incremental Capacity Analysis and Gaussian Process Regression. *Energy* **2020**, *190*, 116467. [CrossRef]
8. Cui, Z.; Wang, C.; Gao, X.; Tian, S. State of Health Estimation for Lithium-Ion Battery Based on the Coupling-Loop Nonlinear Autoregressive with Exogenous Inputs Neural Network. *Electrochim. Acta* **2021**, *393*, 139047. [CrossRef]
9. Cao, M.; Zhang, T.; Wang, J.; Liu, Y. A Deep Belief Network Approach to Remaining Capacity Estimation for Lithium-Ion Batteries Based on Charging Process Features. *J. Energy Storage* **2022**, *48*, 103825. [CrossRef]
10. Severson, K.A.; Attia, P.M.; Jin, N.; Perkins, N.; Jiang, B.; Yang, Z.; Chen, M.H.; Aykol, M.; Herring, P.K.; Fraggedakis, D.; et al. Data-Driven Prediction of Battery Cycle Life before Capacity Degradation. *Nat. Energy* **2019**, *4*, 383–391. [CrossRef]
11. Ansean, D.; Garcia, V.M.; Gonzalez, M.; Blanco-Viejo, C.; Viera, J.C.; Pulido, Y.F.; Sanchez, L. Lithium-Ion Battery Degradation Indicators Via Incremental Capacity Analysis. *IEEE Trans. Ind. Appl.* **2019**, *55*, 2992–3002. [CrossRef]
12. He, J.; Wei, Z.; Bian, X.; Yan, F. State-of-Health Estimation of Lithium-Ion Batteries Using Incremental Capacity Analysis Based on Voltage–Capacity Model. *IEEE Trans. Transp. Electrif.* **2020**, *6*, 417–426. [CrossRef]
13. Zhou, R.; Zhu, R.; Huang, C.-G.; Peng, W. State of Health Estimation for Fast-Charging Lithium-Ion Battery Based on Incremental Capacity Analysis. *J. Energy Storage* **2022**, *51*, 104560. [CrossRef]
14. Richardson, R.R.; Birkl, C.R.; Osborne, M.A.; Howey, D.A. Gaussian Process Regression for In Situ Capacity Estimation of Lithium-Ion Batteries. *IEEE Trans. Ind. Inform.* **2019**, *15*, 127–138. [CrossRef]
15. Richardson, R.R.; Osborne, M.A.; Howey, D.A. Gaussian Process Regression for Forecasting Battery State of Health. *J. Power Sources* **2017**, *357*, 209–219. [CrossRef]
16. Jia, J.; Liang, J.; Shi, Y.; Wen, J.; Pang, X.; Zeng, J. SOH and RUL Prediction of Lithium-Ion Batteries Based on Gaussian Process Regression with Indirect Health Indicators. *Energies* **2020**, *13*, 375. [CrossRef]
17. Barcellona, S.; Cristaldi, L.; Faifer, M.; Petkovski, E.; Piegari, L.; Toscani, S. State of Health Prediction of Lithium-Ion Batteries. In Proceedings of the 2021 IEEE International Workshop on Metrology for Industry 4.0 & IoT (MetroInd4.0 & IoT), Rome, Italy, 7–9 June 2021; pp. 12–17.
18. Lashgari, F.; Petkovski, E.; Cristaldi, L. State of Health Analysis for Lithium-Ion Batteries Considering Temperature Effect. In Proceedings of the 2022 IEEE International Workshop on Metrology for Extended Reality, Artificial Intelligence and Neural Engineering (MetroXRAINE 2022-Proceedings), Rome, Italy, 26–28 October; pp. 40–45. [CrossRef]
19. Weng, C.; Sun, J.; Peng, H. Model Parametrization and Adaptation Based on the Invariance of Support Vectors With Applications to Battery State-of-Health Monitoring. *IEEE Trans. Veh. Technol.* **2015**, *64*, 3908–3917. [CrossRef]
20. Feng, X.; Weng, C.; He, X.; Han, X.; Lu, L.; Ren, D.; Ouyang, M. Online State-of-Health Estimation for Li-Ion Battery Using Partial Charging Segment Based on Support Vector Machine. *IEEE Trans. Veh. Technol.* **2019**, *68*, 8583–8592. [CrossRef]
21. Marri, I.; Petkovski, E.; Cristaldi, L.; Faifer, M. Lithium-Ion Batteries Soh Estimation, Based on Support-Vector Regression and a Feature-Based Approach. In Proceedings of the 18th IMEKO TC10 Conference on Measurement for Diagnostic, Optimisation and Control to Support Sustainability and Resilience 2022, Warsaw, Poland, 26–27 September 2022; pp. 109–113.

22. Marri, I.; Petkovski, E.; Cristaldi, L.; Faifer, M. Comparing Machine Learning Strategies for SoH Estimation of Lithium-Ion Batteries Using a Feature-Based Approach. *Energies* **2023**, *16*, 4423. [CrossRef]
23. Wen, J.; Chen, X.; Li, X.; Li, Y. SOH Prediction of Lithium Battery Based on IC Curve Feature and BP Neural Network. *Energy* **2022**, *261*, 125234. [CrossRef]
24. Tian, Y.; Dong, Q.; Tian, J.; Li, X.; Kukkapalli, V.K.; Kim, S.; Thomas, S.A. Capacity Estimation of Lithium-Ion Batteries Based on Multiple Small Voltage Sections and BP Neural Networks. *Energies* **2023**, *16*, 674. [CrossRef]
25. Li, Y.; Li, K.; Liu, X.; Wang, Y.; Zhang, L. Lithium-Ion Battery Capacity Estimation—A Pruned Convolutional Neural Network Approach Assisted with Transfer Learning. *Appl. Energy* **2021**, *285*, 116410. [CrossRef]
26. Catelani, M.; Ciani, L.; Fantacci, R.; Patrizi, G.; Picano, B. Remaining Useful Life Estimation for Prognostics of Lithium-Ion Batteries Based on Recurrent Neural Network. *IEEE Trans. Instrum. Meas.* **2021**, *70*, 3524611. [CrossRef]
27. Marri, I.; Petkovski, E.; Cristaldi, L.; Faifer, M. Battery Remaining Useful Life Prediction Supported by Long Short-Term Memory Neural Network. In Proceedings of the IEEE International Instrumentation and Measurement Technology Conference (I2MTC), Kuala Lumpur, Malaysia, 22–25 May 2023; pp. 1–6. [CrossRef]
28. Qu, J.; Liu, F.; Ma, Y.; Fan, J. A Neural-Network-Based Method for RUL Prediction and SOH Monitoring of Lithium-Ion Battery. *IEEE Access* **2019**, *7*, 87178–87191. [CrossRef]
29. Tan, Y.; Zhao, G. Transfer Learning with Long Short-Term Memory Network for State-of-Health Prediction of Lithium-Ion Batteries. *IEEE Trans. Ind. Electron.* **2020**, *67*, 8723–8731. [CrossRef]
30. Wang, S.; Takyi-Aninakwa, P.; Jin, S.; Yu, C.; Fernandez, C.; Stroe, D.-I. An Improved Feedforward-Long Short-Term Memory Modeling Method for the Whole-Life-Cycle State of Charge Prediction of Lithium-Ion Batteries Considering Current-Voltage-Temperature Variation. *Energy* **2022**, *254*, 124224. [CrossRef]
31. Wang, S.; Wu, F.; Takyi-Aninakwa, P.; Fernandez, C.; Stroe, D.-I.; Huang, Q. Improved Singular Filtering-Gaussian Process Regression-Long Short-Term Memory Model for Whole-Life-Cycle Remaining Capacity Estimation of Lithium-Ion Batteries Adaptive to Fast Aging and Multi-Current Variations. *Energy* **2023**, *284*, 128677. [CrossRef]
32. Toyota Research Institute. Experimental Data Platform. 2021. Available online: https://data.matr.io/1/ (accessed on 1 September 2023).

energies

MDPI

Article

Battery State of Health Estimation Using the Sliding Interacting Multiple Model Strategy

Richard Bustos [1], Stephen Andrew Gadsden [2,*], Mohammad Biglarbegian [3], Mohammad AlShabi [4] and Shohel Mahmud [1]

[1] College of Engineering and Physical Sciences, University of Guelph, Guelph, ON N1G 2W1, Canada
[2] Department of Mechanical Engineering, McMaster University, Hamilton, ON L8S 4L8, Canada
[3] Department of Mechanical and Aerospace Engineering, Carleton University, Ottawa, ON K1S 5B6, Canada
[4] Department of Mechanical and Nuclear Engineering, University of Sharjah, Sharjah 27272, United Arab Emirates
* Correspondence: gadsden@mcmaster.ca

Abstract: Due to their nonlinear behavior and the harsh environments to which batteries are subjected, they require a robust battery monitoring system (BMS) that accurately estimates their state of charge (SOC) and state of health (SOH) to ensure each battery's safe operation. In this study, the interacting multiple model (IMM) algorithm is implemented in conjunction with an estimation strategy to accurately estimate the SOH and SOC of batteries under cycling conditions. The IMM allows for an adaptive mechanism to account for the decaying battery capacity while the battery is in use. The proposed strategy utilizes the sliding innovation filter (SIF) to estimate the SOC while the IMM serves as a process to update the parameter values of the battery model as the battery ages. The performance of the proposed strategy was tested using the well-known B005 battery dataset available at NASA's Prognostic Data Repository. This strategy partitions the experimental dataset to build a database of different SOH models of the battery, allowing the IMM to select the most accurate representation of the battery's current conditions while in operation, thus determining the current SOH of the battery. Future work in the area of battery retirement is also considered.

Keywords: lithium batteries; Kalman filters; sliding innovation filter; interacting multiple model; state of health; state of charge; battery monitoring system; B005 battery dataset

Citation: Bustos, R.; Gadsden, S.A.; Biglarbegian, M.; AlShabi, M.; Mahmud, S. Battery State of Health Estimation Using the Sliding Interacting Multiple Model Strategy. *Energies* **2024**, *17*, 536. https://doi.org/10.3390/en17020536

Academic Editor: Simone Barcellona

Received: 16 October 2023
Revised: 1 January 2024
Accepted: 4 January 2024
Published: 22 January 2024

1. Introduction

Owing to their high specific energy and high operational voltage, lithium-ion batteries (LiB) have received great attention and are used in many applications [1]. Unfortunately, LiB have a limited operational area mainly bound by two important parameters: voltage and temperature. As such, careful monitoring of a battery's working temperature and voltage is necessary for its optimal and safe operation [2]. If the battery's voltage exceeds its limit, the battery may develop dendrites over time, which increases the battery's internal resistance, resulting in a lower output voltage. Moreover, if the working temperature is substantially increased, the battery may release toxic gases or burst into flames [3].

Another challenge presented by LiB technology is the accurate estimation of its available power or state of charge (SOC). SOC describes the amount of charge available in the battery at any given time during usage. SOC is often represented as a percentage value of available power vs. maximum power, or the available capacity vs. maximum capacity of the battery [3]. The main problem in determining the SOC is the absence of instrumentation that can accurately measure SOC during the battery's operation. This results in an estimation problem where the SOC must be estimated using indirect measurements such as the battery's terminal voltage and current [4].

Different techniques to estimate the LiB's SOC have been proposed in the literature. Some techniques such as neural networks (NN) have been used with great success [5];

however, NN make use of extensive data that must be collected beforehand and are computationally expensive compared to other solutions [5]. Other techniques make use of electrochemical impedance spectroscopy (EIS) data, which requires special instrumentation to be installed in the system [6].

One popular SOC estimation solution is the ampere-hour counting method. The ampere-hour counting method determines SOC based on current measurements and the remaining capacity of the battery [3]. This method's popularity relies on its simplistic approach. If the initial SOC is known, the previous SOC value can be subtracted or added based on the current profile. However, this method comes with many drawbacks. Its accuracy is highly dependent on the initial SOC value, correct current measurements, and accurate battery capacity readings [3]. To ensure proper estimates of the SOC, this method must be frequently calibrated; some calibration techniques include voltage-based corrections using lookup tables [3]. Another method was presented in [7], where the authors were able to jointly estimate the SOC and temperature at the same time. The ability to track the temperature in conjunction with the SOC provides useful insights in terms of battery life management and operational safety.

Furthermore, Kalman filters (KFs) present other estimation techniques that, when combined with the ampere-hour counting method, have proven to be accurate at estimating SOC. KFs provide an accurate and computationally inexpensive solution, but require an accurate battery model for their successful implementation [3]. A linear KF provides an optimal solution to the linear discrete estimation problem. However, due to the battery's nonlinear nature, only modified versions of the KF have been used for SOC estimation. Some KF variations include the extended Kalman filter (EKF) and the unscented Kalman filter (UKF), among others [8,9]. Between these two strategies, the EKF is known to introduce instability in the estimation process due to the linearization process embedded in the algorithm [10]. On the other hand, the UKF has proven to be a more robust strategy [11,12]. Another robust strategy, known as robust fixed-lag smoothing, attempts to overcome model uncertainties or mismatch by utilizing the least favorable model over a finite time horizon [13]. This method is characterized by a dynamic game between two players: one player selects the least favorable model in a prescribed ambiguity set, while the other player selects the fixed-lag smoother, minimizing the smoothing error with respect to the least favorable model. Efficient implementation of the robust fixed-lag smoother may reduce computational burdens and avoid numerical instabilities, which may be helpful for battery applications.

Electrochemical and equivalent circuit models (ECMs) are among the most popular models for batteries. Electrochemical models are based on the underlying physics of the battery using 10–14 partial differential equations, resulting in highly complex and computationally demanding models, but providing high-accuracy information about the battery's state. These types of models are often used for laboratory and battery development research [14–17].

On the other hand, ECMs represent the battery as an electric circuit using voltage sources, resistors, and capacitors. These types of models require low computational power and have low complexity, but are less accurate and yield little information about the battery [18]. Nevertheless, these traits allow for their implementation online.

Some ECMs studied include Rint model, Thevenin model, PNGV model, and Dual Polarity (DP) model [19]. These models can be differentiated by the number of Resistor–Capacitor (RC) branches in the circuit. Adding more RC branches allows the capture of higher-order nonlinearities, resulting in a more accurate model [19]. However, adding more RC branches increases the complexity and computational time of the algorithms.

In summary, a battery monitoring system (BMS) should be implemented to ensure safe operation of LiB. The BMS's main function is the accurate estimation of the battery's current SOC and operating temperature. In addition, the BMS can also track other parameters such as the battery's state of health (SOH); SOH is a measurement of the current health of the battery and is sometimes calculated based on its available maximum capacity [20].

As the battery is subject to aggressive current profiles, excessive cycling, or regular use, its maximum battery capacity degrades over time [20]. Moreover, accurate estimation of the battery's SOH can significantly increase the accuracy of the ampere-hour counting method, since it is dependent on the battery's capacity [20]. Lastly, accurate tracking of the battery's SOH allows for an effective planned retirement of the battery, which ensures that the system continues to operate optimally.

A battery is referred to as due for retirement once its SOH is at 80%, or in other words, when the battery's maximum available capacity is at 80% or less of its designed capacity [20]. Battery retirement can be presented as a fault diagnosis problem, where a SOH value of 80% or lower signals a fault in the battery [21]. A recent paper presented a degradation empirical model-free battery end-of-life prediction framework [22]. This method utilized the KF and Gaussian process regression. It is important to note that the SOH should be rapidly tracked and updated for improved performance and reliability. The authors in [23] introduced a fast capacity estimation method as well as a fast accelerated degradation fault diagnosis strategy for SOH estimation. This article offers insights into the importance of tracking micro-health parameters in batteries, which directly correspond to the overall SOH of the battery or set of batteries.

The multiple model (MM) strategy has been used to detect faults in batteries [24]. In the MM strategy, several models representing different behaviors of the system are generated to make the algorithm resilient against uncertainty [25]. Moreover, [25] presented an interacting multiple model (IMM) strategy where the IMM was combined with the EKF to accurately estimate the SOC of a LiB. The IMM was given allowed two different variations of noise in the battery model to account for the different degrees of parameter shift during the estimation process. Lastly, in [26], a multiple model adaptive estimation (MMAE) technique was used for fault diagnosis. The proposed strategy made use of EIS data and EKFs to generate residual signals that were fed into an MMAE block to detect a fault in the battery.

This paper focuses on the implementation of a MM strategy, i.e., the IMM strategy, to estimate the battery's capacity degradation while accurately estimating the SOC of a battery under cycling conditions [27]. This is a unique contribution to the field of battery monitoring, particularly when utilizing the relatively new sliding innovation filter (SIF). The experimental dataset was partitioned into sections representing a 100% SOH, 75% SOH, 50% SOH, 25% SOH, and 0% SOH, where each section can be identified as a mode to be used within the IMM algorithm. The motivation behind this partition is that the IMM would yield the best matching mode, thus identifying the current SOH of the battery.

The main contribution of this paper is the development of the SIF in conjunction with the IMM (the so-called SIF-IMM) for determining the SOC and SOH of a battery. The IMM algorithm is used for SOH estimation by partitioning the experimental dataset into several SOH modes. This strategy has not been presented in the literature. In addition to introducing this method, the paper compares the performances of SIF-IMM and KF-IMM in estimating SOH.

The remainder of the paper is structured as follows: Section 2 presents the battery and parameter models. Section 3 details the experimental data and estimation algorithms. Section 4 covers the artificial measurements. Section 5 describes the model parameter identification results. Section 6 presents the experimental setup and details the results of the proposed strategy. Section 7 presents the concluding arguments of the work.

2. Battery Models

This section presents the battery model used for the experiment.

Dual Polarity Model

The DP model is an ECM composed of a voltage source, a resistor, and two RC-branches. These elements represent the battery's output voltage, internal resistance, and short-term and long-term transient behaviors, respectively [16]. The battery's transient behaviors are most noticeable at the end of a discharge charge, once the current is cut off. It is evident that there is a quick rise in voltage followed by a slower rise in voltage. These two phenomena have been attributed to the battery's concentration polarization and electrochemical polarization, respectively. In the literature, the DP model has been shown to capture these behaviors by implementing a two RC-branch model [16]. The DP model was selected due to its high accuracy and ability to capture more nonlinearities while remaining computationally efficient. Figure 1 depicts the circuit diagram of the DP model.

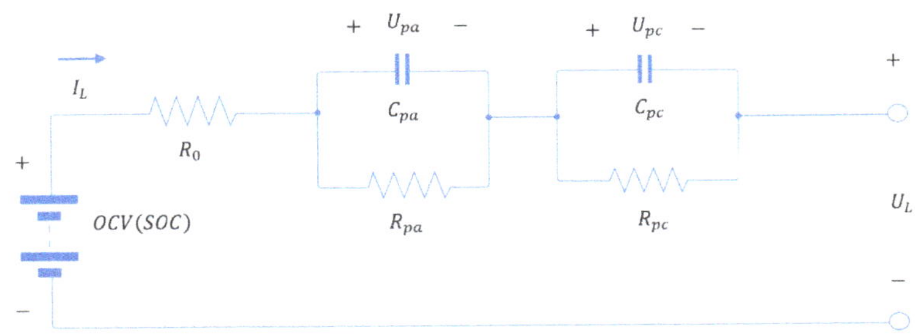

Figure 1. DP model circuit architecture [16].

The circuit can be analyzed by breaking it down into three parts: OCV, resistance, and capacitance. OCV represents the open circuit voltage of the battery, R_o represents the internal resistance of the battery, and R_{pa} and R_{pc} represent the electrochemical polarization resistance and concentration polarization resistance, respectively. Lastly, C_{pa} and C_{pc} characterize the transient response during the transfer of power to/from the battery during the electrochemical and concentration polarization [16].

Lastly, the system's state space representation is described by the following equations:

$$\begin{bmatrix} U_{pa,\,k+1} \\ U_{pc,\,k+1} \end{bmatrix} = \begin{bmatrix} e^{-\frac{\Delta t}{R_{pa}C_{pa}}} & 0 \\ 0 & e^{-\frac{\Delta t}{R_{pc}C_{pc}}} \end{bmatrix} \begin{bmatrix} U_{pa,\,k} \\ U_{pc,\,k} \end{bmatrix} + \begin{bmatrix} R_{pa}\left(1 - e^{-\frac{\Delta t}{R_{pa}C_{pa}}}\right) \\ R_{pc}\left(1 - e^{-\frac{\Delta t}{R_{pc}C_{pc}}}\right) \end{bmatrix} I_L, \tag{1}$$

$$U_{L,\,k+1} = U_{ocv}(SOC_{k+1}) - U_{pa,k+1} - U_{pc,k+1} - I_{L,k}R_0. \tag{2}$$

3. Experimental Data and Estimation Algorithms

This section presents the selected experimental data and the estimation algorithms used for the experiments.

3.1. B005 DATASET

The B005 dataset is part of several datasets released by the Prognostic Center of Excellence (PCoE) and published by the Prognostic Data Repository for the advancement of prognostic algorithms [28]. This dataset provides cycling data of a 2 Ah battery up to a capacity of ~70%. The cycling data includes time, current, voltage, and temperature measurements for each cycle [28]. The B005 dataset was selected to test the proposed strategy as it provides data for each cycle. If the impedance data is counted, there are 615 cycles in total in this dataset. In this study, this dataset provides 340 discharge and charge cycles. The battery has an initial capacity of 1.856 Ah and is cycled to 1.303 Ah. Finally, the data was resampled from 3 s to 0.6 s to allow for improved algorithm adherence.

Figure 2 illustrates the battery capacity degradation over the 615 cycles.

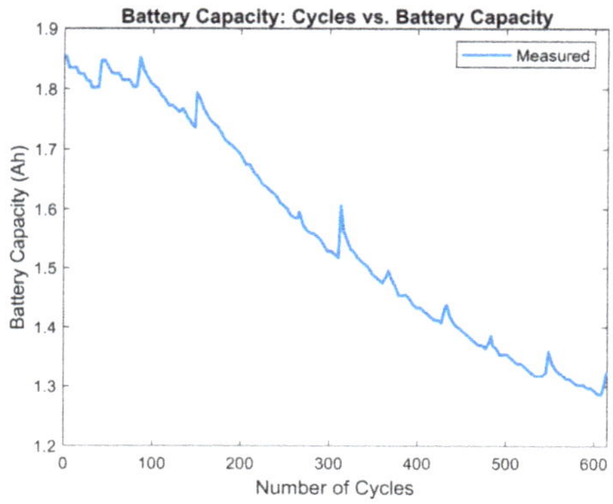

Figure 2. Measured battery capacity at the end of each discharge cycle [29].

3.2. Ampere-Hour Counting

This method is the most popular for estimating SOC, where the SOC is estimated based on an initial value, i.e., SOC_0, and this value is decreased/increased depending on the demanded/supplied current. SOC estimated using this method is given by [3]:

$$SOC = SOC_0 - \frac{1}{C_n}\int_{t_0}^{t} I \, d\tau, \tag{3}$$

where SOC_o is the initial SOC, C_n is the nominal capacity of the battery, and I is the discharge current.

3.3. Kalman Filter

The Kalman filter (KF) yields the optimal solution by minimizing the state estimation error for a known linear system that is subject to white Gaussian noise [30]. The linear system dynamics and the measurement model are described by the following two equations [31]:

$$x_{k+1} = Ax_k + Bu_k + w_k, \tag{4}$$

$$z_{k+1} = Cx_{k+1} + v_{k+1}, \tag{5}$$

where A is the dynamics matrix, B is the input matrix, C is the output matrix, x is the system states, z is the measurement output, u is the input, w is the system noise, and v is the measurement noise.

The KF algorithm is summarized in two stages: prediction and update [30].

(i) Prediction Stage:

$$\hat{x}_{k+1|k} = A\hat{x}_{k|k} + Bu_k \tag{6}$$

$$P_{k+1|k} = AP_{k|k}A^T + Q \tag{7}$$

$$w \sim N(0, Q) \tag{8}$$

$$v \sim N(0, R) \tag{9}$$

(ii) Update Stage:

$$K_{k+1} = P_{k+1|k}C^T\left[CP_{k+1|k}C^T + R\right]^{-1} \tag{10}$$

$$\hat{x}_{k+1|k+1} = \hat{x}_{k+1|k} + K_{k+1}\left(z_{k+1} - C\hat{x}_{k+1|k}\right) \tag{11}$$

$$P_{k+1|k+1} = [I - K_{k+1}C]P_{k+1|k}(I - K_{k+1}C)^T + K_{k+1}RK_{k+1}^T \tag{12}$$

where Q and R are the system and measurement noise covariance matrices, respectively.

3.4. Sliding Innovation Filter

A recently proposed estimation strategy is the novel sliding innovation filter (SIF). The SIF was formulated based on a predictor–corrector estimation method [32]. Like the KF, the SIF first makes predictions of the state estimates and state error covariances using values from previous time steps, and then updates the predictions based on the measurements and a correction term referred to as the SIF gain [32].

The main difference between the SIF and KF algorithms is how the gain was constructed. The SIF gain was built using the measurement matrix, the innovation, and a sliding boundary layer term [32]. On the other hand, the KF makes use of the state error covariance. In other words, the SIF makes use of the gain to drive the state estimates within the defined boundary layer and forces the state estimates to switch about the true trajectory [32]. The following equations describe the SIF algorithm [32].

(i) Update Stage:

$$K_{k+1} = C^{+}\overline{sat}\left(\left|\tilde{z}_{k+1|k}\right|/\delta\right)\hat{x}_{k+1|k+1} \tag{13}$$

$$\hat{x}_{k+1|k+1} = \hat{x}_{k+1|k} + K_{k+1}\tilde{z}_{k+1|k}P_{k+1|k+1} \tag{14}$$

$$P_{k+1|k+1} = (I - K_{k+1}C_{k+1})P_{k+1|k}(I - K_{k+1}C_{k+1})^{T} + K_{k+1}\ R\ K_{k+1}^{T} \tag{15}$$

Note that C^{+} refers to the pseudoinverse of C, $\overline{sat}$ refers to the diagonal of the saturation term (value between -1 and $+1$), and δ is the sliding boundary layer width.

3.5. Interacting Multiple Model (IMM)

The IMM method is most useful for systems with more than one operating mode. The IMM algorithm utilizes a number of models and computes the likelihood values for each model based on the state estimates and the corresponding state error covariance. The likelihood value is also based on the innovation (or measurement error) and provides an indication of how similar the actual system's behavior is compared to the filter model. These likelihood values are then used to determine the operating mode [33]. Note that the IMM is essentially a set of filters that run in parallel, with each filter utilizing a different dynamic system and/or measurement model. The IMM algorithm can be described in a set of five stages. These five stages are presented below [33].

(i) Calculation of the mixing probabilities $\mu_{i|j,k|k}$:

$$\mu_{i|j,k|k} = \frac{1}{\bar{c}_j}\, p_{ij}\mu_{i,k} \tag{16}$$

$$\bar{c}_j = \sum_{i=1}^{r} p_{ij}\mu_{i,k} \tag{17}$$

where $\mu_{i|j,k|k}$ refers to the mixing probabilities between modes i and j at time k, p_{ij} refers to the mixing parameter (predefined value), and $\mu_{i,k}$ refers to the mode probability.

(ii) Mixing Stage:

$$\hat{x}_{0j,k|k} = \sum_{i=1}^{r} \hat{x}_{i,k|k}\mu_{i|j,k|k} \tag{18}$$

$$P_{0j,k|k} = \sum_{i=1}^{r} \mu_{i|j,k|k}\left\{P_{i,k|k} + \left(\hat{x}_{i,k|k} - \hat{x}_{0j,\,k|k}\right)\left(\hat{x}_{i,k|k} - \hat{x}_{0j,\,k|k}\right)^{T}\right\} \tag{19}$$

where $\hat{x}_{0j,k|k}$ and $P_{0j,k|k}$ are the state estimates and state error covariances used as initial conditions for the filters (KF or SIF).

(iii) Mode-Matched Fitlering:

$$\Lambda_{j,k+1} = \mathcal{N}\left(z_{k+1}; \hat{z}_{j,k+1|k}, \ S_{j,k+1}\right) \tag{20}$$

$$\Lambda_{j,k+1} = \frac{1}{\sqrt{\left|2\pi S_{j,k+1}\right|_{Abs}}} \exp\left(\frac{-\frac{1}{2}\, e^T_{j,z,k+1} e_{j,z,k+1|k}}{S_{j,k+1}}\right) \tag{21}$$

where $\Lambda_{j,k+1}$ refers to the likelihood value generated based on the measurement, estimate measurement (from the corresponding filter), and the innovation covariance matrix (from the corresponding filter). The likelihood value is then used to update the mode probability (described in the next stage).

(iv) Mode Probability Update:

$$\mu_{j,k} = \frac{1}{c} \Lambda_{j,k+1} \sum_{i=1}^{r} p_{ij} \mu_{i,k} \tag{22}$$

$$c = \sum_{j=1}^{r} \Lambda_{j,k+1} \sum_{i=1}^{r} p_{ij} \mu_{i,k} \tag{23}$$

(v) State Estimate and Covariance Combination

$$\hat{x}_{k+1|k+1} = \sum_{j=1}^{r} \mu_{j,k+1} \hat{x}_{j,k+1|k+1} \tag{24}$$

$$P_{k+1|k+1} = \sum_{j=1}^{r} \mu_{j,k+1} \left\{ P_{j,k+1|k+1} + \left(\hat{x}_{j,k+1|k+1} - \hat{x}_{k+1|k+1}\right)\left(\hat{x}_{j,k+1|k+1} - \hat{x}_{k+1|k+1}\right)^T \right\} \tag{25}$$

Note that for this final stage (v), the overall state estimates and state error covariance are used outside of the IMM process, and are used for output purposes only.

In the proposed strategy, the IMM is used to track the correct battery capacity, thus estimating the SOH of the battery. Meanwhile, the KF and SIF are used to estimate the SOC of the battery during cycling. The integration of these algorithms with the IMM resulted in strategies referred to as KF-IMM and SIF-IMM. The figure shown in Appendix A helps to further illustrate the overall architecture of the proposed strategies.

4. Artificial Measurements

As mentioned in Section 3.4, the SIF gain was derived based on the measurement matrix, meaning that it is dependent on the availability of individual measurements for each state estimate [32]. When individual measurements for each state are not available, which is often the case for LiBs and most types of batteries, generating artificial measurements is necessary to ensure that the SIF is effective [32]. The SIF strategy relies on the measurements in order to formulate good estimates of the states through the use of a full identity measurement matrix. This is one of the main disadvantages of this method. This section presents how artificial measurements were generated for the battery and parameter model.

4.1. State Measurement Equations

LiBs only offer measurements of current, voltage, and temperature. To generate artificial measurements for each state of the battery model, the model equations were rearranged as followed:

$$\widehat{U}_{pa,k+1} = OCV(SOC_k) - U_{L,k+1} - R_{0,k} I_{s,k} - U_{pc,k} \tag{26}$$

$$\widehat{U}_{pc,k+1} = OCV(SOC) - U_{L,k+1} - R_{o,k} I_{s,k} - U_{pa,k} \tag{27}$$

$$\widehat{SOC}_{k+1} = OCV^{-1}\left(U_{L,k+1} + U_{pa,k} + U_{pc,k} + R_{o,k} I_{s,k}\right) \tag{28}$$

where $\widehat{U}_{pa}$, $\widehat{U}_{pc}$, and $\widehat{SOC}$ are the measurements for each state of the battery model, and $OCV^{-1}(.)$ is the inverse function of $OCV(SOC)$ found in (27).

4.2. Parameter Measurement Equations

To generate artificial measurements for the parameters of interest, $Batt_{cap}$ and R_0, Equations (2) and (3) were rearranged as follows:

$$\widehat{R}_o = \frac{1}{I_{s,k}}[OCV(SOC) - U_L - U_{pa} - U_{pc}] \tag{29}$$

$$\widehat{Batt}_{Cap} = \frac{\Delta t \times I_{s,k}}{abs(3.6 \times \Delta SOC_k)} \tag{30}$$

where $\widehat{R}_0$ and $\widehat{Batt}_{cap}$ represent the artificial measurements for R_0 and $Batt_{cap}$.

5. Model Parameter Identification

As mentioned before, the IMM makes use of several models. These models were identified by breaking the B005 dataset into four regions, resulting in five different modes to be identified corresponding to 100%, 75%, 50%, 25%, and 0% SOH. Furthermore, the B005 dataset contains 340 discharge and charge cycles; however, only the discharge cycles provide battery capacity measurements. Since this paper focuses on battery capacity estimation, the models were derived using discharge cycles. To derive each battery model, the following parameters needed to be identified (for each of the selected cycles): $OCV(SOC)$, R_0, R_{pa}, C_{pa}, R_{pc}, C_{pc}. This section presents the model's parameter identification results using the nonlinear least squares (NLLS) algorithm for each of the modes selected: 100%, 75%, 50%, 25%, and 0% SOH.

5.1. Least Squares Setup

To use the NNLS algorithm, a relationship between the measurable data and the parameters must be established. Making use of the battery model, the following relationship can be derived [29]:

$$OCV(SOC) = \alpha_0 + \alpha_1 SOC + \alpha_2 SOC^2 + \alpha_3 SOC^3 + \alpha_4 SOC^4 + \alpha_5 SOC^5 \tag{31}$$

$$U_{pa} = I_L R_{pa}\left(1 - e^{-\frac{t}{R_{pa}C_{pa}}}\right) \tag{32}$$

$$U_{pc} = I_L R_{pc}\left(1 - e^{-\frac{t}{R_{pc}C_{pc}}}\right) \tag{33}$$

$$U_L = OCV(SOC) - I_L R_0 - I_L R_{pa}\left(1 - e^{-\frac{t}{R_{pa}C_{pa}}}\right) - I_L R_{pc}\left(1 - e^{-\frac{t}{R_{pc}C_{pc}}}\right) \tag{34}$$

$$\theta = \left[\alpha_0, \alpha_1, \alpha_2, \alpha_3, \alpha_4, \alpha_5,\ R_0,\ R_{pa}, \frac{1}{R_{pa}C_{pa}}, R_{pc}, \frac{1}{R_{pc}C_{pc}}\right] \tag{35}$$

where t represents the time vector, $OCV(SOC)$ is the OCV curve approximated to a 5th order polynomial, and θ represents the parameter vector, i.e., a vector consisting of all the parameters. Here, the capacitance values are estimated using a fraction to account for their higher magnitude [29,34].

Lastly, MATLAB®'s R2023a 'lsqcurvefit' function was used to perform the parameter estimation process. This function makes use of boundaries and initial conditions which are detailed in Table 1.

Table 1. NLLS boundaries and initial guess.

Parameters	R_0	R_{pa}	C_{pa}	R_{pc}	C_{pc}
Unit	Ω	Ω	1/(ΩF)	Ω	1/(ΩF)
LB	0.001	0.01	0.0001	0.01	0.01
UB	0.500	0.500	0.002	0.500	0.100
Guess	0.020	0.100	0.001	0.100	0.010

5.2. Least Squares Results

5.2.1. The 100% SOH Model

As indicated before, 100% SOH represents a battery with an available capacity equal to its designed capacity. Table 2 shows the results of the parameter identification process for the 100% SOH mode. The terminal voltage's RMSE was 0.063 V. Figure 3 depicts the terminal voltage's error plot between the model and the measured terminal voltage. Finally, Figure 4 illustrates the generated model's terminal voltage plot vs. the measured terminal voltage.

Table 2. NLLS parameter estimation results for 100% SOH model.

RC Parameters	Value	OCV (SOC)	Value
R_0	0.0700	α_1	0.6996
R_{pa}	0.1070	α_2	17.4679
C_{pa}	1329.29	α_3	−62.4061
R_{pc}	0.0401	α_4	76.7998
C_{pc}	19325	α_5	−31.7285

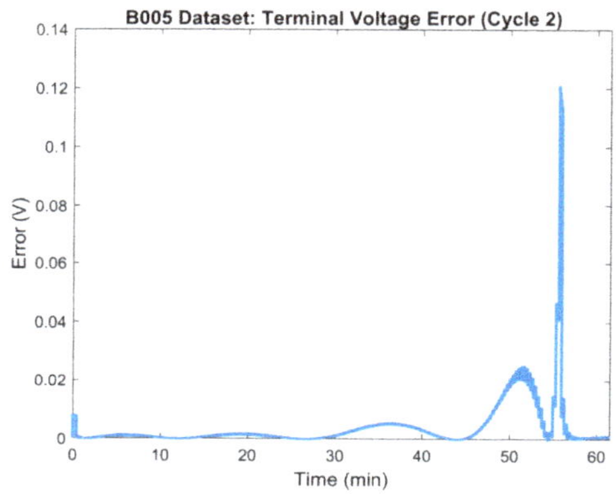

Figure 3. 100% SOH: LS model error—terminal voltage.

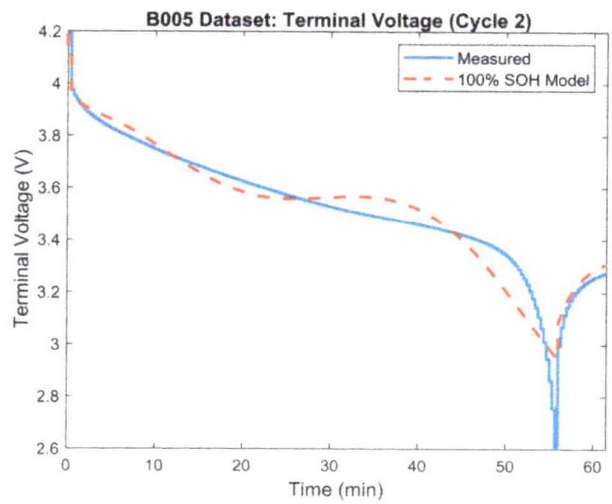

Figure 4. Terminal voltage: model vs. measured voltage for 100% SOH model.

5.2.2. The 75% SOH Model

The 75% SOH of the battery was determined to occur when the battery reaches a capacity of 1.83 Ah which, based on the data, happens at the ~98th cycle. Table 3 illustrates the results of the parameter identification process for the 75% SOH mode. The terminal voltage's RMSE was 0.032 V. Figure 5 depicts the terminal voltage's error plot between the model and the measured terminal voltage. Figure 6 shows the generated model's terminal voltage plot vs. the measured terminal voltage.

Table 3. NLLS parameter estimation results for 75% SOH model.

RC Parameters	Value	OCV (SOC)	Value
R_0	0.07	α_1	5.8933
R_{pa}	0.10	α_2	−23.1629
C_{pa}	1428.60	α_3	42.4898
R_{pc}	0.04	α_4	−35.0472
C_{pc}	53333	α_5	10.7280

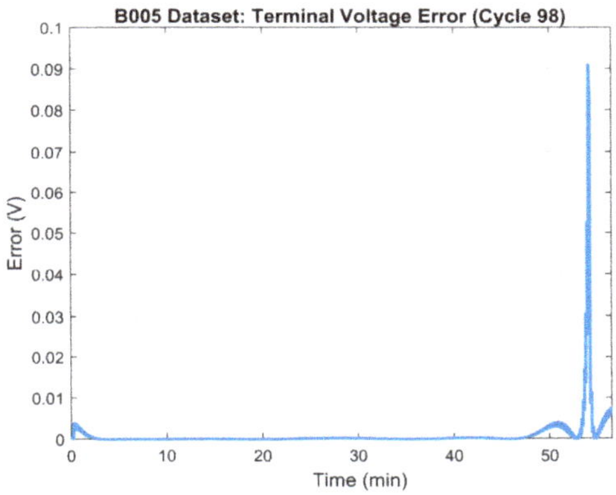

Figure 5. 75% SOH: LS model error—terminal voltage.

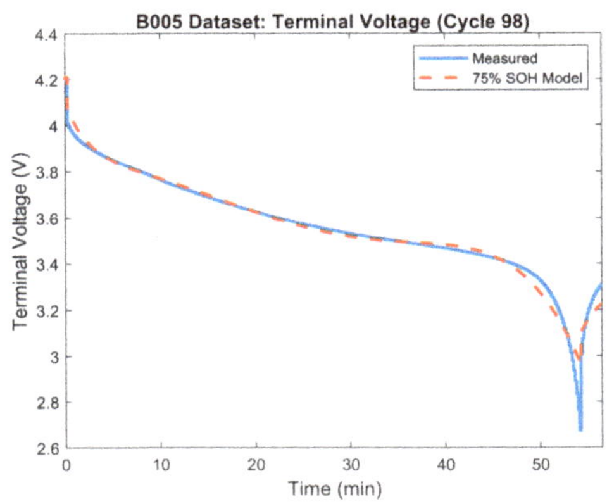

Figure 6. Terminal voltage: model vs. measured voltage for 75% SOH model.

5.2.3. The 50% SOH Model

The 50% SOH of the battery was determined to occur when the battery reaches a capacity of 1.65 Ah which, based on the data, happens at the ~225th cycle. Table 4 illustrates the results of the parameter identification process for the 50% SOH mode. The terminal voltage's RMSE was 0.032 V. Figure 7 shows the terminal voltage's error plot between the model and the measured terminal voltage, and Figure 8 illustrates the generated model's terminal voltage plot vs. the measured terminal voltage.

Table 4. NLLS parameter estimation results for 50% SOH model.

RC Parameters	Value	OCV (SOC)	Value
R_0	0.08	α_1	5.0727
R_{pa}	0.137	α_2	−17.8196
C_{pa}	428.730	α_3	29.3507
R_{pc}	0.078	α_4	−21.1457
C_{pc}	9500.530	α_5	5.3372

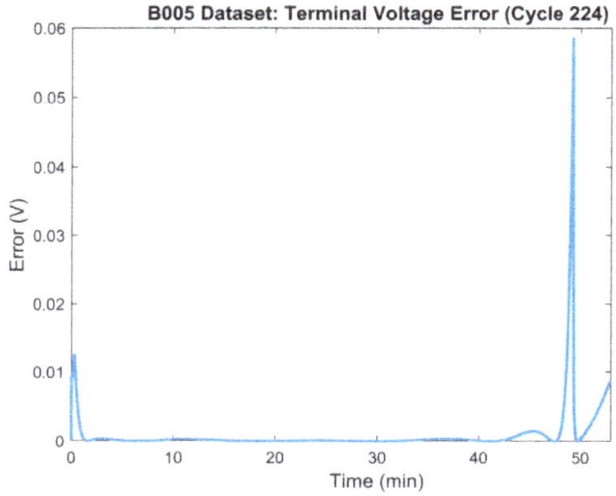

Figure 7. 50% SOH: LS model error—terminal voltage.

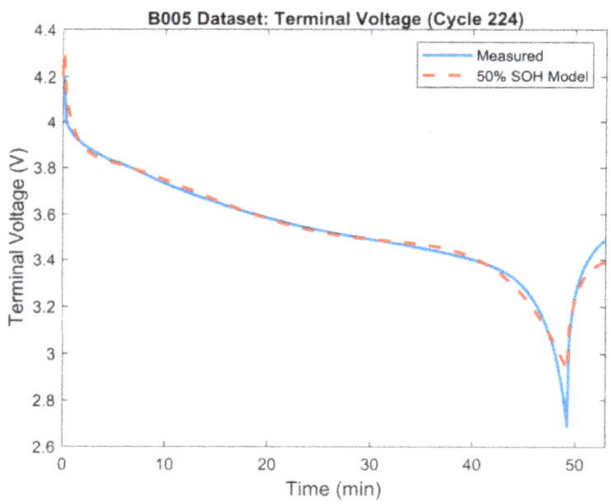

Figure 8. Terminal voltage: model vs. measured voltage for 50% SOH model.

5.2.4. The 25% SOH Model

The 25% SOH of the battery was determined to occur when the battery reaches a capacity of 1.48 Ah which, based on the data, happens at the ~375th cycle. Table 5 illustrates the results of the parameter identification process for the 25% SOH mode. The terminal voltage's RMSE was 0.034 V. The terminal voltage's error plot between the model and the measured terminal voltage is depicted in Figure 9. The generated model's terminal voltage plot vs. the measured terminal voltage in shown in Figure 10.

Table 5. NLLS parameter estimation results for 25% SOH model.

RC Parameters	Value	OCV (SOC)	Value
R_0	0.08	α_1	3.1583
R_{pa}	0.19	α_2	−5.2627
C_{pa}	401.82	α_3	−2.0509
R_{pc}	0.0277	α_4	12.8807
C_{pc}	86,473.20	α_5	−7.9551

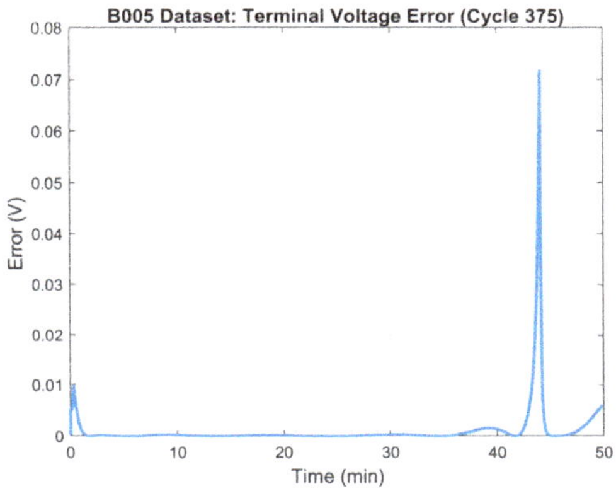

Figure 9. 25% SOH: LS model error—terminal voltage.

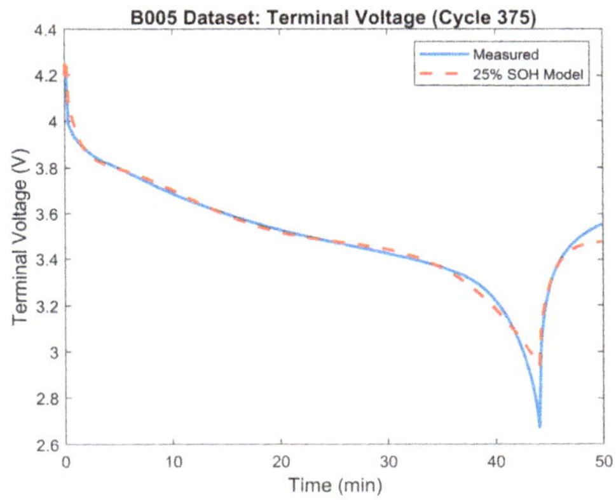

Figure 10. Terminal voltage: model vs. measured voltage for 25% SOH model.

5.2.5. The 0% SOH Model

The 0% SOH of the battery occurs when the battery reaches a capacity of 1.30 Ah which, based on the data, happens at the ~588th cycle. Table 6 illustrates the results of the parameter identification process for the 0% SOH mode. The terminal voltage's RMSE was 0.042 V. Figure 11 shows the terminal voltage's error plot between the model and the measured terminal voltage. The generated model's terminal voltage plot vs. the measured terminal voltage is illustrated in Figure 12.

Table 6. NLLS parameter estimation results for 0% SOH model.

RC Parameters	Value	OCV (SOC)	Value
R_0	0.080	α_1	2.4549
R_{pa}	0.202	α_2	−2.8347
C_{pa}	401.510	α_3	−5.1815
R_{pc}	0.001	α_4	14.5322
C_{pc}	86,963.20	α_5	−8.2310

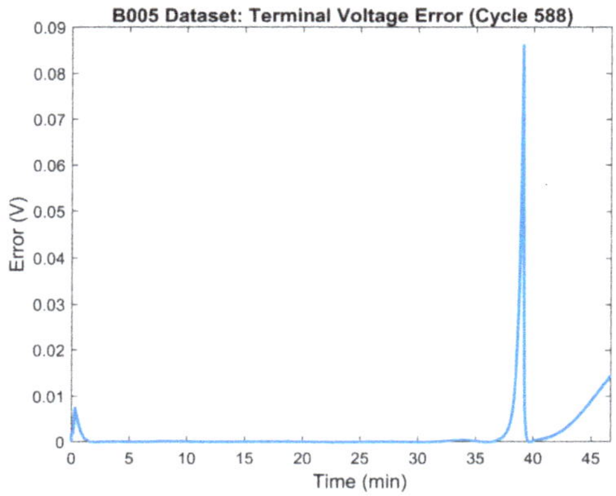

Figure 11. 0% SOH: LS model error—terminal voltage.

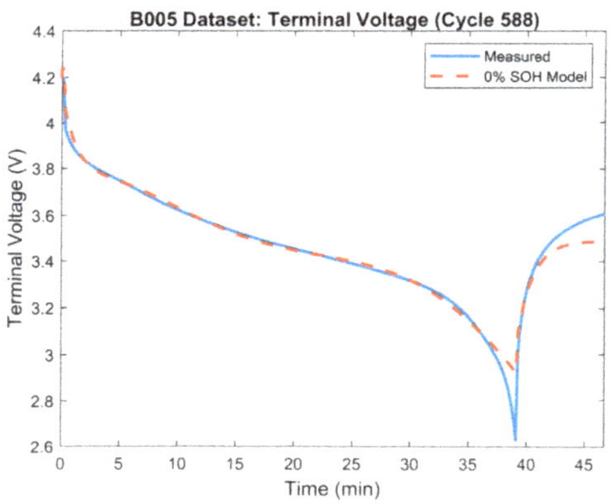

Figure 12. Terminal voltage: model vs. measured voltage for 0% SOH model.

6. Simulation Setup and Results

6.1. Simulation Setup

Mode Matching

In this experiment, the IMM was used for tracking the battery capacity of the LiB. The IMM was combined with the KF and SIF, resulting in the KF-IMM and SIF-IMM algorithms. The initial mode probabilities were set to 95% normal and 5% faulty. The IMM mixing parameter was defined as $p_{ij} = \begin{bmatrix} 0.95 & 0.05 \\ 0.05 & 0.95 \end{bmatrix}$. Initial estimates were set to zero.

The IMM has access to five different battery models representing different stages of the SOH of the battery. As the battery ages, the IMM selects the mode that best matches the battery's current state and by doing so, the battery's current SOH can be determined. Table 7 summarizes the initial conditions used for this experiment.

Table 7. Mode matching experiment: initial conditions.

Variables	Values
V_{pa}	0
V_{pc}	0
SOC	100%
R_0	0.07
$Batt_{Cap}$	2
$Q_{100,75,25\% SOH}$	$\text{diag}(5 \times 10^{-9}, 5 \times 10^{-9}, 1 \times 10^{-8})$
$Q_{50\%\ SOH}$	$\text{diag}(5 \times 10^{-9}, 5 \times 10^{-9}, 1 \times 10^{-7})$
$Q_{0\%\ SOH}$	$\text{diag}(5 \times 10^{-8}, 5 \times 10^{-8}, 1 \times 10^{-6})$
$R_{100-0\%\ SOH}$	$\text{diag}(5 \times 10^{-2}, 5 \times 10^{-2}, 5 \times 10^{-2})$
$Delta_{100\%\ SOH}$	$\text{diag}(5 \times 10^{1}, 5 \times 10^{1}, 1 \times 10^{2})$
$Delta_{75,25,0\%\ SOH}$	$\text{diag}(1 \times 10^{1}, 1 \times 10^{1}, 1 \times 10^{2})$
$Delta_{50\%\ SOH}$	$\text{diag}(9, 9, 1 \times 10^{2})$
p	0.9999
μ	0.2

6.2. Simulation Results

This section presents the results obtained from the proposed strategy.

Mode Matching

Five different modes were identified in the B005 dataset and served as the database for the IMM algorithm. At the end of each time step of each cycle, the parameter values of each mode were multiplied by the probability of each mode and combined to determine the estimated parameter values of the proposed IMM strategy. The following cycles were chosen to demonstrate the results of the proposed strategy at each stage of SOH: 106, 278, 441, 596. These cycles correspond to 75% SOH, 50% SOH, 25% SOH, and 0% SOH, respectively.

At cycle 106, the battery's measured capacity is ~1.81 Ah. Therefore, the KF-IMM and SIF-IMM should select, for the most part, the 75% SOH mode to better match the current state of the battery. Figure 13 illustrates the estimation results of the KF-IMM and SIF-IMM. Part (a) shows the terminal voltage estimation. Part (b) depicts the SOC estimation and part (c) illustrates the battery capacity estimation at the 106th cycle.

The terminal voltage's RMSE values for the KF-IMM and SIF-IMM were 0.0249 and 0.0255, respectively; both of these values show strong accuracy. Furthermore, both SOC estimation results showed a good profile. Moreover, the estimated battery capacity for both algorithms remained, for the most part, at ~1.83 Ah.

Lastly, Figure 14 illustrates the mode probability of each algorithm at cycle 106. Both algorithms switch between various modes to account for the sudden drop in voltage at the beginning of the discharge process. This switching can also be observed when the current is cut off, and the battery starts to recover after a period of ~53 min. Most importantly, it is evident that both filters select the 75% SOH mode as the better match, which reflects the measured battery capacity in the B005 dataset.

At cycle 278, the battery's measured capacity is ~1.56 Ah. Therefore, the KF-IMM and SIF-IMM should select, for the most part, the 50% SOH mode. Figure 15 illustrates the terminal voltage, SOC, and battery capacity estimation results of the KF-IMM and SIF-IMM at the 278th cycle.

The terminal voltage's RMSE values for the KF-IMM and SIF-IMM were 0.0377 and 0.0257, respectively; both values suggest high accuracy. Furthermore, both SOC estimation results showed a good constant discharge profile, with the KF-IMM fully discharging the battery. In terms of battery capacity estimates, the KF-IMM had a mean value of 1.56 Ah, which matches the measured value from the dataset. The SIF-IMM has a higher mean value at 1.65 Ah. This higher value in battery capacity enforces the discrepancy in the SOC discharge profiles between both algorithms.

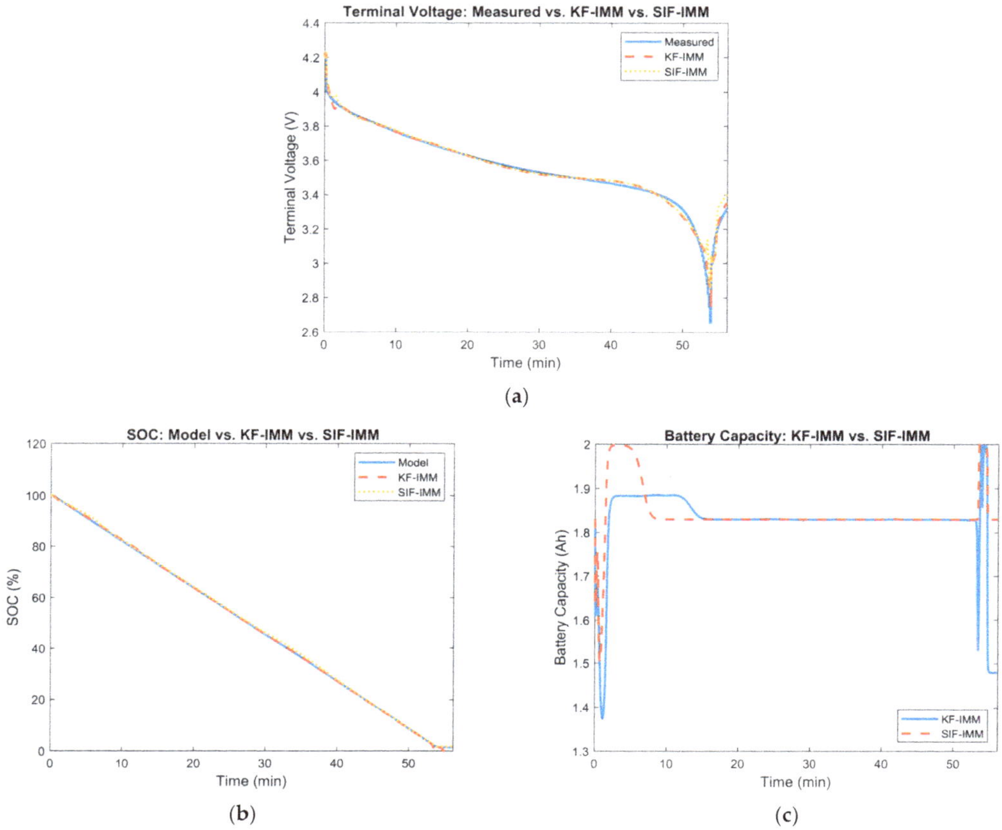

Figure 13. (**a**) Measured terminal voltage vs. KF-IMM and SIF-IMM estimation at cycle 106. (**b**) KF-IMM and SIF-IMM SOC estimation at cycle 106. (**c**) KF-IMM and SIF-IMM battery capacity estimation at cycle 106.

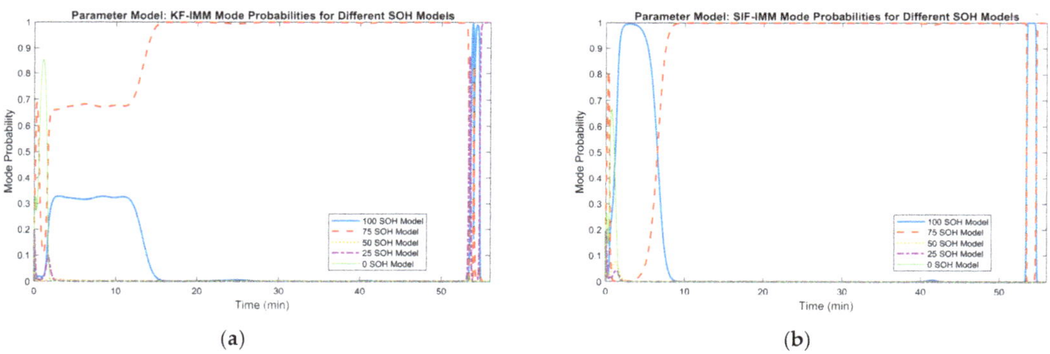

Figure 14. (**a**) KF-IMM mode probability at cycle 106. (**b**) SIF-IMM mode probability at cycle 106.

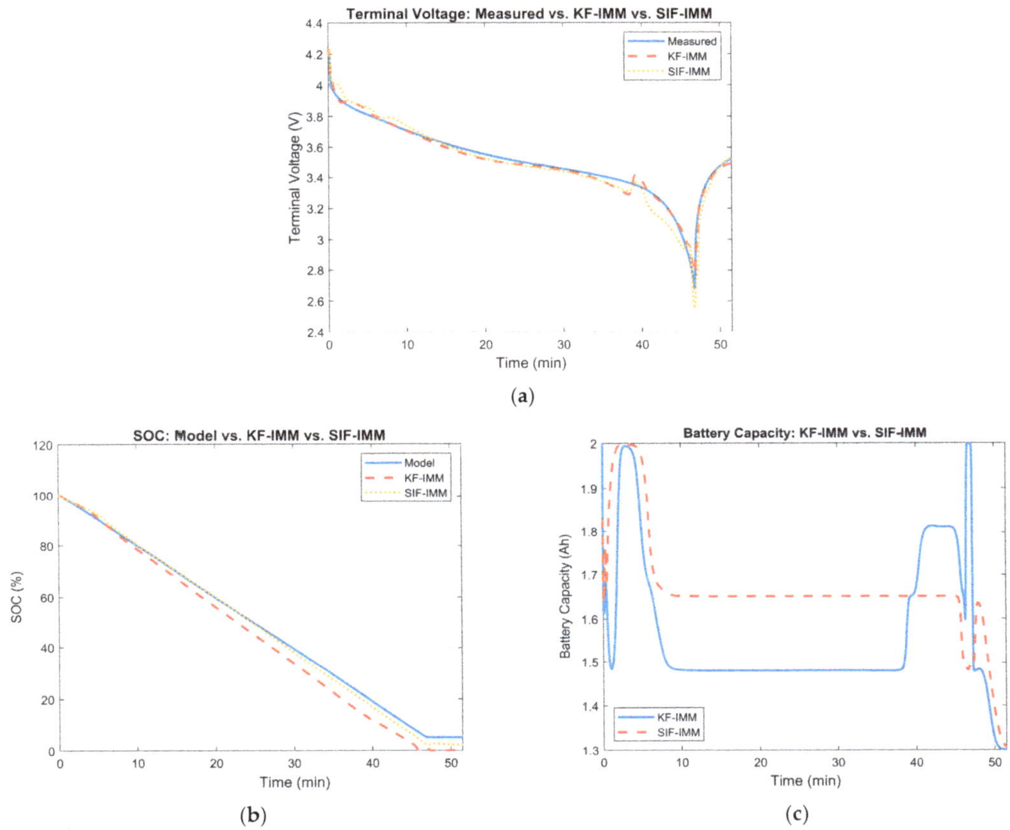

Figure 15. (**a**) Measured terminal voltage vs. KF-IMM and SIF-IMM estimation at cycle 278. (**b**) KF-IMM and SIF-IMM SOC estimation at cycle 278. (**c**) KF-IMM and SIF-IMM battery capacity estimation at cycle 278.

Figure 16 demonstrates the mode probability of each algorithm at cycle 278. In part (a), the KF-IMM selects the 25% SOH model as the predominant mode for most of the cycle. Looking at the same window of time in part (b) of Figure 15, the lower battery capacity allows for a faster discharge rate, which is more advantageous towards the end of the cycle when compared to the SIF-IMM results. However, the sudden jump in voltage near the 40 min mark, and the selection of the 75% SOH mode afterwards, can be a consequence of the selection of the 25% SOH mode.

On the other hand, the SIF-IMM method chooses the 50% SOH mode for most of the cycle. This mode selection reflects a better estimate of the current SOH of the battery based on the measured battery capacity.

At cycle 411, the battery's measured capacity is ~1.42 Ah. Therefore, the KF-IMM and SIF-IMM should select, for the most part, the 25% SOH mode to match the current state of the battery. The estimation results of the KF-IMM and SIF-IMM are shown in Figure 17. Part (a) shows the terminal voltage results. Part (b) depicts the SOC estimation and part (c) illustrates the battery capacity estimation at the 411th cycle.

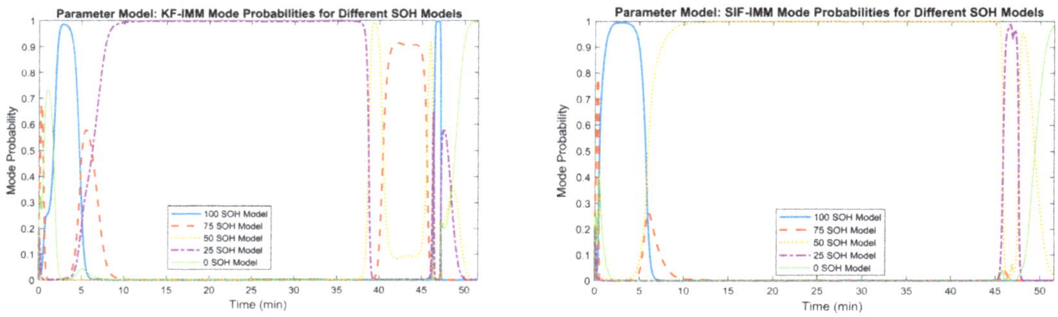

(a) (b)

Figure 16. (a) KF-IMM mode probability at cycle 278. (b) SIF-IMM mode probability at cycle 278.

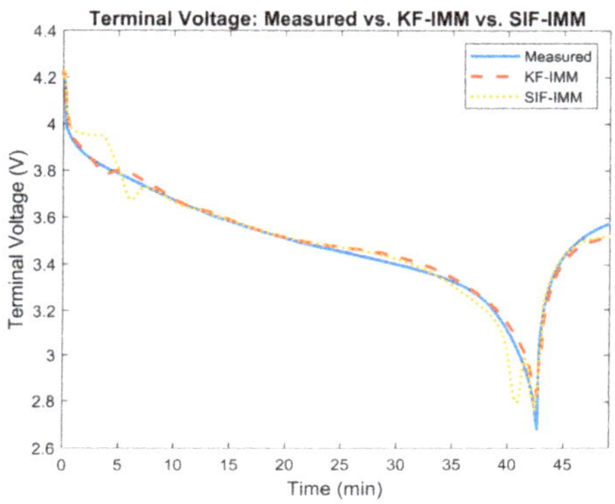

(a)

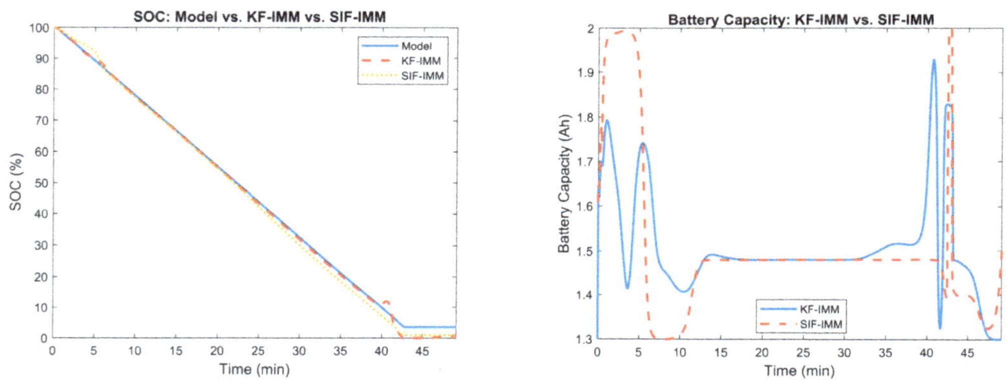

(b) (c)

Figure 17. (a) Measured terminal voltage vs. KF-IMM and SIF-IMM estimation at cycle 411. (b) KF-IMM and SIF-IMM SOC estimation at cycle 411. (c) KF-IMM and SIF-IMM battery capacity estimation at cycle 411.

The terminal voltage's RMSE values for the KF-IMM and SIF-IMM were 0.0373 and 0.0239, respectively, confirming their accuracy. In part (b), both SOC estimation curves have a good overall profile that reflects a full discharge of the battery. Moreover, in terms of battery capacity, the mean value for both algorithms was ~1.51 Ah. These values are close to the measured value of 1.42 Ah in the dataset. The corresponding mode probabilities for this case are shown in Figure 18.

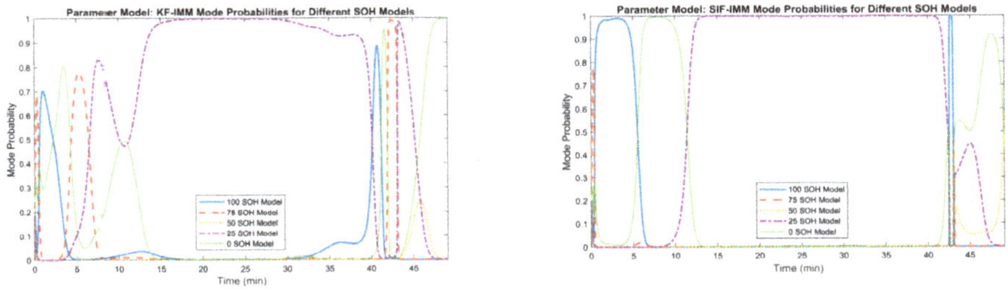

(a) (b)

Figure 18. (**a**) KF-IMM mode probability at cycle 411. (**b**) SIF-IMM mode probability at cycle 411.

At cycle 596, the battery's measured capacity is ~1.3 Ah. Therefore, the KF-IMM and SIF-IMM should select, for the most part, the 0% SOH mode. Figure 19 illustrates the estimation results of the KF-IMM and SIF-IMM. Part (a) shows the terminal voltage estimation. Part (b) depicts the SOC estimation and part (c) illustrates the battery capacity estimation at the 596th cycle.

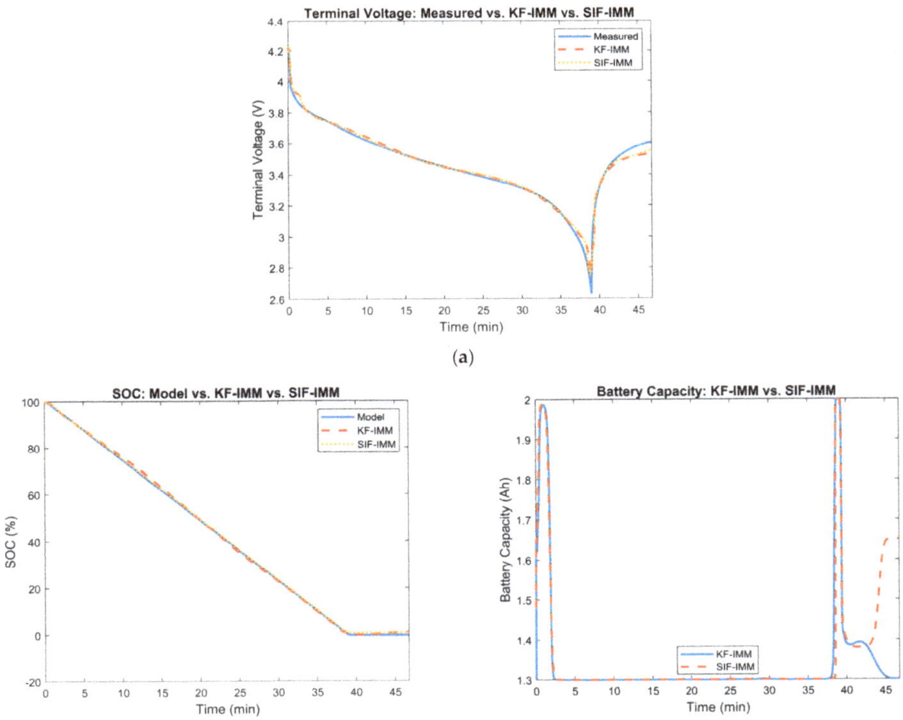

(b) (c)

Figure 19. (**a**) Measured terminal voltage vs. KF-IMM and SIF-IMM estimation at cycle 596. (**b**) KF-IMM and SIF-IMM SOC estimation at cycle 596. (**c**) KF-IMM and SIF-IMM battery capacity estimation at cycle 596.

The terminal voltage's RMSE values for the KF-IMM and SIF-IMM were 0.0311 and 0.0251, respectively; both values show strong accuracy. Both algorithms' SOC curves show a good full discharge profile. Moreover, the estimated battery capacity for both algorithms remained, for the most part, at ~1.33 Ah.

The mode probability of each algorithm at cycle 596 is shown in Figure 20. Both algorithms select the correct mode, which reflects the current SOH of the battery in the B005 dataset.

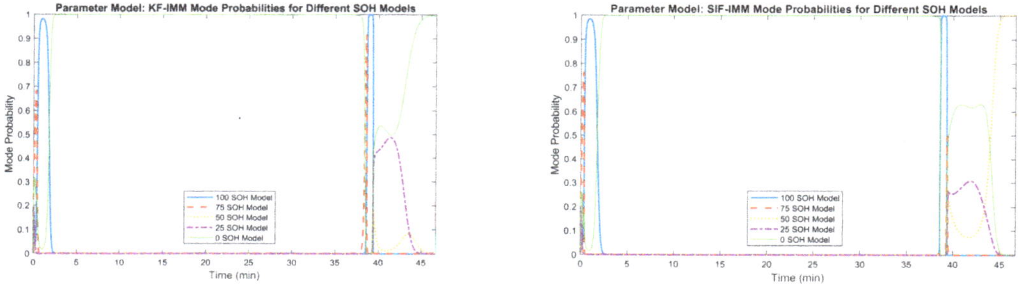

(a) (b)

Figure 20. (**a**) KF-IMM mode probability at cycle 596. (**b**) SIF-IMM mode probability at cycle 596.

To analyze the overall estimation results of the KF-IMM and SIF-IMM, the mean battery capacity for each cycle was generated, and the mode chosen by the IMM-selected models was taken at the end of each cycle. Figure 21 shows the most frequently selected mode for each cycle. Part (a) depicts the most-selected mode of the KF-IMM and part (b) illustrates the most-selected mode of the SIF-IMM. Figure 22 depicts the mean battery capacity of both algorithms for each cycle compared to the measured battery capacity.

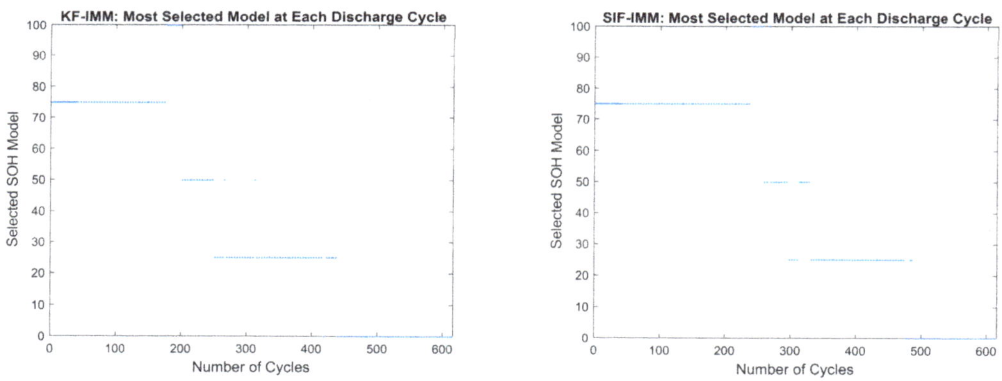

(a) (b)

Figure 21. (**a**) KF-IMM's most selected model for each cycle. (**b**) SIF-IMM's most selected model for each cycle.

From Figure 21 it is evident that neither algorithm made use of the 100% SOH model at the beginning of the experiment. The reason is evident if one were to look at the starting capacity of the battery, which is closer to the 75% SOH model. Based on the break points for each identified SOH region, the algorithms should ideally switch between modes after cycles 98, 225, 375, and 588. Since the battery's capacity starts at 75% SOH, the first switch should occur after the 225th cycle. However, the KF-IMM switches to the 50% SOH model earlier than the SIF-IMM. This early jump reinforces the results from Figures 15 and 16.

Furthermore, neither algorithm makes consistent use of the 50% SOH model and both algorithms make an early jump to the 25% SOH model. Finally, both algorithms switch to the 0% SOH model before the expected 588th cycle. Taking a holistic view on Figure 21, both algorithms show the desired downward step trend.

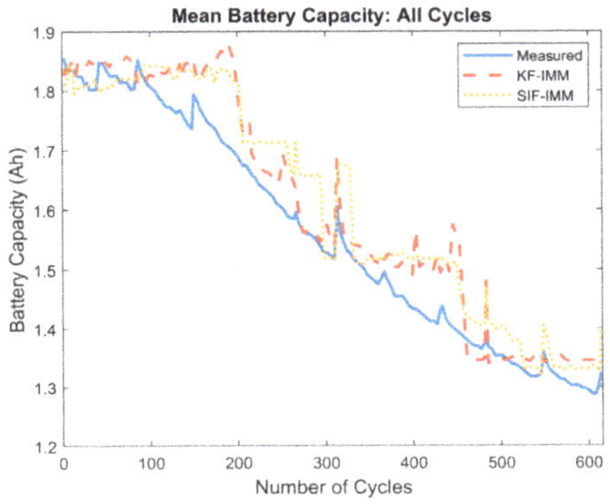

Figure 22. Mean battery capacity across all cycles.

It can be seen from Figure 22 that both algorithms show a good battery capacity trend. The battery capacity RMSE values for the KF-IMM and SIF-IMM were 0.060 and 0.065, respectively. When compared to the difference between the healthy and retired battery capacity values, the RMSE values correspond to an error of ~9% in determining the current SOH of the battery. After further inspection of Figure 22, the early switches between modes at cycles ~190 and ~440 can be justified.

7. Conclusions

This paper presented a strategy that utilizes the interacting multiple model (IMM) algorithm integrated with the Kalman filter (KF) and sliding innovation filter (SIF) methods for mode-matched filtering. As demonstrated by the results, the proposed strategy accurately estimates the state of charge (SOC) and state of health (SOH) of a lithium-ion battery (LiB) under cycling conditions. More specifically, the KF-based methods worked more accurately under 'normal' operating conditions (e.g., when the dynamic models closely resembled the actual operating models). However, when the SOH started to reduce or the battery began to degrade (essentially when there was model mismatch), the SIF-based methods yielded better estimates.

In the proposed strategy, the IMM is used to track the correct battery capacity, thus estimating the SOH of the battery. Meanwhile, the KF and SIF are used to estimate the SOC of the battery during cycling. The integration of these algorithms with the IMM resulted in strategies called KF-IMM and SIF-IMM. Moreover, the proposed strategy was evaluated using experimental data and was found to be successful in tracking the SOH of the battery. In the future, the proposed strategy will be implemented on a real-world BMS for further testing and verification. Lastly, this strategy has the potential to be used for predicting battery retirement, the results of which will be explored further in future studies.

Author Contributions: Investigation, R.B. and S.A.G.; writing—original draft preparation, R.B.; writing—review and editing, S.A.G., M.B., M.A. and S.M. All authors have read and agreed to the published version of the manuscript.

Funding: This research received no external funding.

Data Availability Statement: Publicly available datasets were analyzed in this study. The B005 dataset can be found here: https://ti.arc.nasa.gov/tech/dash/groups/pcoe/prognostic-data-repository/, accessed on 20 August 2022. B. Saha and K. Goebel (2007). The "Battery Data Set" is provided by the NASA Ames Prognostics Data Repository (http://ti.arc.nasa.gov/project/prognostic-data-repository, accessed on 20 August 2022), NASA Ames Research Center, Moffett Field, CA.

Conflicts of Interest: The authors declare no conflict of interest.

Appendix A

In the proposed strategy, the IMM is used to track the correct battery capacity, thus estimating the SOH of the battery. Meanwhile, the KF and SIF are used to estimate the SOC of the battery during cycling. The integration of these algorithms with the IMM resulted in strategies referred to as KF-IMM and SIF-IMM. The following figure helps further illustrate the overall architecture of the proposed strategies.

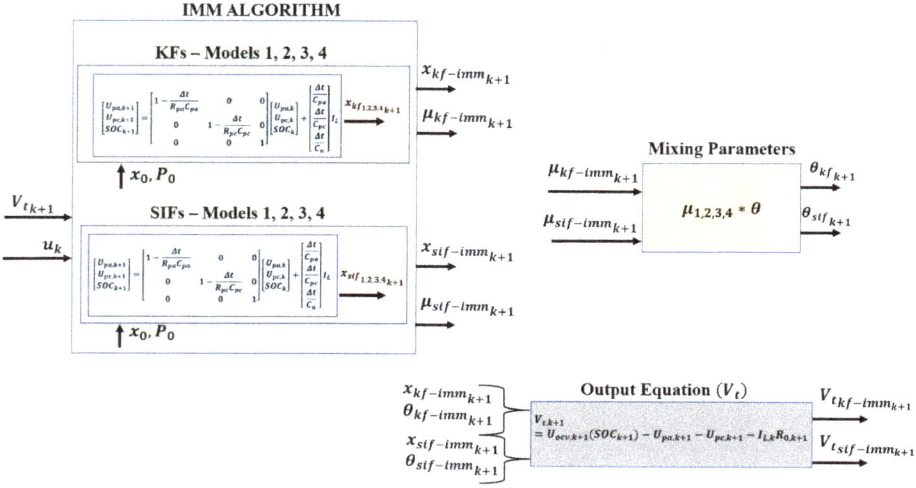

Figure A1. Overall architecture of IMM-KF and IMM-SIF strategies.

References

1. Ren, G.; Ma, G.; Cong, N. Review of Electrical Energy Storage System for Vehicular Applications. *Renew. Sustain. Energy Rev.* **2015**, *41*, 225–236. [CrossRef]
2. Thackeray, M.M.; Wolverton, C.; Isaacs, E.D. Electrical Energy Storage for Transportation—Approaching the Limits of, and Going beyond, Lithium-Ion Batteries. *Energy Environ. Sci.* **2012**, *5*, 7854. [CrossRef]
3. Lu, L.; Han, X.; Li, J.; Hua, J.; Ouyang, M. A Review on the Key Issues for Lithium-Ion Battery Management in Electric Vehicles. *J. Power Sources* **2013**, *226*, 272–288. [CrossRef]
4. Zhang, S.; Sun, H.; Lyu, C. A Method of SOC Estimation for Power Li-Ion Batteries Based on Equivalent Circuit Model and Extended Kalman Filter. In Proceedings of the 2018 13th IEEE Conference on Industrial Electronics and Applications (ICIEA), Wuhan, China, 31 May–2 June 2018; IEEE: Piscataway, NJ, USA, 2018; pp. 2683–2687.
5. Vidal, C.; Malysz, P.; Kollmeyer, P.; Emadi, A. Machine Learning Applied to Electrified Vehicle Battery State of Charge and State of Health Estimation: State-of-the-Art. *IEEE Access* **2020**, *8*, 52796–52814. [CrossRef]
6. Meddings, N.; Heinrich, M.; Overney, F.; Lee, J.-S.; Ruiz, V.; Napolitano, E.; Seitz, S.; Hinds, G.; Raccichini, R.; Gaberšček, M.; et al. Application of electrochemical impedance spectroscopy to commercial Li-ion cells: A review. *J. Power Sources* **2020**, *480*, 228742. [CrossRef]
7. Zhang, R.; Li, X.; Sun, C.; Yang, S.; Tian, Y.; Tian, J. State of Charge and Temperature Joint Estimation Based on Ultrasonic Reflection Waves for Lithium-Ion Battery Applications. *Batteries* **2023**, *9*, 335. [CrossRef]
8. Kandepu, R.; Foss, B.; Imsland, L. Applying the Unscented Kalman Filter for Nonlinear State Estimation. *J. Process Control* **2008**, *18*, 753–768. [CrossRef]
9. Partovibakhsh, M.; Liu, G. An Adaptive Unscented Kalman Filtering Approach for Online Estimation of Model Parameters and State-of-Charge of Lithium-Ion Batteries for Autonomous Mobile Robots. *IEEE Trans. Control Syst. Technol.* **2015**, *23*, 357–363. [CrossRef]
10. Yadkuri, F.F.; Khosrowjerdi, M.J. Methods for Improving the Linearization Problem of Extended Kalman Filter. *J. Intell. Robot. Syst.* **2015**, *78*, 485–497. [CrossRef]

11. Sanjurjo, E.; Naya, M.Á.; Blanco-Claraco, J.L.; Torres-Moreno, J.L.; Giménez-Fernández, A. Accuracy and Efficiency Comparison of Various Nonlinear Kalman Filters Applied to Multibody Models. *Nonlinear Dyn.* **2017**, *88*, 1935–1951. [CrossRef]
12. Foletto, T.C.; Moreno, U.F.; Lopes, A.M. Comparison of Packet Loss Compensation in Nonlinear Networked Control Systems Using the Methods Extended Kalman Filter and Unscented Kalman Filter. In Proceedings of the ETFA2011, Toulouse, France, 5–9 September 2011; IEEE: Piscataway, NJ, USA, 2011; pp. 1–7.
13. Yi, S.; Zorzi, M. Robust fixed-lag smoothing under model perturbations. *J. Frankl. Inst.* **2023**, *360*, 458–483. [CrossRef]
14. Wang, D.; Huang, H.; Tang, Z.; Zhang, Q.; Yang, B.; Zhang, B. A Lithium-Ion Battery Electrochemical–Thermal Model for a Wide Temperature Range Applications. *Electrochim. Acta* **2020**, *362*, 137118. [CrossRef]
15. Tang, S.-X.; Camacho-Solorio, L.; Wang, Y.; Krstic, M. State-of-Charge Estimation from a Thermal–Electrochemical Model of Lithium-Ion Batteries. *Automatica* **2017**, *83*, 206–219. [CrossRef]
16. Ghalkhani, M.; Bahiraei, F.; Nazri, G.-A.; Saif, M. Electrochemical–Thermal Model of Pouch-Type Lithium-Ion Batteries. *Electrochim. Acta* **2017**, *247*, 569–587. [CrossRef]
17. Wang, L.; Niu, J.; Zhao, W.; Li, G.; Zhao, X. Study on Electrochemical and Thermal Characteristics of Lithium-Ion Battery Using the Electrochemical-Thermal Coupled Model. *Int. J. Energy Res.* **2019**, *43*, 2086–2107. [CrossRef]
18. Barcellona, S.; Piegari, L. Lithium Ion Battery Models and Parameter Identification Techniques. *Energies* **2017**, *10*, 2007. [CrossRef]
19. Zhang, X.; Zhang, W.; Lei, G. A Review of Li-Ion Battery Equivalent Circuit Models. *Trans. Electr. Electron. Mater.* **2016**, *17*, 311–316. [CrossRef]
20. Davide, A. *Battery Management Systems for Large Lithium-Ion Battery Packs*; Artech: Morristown, NJ, USA, 2010.
21. Marcicki, J.; Onori, S.; Rizzoni, G. Nonlinear Fault Detection and Isolation for a Lithium-Ion Battery Management System. In Proceedings of the ASME 2010 Dynamic Systems and Control Conference, Cambridge, MA, USA, 1 January 2010; ASMEDC: New York, NY, USA, 2010; Volume 1, pp. 607–614.
22. Meng, J.; Yue, M.; Diallo, D. A Degradation Empirical-Model-Free Battery End-Of-Life Prediction Framework Based on Gaussian Process Regression and Kalman Filter. *IEEE Trans. Transp. Electrif.* **2022**, *9*, 4898–4908. [CrossRef]
23. Xu, J.; Sun, C.; Ni, Y.; Lyu, C.; Wu, C.; Zhang, H.; Yang, Q.; Feng, F. Fast Identification of Micro-Health Parameters for Retired Batteries Based on a Simplified P2D Model by Using Padé Approximation. *Batteries* **2023**, *9*, 64. [CrossRef]
24. Singh, A.; Izadian, A.; Anwar, S. Nonlinear Model Based Fault Detection of Lithium Ion Battery Using Multiple Model Adaptive Estimation. *IFAC Proc. Vol.* **2014**, *47*, 8546–8551. [CrossRef]
25. Xia, X.; Wei, Y. Lithium-Ion Batteries State-of-Charge Estimation Based on Interactive Multiple-Model Extended Kalman Filter. In Proceedings of the 2016 22nd International Conference on Automation and Computing (ICAC), Colchester, UK, 7–8 September 2016; IEEE: Piscataway, NJ, USA, 2016; pp. 204–207.
26. Singh, A.; Izadian, A.; Anwar, S. Fault Diagnosis of Li-Ion Batteries Using Multiple-Model Adaptive Estimation. In Proceedings of the IECON 2013—39th Annual Conference of the IEEE Industrial Electronics Society, Vienna, Austria, 10–13 November 2013; IEEE: Piscataway, NJ, USA, 2013; pp. 3524–3529.
27. Bustos, R. State of Health Estimation of Lithium-Ion Batteries Using Dual Filters and the IMM Strategy. Ph.D. Thesis, University of Guelph, Guelph, ON, Canada, 2022.
28. Hasib, S.A.; Islam, S.; Chakrabortty, R.K.; Ryan, M.J.; Saha, D.K.; Ahamed, M.H.; Moyeen, S.I.; Das, S.K.; Ali, M.F.; Islam, M.R.; et al. A Comprehensive Review of Available Battery Datasets, RUL Prediction Approaches, and Advanced Battery Management. *IEEE Access* **2021**, *9*, 86166–86193. [CrossRef]
29. Tian, N.; Wang, Y.; Chen, J.; Fang, H. On Parameter Identification of an Equivalent Circuit Model for Lithium-Ion Batteries. In Proceedings of the 2017 IEEE Conference on Control Technology and Applications (CCTA), Mauna Lani Resort, HI, USA, 27–30 August 2017; IEEE: Piscataway, NJ, USA, 2017; pp. 187–192.
30. Kalman, R.E. A New Approach to Linear Filtering and Prediction Problems. *J. Basic Eng. Mar* **1960**, *82*, 35–45. [CrossRef]
31. Gadsden, S.A.; Habibi, S.; Kirubarajan, T. Kalman and Smooth Variable Structure Filters for Robust Estimation. *IEEE Trans. Aerosp. Electron. Syst.* **2014**, *50*, 1038–1050. [CrossRef]
32. Gadsden, S.A.; Al-Shabi, M. The Sliding Innovation Filter. *IEEE Access* **2020**, *8*, 96129–96138. [CrossRef]
33. Bustos, R.; Gadsden, S.A.; Al-Shabi, M.; Mahmud, S. Lithium-Ion Battery Health Estimation Using an Adaptive Dual Interacting Model Algorithm for Electric Vehicles. *Appl. Sci.* **2023**, *13*, 1132. [CrossRef]
34. Wei, X.; Yimin, M.; Feng, Z. Lithium-Ion Battery Modeling and State of Charge Estimation. *Integr. Ferroelectr.* **2019**, *200*, 59–72. [CrossRef]

MDPI
St. Alban-Anlage 66
4052 Basel
Switzerland
www.mdpi.com

Energies Editorial Office
E-mail: energies@mdpi.com
www.mdpi.com/journal/energies

www.ingramcontent.com/pod-product-compliance
Lightning Source LLC
Chambersburg PA
CBHW040805250926
51850CB00007BA/1090
9783725806058